Blue Economy of the Indian Ocean

Blue Economy of the Indian Ocean

Resource Economics, Strategic Vision, and Ethical Governance

Ranadhir Mukhopadhyay,
Victor J. Loveson,
Sridhar D. Iyer,
P.K. Sudarsan

CRC Press
Taylor & Francis Group
Boca Raton London New York

CRC Press is an imprint of the
Taylor & Francis Group, an **informa** business

First edition published 2021
by CRC Press
6000 Broken Sound Parkway NW, Suite 300, Boca Raton, FL 33487-2742

and by CRC Press
2 Park Square, Milton Park, Abingdon, Oxon OX14 4RN

ISBN: 978-0-367-33467-3 (hbk)
ISBN: 978-0-429-32600-4 (ebk)

Typeset in Times
by Newgen Publishing UK

We dedicate this book
To the departed soul of our mentor and former co-author

Prof. Anil Kumar Ghosh
(8 November 1940–21 November 2018)

Who left for the heavenly abode while working on an earlier book

Contents

Preface

The world has deviated far from a sustainable paradigm, as pollution in air, rivers, and oceans has reached an agonizing limit. Water resources are dwindling, economic disparity in society has increased, poverty is rampant, and dignity of living has nosedived. This is because the world has transformed into an economy wherein respect, empathy, and responsibility are replaced by greed, hollow competitions, and cutting corners. The threats from climate change and greed-based commercial exploitation on humanity are looming large, as is the possibility of military intervention. This is despite the fact that Earth has enough resources to take care of the genuine needs of all her inhabitants.

Against this background, the blue economy is emerging fast as an alternative to bring about a balance among the incoherencies. This upcoming concept of economy is expected to shift the society from scarcity to abundance "with what is locally available," and by tackling issues that cause environmental and related problems. Such initiatives are more important to the Indian Ocean region (IOR) as it encompasses the world's poorest and most densely populated countries. This is notwithstanding the fact that globally the Indian Ocean hosts major international shipping lanes (>100,000 ship transits annually) that transport 66% of crude oil, 55% of container shipment, and 33% of bulk cargo.

While exploring the immense socioeconomic potential of the blue economy in the Indian Ocean, the authors of this book regret that this ocean has not received as much attention as it truly deserves. The unique geography of this ocean does not have pole-to-pole water connection like other oceans. This "disconnect" contributes enormously to the generation of southwest and northeast monsoons that influence the economy of a major part of the IOR countries. Moreover, climate change is an issue that needs special attention, as a vast tract of underdeveloped countries in the IOR would receive the maximum brunt of such changes. Since its first conceptualization in the last decade of the twentieth century and consolidation by the first decade of this century, the blue economy has taken the world by storm. With questions mounting on the failure of the existing governance setup to arrest unabashed and uneven social and economic growth, to seize the damaging climate change impact, and to halt increasing domination over nature by humankind, an urgent need is felt to relook at the way success and development are presently perceived. This book discusses the shortcomings of the existing dispensation and suggests ways to a new world.

The concept of the *blue economy* has emerged recently also as an important approach to drive sustainable development of coasts and oceans. Such sustainable economic development of the *ocean* is only possible if it contributes to long-term prosperity and resilience. The *blue economy* covers a wide range of interlinked established and emerging sectors. In addition, the skill development under this economy could help the large numbers of unemployed youth in the region by educating them enough to take up new types of jobs and/or for entrepreneurship. This book traces this emerging multidimensional, multilateral, interdisciplinary, and dynamic economic model that goes beyond national boundaries.

The book starts by discussing the concept, rationale, and potentials of the blue economic order. This line of economy offers an approach to sustained growth, enhanced social inclusion, improved human welfare, and helps create opportunities for employment, while maintaining the healthy functioning of the Earth's ecosystems (Chapter 1). The geographical significance of the Indian Ocean is briefly traced, and also the culture, civilization, and human migration in the IOR. A possible blue economy vision for the entire Indian Ocean is also drawn.

All the facets of resources that can be conceivably procured from the Indian Ocean under the blue economy architecture are discussed next (Chapter 2). These include living and non-living resources, including minerals and energy. In addition, the cultural linkages that the ocean facilitates and services that the ocean offers are specially spelt out. Further, the question as to why the IOR should support a blue economy paradigm is also discussed.

In the next two chapters this book discusses the blue economy scenario of ten countries each from the Indian Ocean (Chapter 3), and those from the Atlantic and Pacific oceans (Chapter 4). The Indian Ocean countries comprise South Africa, Mauritius, Seychelles, Kenya, Oman, Bangladesh, Sri Lanka, Indonesia, Singapore, Thailand, and Australia. A similar discussion from other oceans pertaining to China, the USA, Japan, Canada, New Zealand, Brazil, Pacific Islands (Papua New Guinea and Fiji Islands), the United Kingdom, Mauritania, and Nigeria are included. The plan, program, preparedness, and implementation of each of these nations vis-à-vis the blue economy are analyzed. The areas of success and shortcoming are tabulated. Furthermore, the contribution of the blue economy to the countries' respective Gross Domestic Product (GDP) is assessed. How prepared these nations are culturally (through ethics, education, customs, and mindset, among others) to enact and execute the alternative blue economy concept are examined.

The three major threats to the blue economy, namely change in climate, strategic geopolitics, and greediness of humans, are discussed in Chapter 5. We feel the impact of change in climate on the blue economy outcome will be enormous. Further, with countries like China, France, India, Japan, Russia, the UK, and the USA focusing on acquiring commercial ports and military facilities, the Indian Ocean has become the platform of prime strategic importance. China, through her One Belt One Road (OBOR) strategy, is expanding her tentacles to establish military-cum-commercial bases in many IOR countries.

In fact, the blue economy up till now has been recognized and described to have only the first two threats (climate and geopolitics), but the near-unlimited aspirations and human mindset have been detailed in this book as the "third major threat." Such threat comes from "within." It is the mindset, belief, and the perception of human beings on how success and progress are defined, so far. A discussion shows that socioeconomic development for a long time has been ill-defined in many countries and the concept of genuine progress has been deliberately ignored for commercial benefit. Erosion of values and ethics is a real challenge. Despite this gloomy picture the IOR must remain as a more balanced theater of cooperation, growth, and peace. We trace the mechanism of such a possibility, following the teachings and thoughts of four visionary philosophers: Lord Buddha, Rabindranath Tagore, Swami Vivekananda, and Mohandas Karamchand Gandhi (the BTVG doctrine).

For a successful blue economy revolution, developments in science and technology (S&T) would remain an important component. The nature of such S&T innovations in the areas of food, minerals, energy, infrastructure, trade, commerce, blue carbon, recreation, navigation, capacity building, education, scientific research, and employment generation are detailed in Chapter 6. The "canvas of opportunities" has been drawn with clear projections of growth for some of these sectors in the years 2030 and 2050, and also the financial contribution of these sectors to the national GDPs of the respective countries (Chapter 7).

In a bid to take the blue economy revolution to its logical sustainable stage, in Chapter 8 we propose the ethical blue economy (EBE), an extended and revised approach of the existing blue economy paradigm. The EBE model calls for rational, critical, and time-bound reforms in various sectors of governance, such as education, health, energy, development, labor, communication, social righteousness, equal opportunity, jobs and employment, pluralism, inclusiveness, local government, law and justice. A detailed roadmap to reach EBE targets is deliberated. A possible transformation of the Indian Ocean into a hub of excellence in human values, ethics, and sustainable economy following the BTVG doctrine is envisioned.

The challenge to make the Indian Ocean an economically beneficial, zero-waste, recyclable, and environmentally sustainable entity is "very real." All such activities must be in balance with the ocean's long-term carrying capacity and its health. Sharing of information among the IOR countries is required to deepen economic and security cooperation with maritime neighbors and island states; and promote cooperation in low-carbon shipping, regional fisheries, and ocean surveillance, amongst others. This book travels through all these uncertainties to lay a roadmap of do-able actions.

It appears the centuries old complementary relationship between human and nature is under extreme threat, and needs to be re-invented. The present book discusses all these issues in the light of the increasing population and their growing aspirations, and the threat and security perspectives in the region. It also draws strength essentially from the BTVG doctrine of ethical governance to make business in the IOR not only sustainable but also responsible and equitable.

The marriage of science, innovation, and entrepreneurship with complete integrity could create the EBE model, which would have the potential to transform the society at large. This model would help draw an architecture mixing economic development with ethical practices in every sphere of life. The components of this model would include food security, harnessing minerals/energy/medicine, mitigation resilience to climate change, enhanced trade and investments, improved maritime connectivity, increased diversification, recreation and tourism, job creation, poverty alleviation, harmony, peace, and socioeconomic growth.

The change in climate is real and is considered as the biggest manifestation of the failure of the market economy, as the inflated materialistic demands are fast exceeding the carrying capacity of this very world. For the question "how to deal with this catastrophe," the answer probably lies in "thinking differently" and quickly opting for a course correction. This could be achieved probably when EBE, as proposed in this book, replaces the present-day market economy. This book explores such a difficult terrain.

In summary, the important requirement of the present-day economy is to reinvent the relation of the individual to nature. In fact, our education, research, teaching, and learning must understand that superiority of human beings does not rest in power of possession, but in power of unison and sharing. Toward this, an ideal comprehension encompassing an integrated balance among humans, nature, land, ocean, and economic growth is much desired.

Our efforts to write this book will be suitably rewarded, if it helps in such integration of thoughts and actions. Additionally, the book must remain useful in informing and empowering people, students, teachers, diplomats, negotiators, and policy makers of the entire IOR, and beyond. As the book explores establishing a strong sustainable economy through responsible activities, it becomes a step toward superior stewardship of our ocean and blue resources.

We acknowledge the support and efforts of the following institutions and personalities in making this book. While RM, VJL, and SDI abundantly express thanks to the CSIR-National Institute of Oceanography, PKS expresses gratitude toward the Goa University for unflinching support in making this study possible. The authors pursued research and teaching for several decades in these institutions. The Tapovan-Centre for Contemporary Research with which the authors are attached presently is also acknowledged.

We thank CRC Press of Taylor & Francis for publishing this book. We acknowledge the support and patience of Irma Britton (Acquisition Editor) and Rebecca Pringle (Editorial Assistant) of Taylor & Francis for constantly supervising the pre-production and that of Flora Kenson (Project Manager) during printing stages of the book. We are also indebted to the reviewers of our original proposal who encouraged us to go ahead with writing the book.

Our wives (Julie, Latha Lavanya, Kamakshi, and Ambily) and children (Ayan, Bhuvaneswari, Sainandan, Kavya, and Hridya) have been a constant source of encouragement. Poornima Dhawaskar helped in drawing most of the figures on her computer. Mohammed Irfan Shaikh designed the cover image of the book. Divakar Naidu, a colleague of ours, went through some portions of the manuscript and made suggestions.

The image on the book cover depicts four fundamental issues of the blue economy—utterly dismayed by the upturned civilization (reflecting completely de-railed present progress due to rampant greed-based irrational growth), a woman (representing ethical sustainability) is voyaging toward the ocean (the last frontier of knowledge and resources) on a wooden boat (signifying a simple change in mindset can bring in the viable blue economy; it is not necessary to have sophisticated technology). More appropriately we dedicate this book to the loving memory of our mentor, co-author and friend the late Professor Anil Kumar Ghosh of the University of Calcutta, who was an uncompromising soldier of sustainability, not only for the environment but also in mindset.

25 March 2020
Ranadhir Mukhopadhyay, Victor J. Loveson, Sridhar D. Iyer
CSIR-National Institute of Oceanography, Dona Paula 403004, Goa
P.K. Sudarsan
Goa Business School. Goa University, Taleigao 403206, Goa

Authors

Ranadhir Mukhopadhyay is the former Director of Mauritius Oceanography Institute (Mauritius), and former Chief Scientist of CSIR-National Institute of Oceanography (Goa). A trained Earth scientist, Dr. Mukhopadhyay has also contributed in HR management, strategic leadership, and in developing an ethical governance model. He has 35 years of research, and 20 years of teaching experience. With doctoral degrees both in natural science (Geology, 1988) and in social science (Public Administration, 2019), he has authored four books so far. ranadhir@nio.org, mranadhir@yahoo.com, ppb1958@gmail.com

Victor J. Loveson is a Professor and Chief Scientist at the CSIR- National Institute of Oceanography, Goa, India. With a PhD in Applied Geology, he has about three decades of experience in mineral resources mapping, Quaternary research, sea level studies, paleoclimate reconstruction, coastal disaster management and Marine Archaeology. Dr. Loveson has edited three books and authored four-dozens of research papers and has been awarded with the ISCA young scientist award, CSIR Golden Jubilee Whittaker award and the INSA-Royal Society exchange scholar fellowship.

Sridhar D. Iyer has a doctoral degree in Geology and is a Professor and former Chief Scientist with CSIR-National Institute of Oceanography, Goa. He has 35 years of research experience in Ocean Science, and two decades of teaching experience. He has authored two books. iyer@nio.org

P.K. Sudarsan, with a doctoral degree in Economics, and several decades of teaching experience behind him, is now the Vice-Dean of the Goa Business School in Goa University, Taleigao (Goa). Formerly, he was the head of the Department of Economics at Goa University. sudha@unigoa.ac.in

1 Introduction

1.1 CONCEPT AND RATIONALE

The way science without ethics is considered perilous, growth in the economy in a region without taking cognizance of the surrounding ecosystem remains unsustainable. This is quite evident from the fact that the most significant global problem of today has not been the change of climate or loss of biodiversity (these are only the symptoms), but is the non-inclusion of indigenous knowledge, native culture, and local values into the realm of economic development and societal progress. Hence, one must close the gulf that got accumulated over the centuries between sustainability and growth by reinventing the relation of the individual to nature.

The need for a plural, inclusive, and complementary economic order emanates from the above philosophy. Such new order, which is defined globally as the blue economy, seeks to promote economic growth, social inclusion, and the preservation and improvement of livelihoods while ensuring at the same time environmental sustainability of the oceans and coastal areas. This would balance the socioeconomic development through ocean-related activities with the environmental and ecosystems degradation.

The reasons for emphasizing the oceans in this book lie in the fact that almost three-fourths of planet Earth is covered by water, of which oceans hold about 96.5%. Oceans support more than 95% of the biosphere by generating oxygen, absorbing in the process about 30–40% of atmospheric carbon dioxide, recycling nutrients and regulating global climate and temperature. Oceans also account for a considerable part of the world population with respect to the availability of food and providing livelihoods.

A sustainable approach is most sought after, as scientific data find that ocean resources are limited and that the health of the oceans has drastically declined due to anthropogenic activities. These changes are already being profoundly felt, affecting human well being, and the impacts are likely to be amplified in the future, especially in view of projected population growth. Hence, it is necessary to use the resources available in the oceans in a judicious manner for economic growth, enhancing in the process the living style of people and creating jobs, while preserving the health of the ecosystem.

To meet the ever-increasing demand for food—especially for protein—harvesting and trading of marine living resources are necessary. This seafood harvesting includes primary fisheries (captured fish catch from rivers and oceans) and secondary fisheries (aquaculture, mariculture). For example, the total fish production from oceans (through both captured and aquaculture methods) shows a steady increase from 101 MT (million tons) in 2009 to 108 MT in 2014, with per-capita fish intake rising from 18.1 kg to 20.1 kg during the same period (FAO, 2016, 2018). In addition, a whole lot of fishing-related economic and livelihood activities include net and gear making, ice production and supply, boat construction and maintenance, manufacturing of fish processing equipment, packaging, marketing and distribution, besides creating employment opportunities. In addition to food, about 80% of global trade is accomplished through oceans. Again, the advanced and sophisticated marine biotechnology and bio-prospecting methods for pharmaceutical products and chemical applications could well be the future of the blue economy by contributing toward health care, cosmetic, enzyme, nutraceutical, and other industries.

Two decades after the Rio de Janeiro Earth Summit in 1992, the United Nations Conference on Sustainable Development (UNCSD), held at the same city between June 20 and 22, 2012 (Rio+20 Conference) maintained its priority focus on poverty alleviation on an urgent basis by advancing the "green economy" concept. However, considering the strong lobbying by coastal and island countries to emphasize oceans and seas in a humane endeavor, the concept of the green economy in a blue world was accepted as the "blue economy." This type of economy offers an approach for sustained fiscal growth, enhanced social inclusion, improved human welfare, and creating opportunities for employment and decent work for all, while maintaining the healthy carrying capacity of the Earth's ecosystems (UNCSD, 2017). Accordingly, the major components of the blue economy are identified as: food security, harnessing minerals/energy/medicine, resilience to climate change, enhanced trade and investments, improved maritime connectivity, increased diversification, recreation and tourism, job creation, poverty alleviation, and socioeconomic growth (Pauli, 2010; Fig. 1.1).

With the dismantling of the socialist economy three decades ago, and the questions on the continuation of the capitalist economy that are now being raised, it is time to choose a new innovative path of development and socioeconomic prosperity. The new course must be sustainable, respectful, and responsible. The challenges for any new economic order must be to be able to: invest less to innovate more, generate multiple cash flow, create jobs, build up social capital, and stimulate entrepreneurship. While doing so, the focus must be on thinking positive, behaving responsibly, and encouraging creative learning. The blue economy fits this bill squarely, as it aptly represents inclusive growth. Hence, it needs to be ensured that the concept and rationale of the blue economy are ingrained in all stages of our life for any future economic activities, right from school education to project conceptualization, planning, and infrastructure development, to project implementation.

The concept of the blue economy is however accepted differently in the world and by the Indian Ocean Region countries (IOR). In some countries, the blue economy plays an important role such as in the USA, China, South Korea, the EU, Australia,

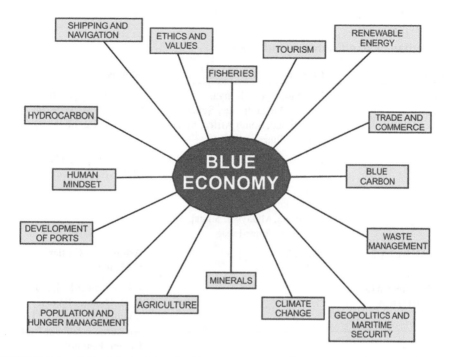

FIGURE 1.1 Components of the Blue Economy

Canada, France, the United Kingdom, Ireland, and the Philippines. These countries have made significant progress by developing a national accounting framework mechanism that connects gain from the blue economy with their respective national gross domestic product (GDP) calculations.

Among the IOR countries, Mauritius and Seychelles have made very good progress. In Mauritius the ocean economy (the blue economy) is identified as one of the pillars of its economic development, which could lead to its transformation into a high-income economy by 2025. At present, the blue economy contributes to around 10–11% of Mauritius's GDP. In Seychelles, there is a separate Ministry of Finance, Trade, and Blue Economy. A budget roughly equivalent to US$58.8 million was allocated for the year 2017.

Against this background, it would be ideal to explore the immense potential of the blue economy of the Indian Ocean (Mohanty et al., 2015). The reason is buttressed by the fact that this ocean has not received as much attention as it deserves. The unique geography of the Indian Ocean denies pole-to-pole water connection, unlike other oceans. This "disconnect" contributes enormously to the generation of southwest and northeast monsoons (see Section 1.3) that influence the economy of a majority of the IOR countries. In addition, the largely poor IOR countries are expected to receive the greatest brunt of climate change impact (rise in sea level, increased frequency of cyclone, storms, pollution, and degradation of the coastal zone). The IOR countries (comprising 32 countries, 12 islands) therefore should have special and requisite apparatus to deal both with poverty and climate change impact

TABLE 1.1
The Indian Ocean Countries and Islands

Continents	Countries	Islands
AFRICA 11 + 6	Djibouti, Egypt, Kenya, Eretrea, Mozambique, Somalia, Somaliland, South Africa, Sudan, Yemen, Tanzania	Mauritius, Seychelles, Comoros, Madagascar, Mayotte, Réunion
ASIA 20 + 4	Bahrain, Iran, Iraq, Israel, Jordan, Kuwait, Oman, Palestine, Qatar, Saudi Arabia, UAE, Pakistan, India, Bangladesh, Myanmar, Indonesia, Malaysia, Singapore, Thailand, Timor-Leste	Chagos & Diego Garcia, Maldives, Sri Lanka, Cocos
OCEANIA 1 + 2	Australia	Ashmore & Cartier, Christmas
Southern Indian Ocean / ANTARCTICA 0 + 6		Prince Edward, Heard & McDonald, French Southern & Antarctic Lands, Amsterdam & Saint Paul, Crozet, Kerguelen

Note: Total countries and islands bordering the Indian Ocean. IOR = 32 countries + 18 islands.

(Table 1.1). In this regard, skill development could be a major boost to help the large numbers of unemployed youth in the region, and by making them capable and educated enough for the new types of job requirement and entrepreneurship.

The challenge now is to make the Indian Ocean an economically beneficial, zero-waste, recyclable, and environmentally sustainable entity. All such activities must be in balance with the ocean's long-term carrying capacity and its health. Sharing of information with maritime neighbors and island states is required to deepen economic and security cooperation to thwart any threats, and promote cooperation in low-carbon shipping, regional fisheries, and ocean surveillance, amongst others. Hence, a strategic perspective of the blue economy for the Indian Ocean is the need of the hour.

1.2 POTENTIALS OF THE BLUE ECONOMY

Potential economic activities in the ocean, particularly in terms of procuring living, non-living, and energy resources from the ocean, are now regulated by Article 76 of the United Nations Convention on the Law of the Sea (UNCLOS) that came into operation by 1994 (Table 1.2). Article 76 defines the rights and responsibilities of nations with respect to their use of the world's oceans, establishing guidelines for businesses, the environment, and the management of marine natural resources. Later following the adoption of resolution No. 70/226 by the UN General Assembly on

TABLE 1.2
Maritime Zone: Rights and Responsibilities

Maritime Zones	Distance from Coast	Rights for Resources by the Country
Coastal Water	0–3	Airspace, Water Column, Seabed, Sub-seabed
Territorial Sea	3–12	Airspace, Water Column, Seabed, Sub-seabed
Contiguous Zone	12–24	Airspace, Water Column, Seabed, Sub-seabed
Exclusive Economic Zone	24–200	Water Column, Seabed, Sub-seabed
Extended Continental Shelf	200–maximum 350, at places	Seabed, Sub-seabed
International Water	➢ No nation has any right beyond 200, and beyond 350 (for those countries with extended Continental Shelf)	

Notes: Distance in nautical miles (1 nm = ~1.85 km), Sub-seabed = beneath seafloor, Zones demarcated as per Article 76 of UNCLOS III.

December 22, 2015 a high-level UN conference to support the implementation of Sustainable Development Goal 14 (SDG 14) using the oceans, seas, and marine resources was convened (World Bank, 2017). To reap benefits particularly to Small Island Developing States (SIDS) and Least Developed Countries (LDC) from sustainable management of marine resources, including fisheries, aquaculture, and tourism, a host of highly esteemed institutions were included as members of working groups.

Some of these institutions are the World Bank Group (WBG), UN Department of Economic and Social Affairs (DESA), UN Environment, Food, and Agriculture Organization (FAO), International Maritime Organization (IMO), UN Office of Legal Affairs/Division for Ocean Affairs and the Law of the Sea (OLA/DOALOS), UN Office of the High Representative for the Least Developed Countries, Landlocked Developing Countries, and Small Island Developing States (OHRLLS), UN Conference on Trade and Development (UNCTAD), United Nations Development Programme (UNDP), United Nations Industrial Development Organization (UNIDO), United Nations World Tourism Organization (UNWTO), International Seabed Authority (ISA), International Union for Conservation of Nature (IUCN), World Trade Organization (WTO), International Council for Science (ICSU), Organisation for Economic Co-operation and Development (OECD), World Ocean Council (WOC), World Wide Fund for Nature (WWF), Conservation International, Ocean Policy Research Institute (OPRI) of Sasakawa Peace Foundation, and National Oceanic and Atmospheric Administration (NOAA) of US Department of Commerce.

The reports of these working groups identified that understanding and better managing the many aspects of oceanic sustainability, ranging from sustainable fisheries to ecosystem health to pollution, will be the first challenge. The second significant issue is the realization that the sustainable management of ocean resources requires collaboration across nation-states and across the public-private sectors, and on a scale that has not been achieved before. This realization underscores the challenge facing the SIDS and LDC, as they turn to better managing their blue economies (Attri and Bohler-Mulleris, 2018).

As listed earlier, the blue economy has diverse components, such as fisheries, tourism, and maritime transport of oil, gas, and bulk material, but also includes new and emerging activities (Fig. 1.1). As many as ten potentials of the blue economy for a sustainable ocean, commerce, and trade are identified (Table 1.3). For example, offshore renewable energy, aquaculture, seabed minerals, and marine biotechnology and bio-prospecting are now included in the blue economy. A number of services provided by ocean ecosystems, and for which markets do not exist, also contribute significantly to natural, economic, and other human activities. These include, for instance, carbon sequestration, coastal protection, waste disposal, and biodiversity. The crowded shipping lanes in the Indian Ocean hint at the enormous dependence of the IOR economy on this ocean (Fig. 1.2).

The main pillars of the blue economy administration as identified by the concept founder Gunter Pauli are clear vision, innovative technology, blue management, good governance, and regulatory reforms. According to Pauli (2010), a marriage of science, innovation, and entrepreneurship could create a new business model that can transform society. The sequential steps to carry out the blue economy model are: (a) estimate the size of the opportunity, (b) comprehend the nature of risks involved, (c) understand the mechanism of sustainable investment in the ocean, (d) assess the nature of necessary investment framework, and (e) make capital available to scale up blue industries.

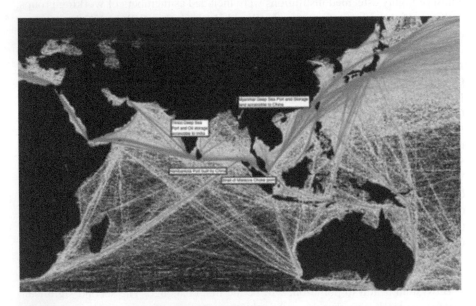

FIGURE 1.2 The Shipping Lanes in the Indian Ocean

TABLE 1.3
Potentials of Blue Economy for a Sustainable Ocean, Commerce, and Trade

Sl.	Areas	Components
01	Fisheries	• Modernize technology to increase fish-catch (mariculture, cage culture, aquaculture) • Use myctophids as food for livestock, poultry, and aquatics
02	Mineral and Oil & Gas	• Develop technology to harness gas hydrate, Mn nodules, hydrothermal sulfides • Collaborate to develop deep-sea technology: AUVs, ROVs, mining equipment • Develop offshore oil & gas platform, use new technology to increase yield
03	Shipping, Trade, Port expansion	• Develop shipyard, repair old ships, and fabricate new ships • Make some ports as hub of international shipping activities • Trade and investments, improve maritime connectivity, develop new shipping lanes
04	Coastal Threats/ Pollution	• Response to oil spill, pollution, ballast water invasion, MARPOL enforcement • Algal blooms and hypoxic zones • Rehabilitation center for extreme events; SOP for relief mechanism and infrastructure
05	Eco-Tourism, Marine Protected Areas	• Form travel and tourism partnership among IOR countries • Form marine park to give impetus to tourism and adventure sports • Resilience to climate change, enhanced recreation and tourism • Job creation, poverty alleviation, and socioeconomic growth
06	Food Processing	Collaborate in food processing technology, Antarctic krill, development of lobster and crab fattening devices—it's a huge opportunity for innovation and employment
07	Energy from Ocean	• Generate energy from currents, tides, winds, waves, and also from algae biofuel, solar insolation, and gas hydrate • Encourage high-end research on sophistication and economization of renewable energy generating technology
08	Medicine from Sea, and Aquatic Disease	• Develop marine pharmaceutical industry, conduct study on bio-active molecules available in seawater for making medicines • Share information and collaborate aquatic disease control program and aquatic animal health
09	Capacity Building, Outreach, Awareness	• Conduct IMO III Training for Coast Guard, IMOOPRC level II and level III training • Conduct workshops and hand-holding on pollution management (oil spill, tar-balls, ballast water invasion), soft coastal protection, CRZ, policy-legal issues, blue carbon, acidification & hypoxia, aquaculture and shellfish, ocean energy, technology

(continued)

TABLE 1.3 (Continued)
Potentials of Blue Economy for a Sustainable Ocean, Commerce, and Trade

Sl.	Areas	Components
10	Marine Scientific Research	IOR countries are advised to collaborate on: • Ocean monitoring and weather forecasting • Ocean acidification • Sea-level rise • Integrated waste management • Land-based sources of marine pollution, e.g., nutrients and other pollutants that cause algal blooms, hypoxic zones • Impact of climate change: ocean acidification on marine tropho-dynamics • Desalination: develop a feasible membrane to generate power from salinity gradient from sea water and river water (up to 10 million litres per day) • Low temperature thermal desalination for power generation • Climate modeling • Understand complete dynamics of ENSO and monsoon • Modeling impact of polar climate on tropical monsoon • Marine ecology • Long-term satellite mapping of coral reefs and PFZ • Explore the world of biotechnology to develop eco-friendly mechanism to transform waste → wealth → reuse → safe disposal of industrial by-products and wastewater

Source: Compiled from IPCC (2014); FAO (2017), World Bank (2017).
Notes: MARPOL = The International Convention for the Prevention of Pollution from Ships; SOP = Standard Operating Procedure; IMO III = International Maritime Organization; IMOOPRC = International Convention on Oil Pollution Preparedness, Response and Co-operation; CRZ = Coastal Regulation Zone; ENSO = El Niño and the Southern Oscillation; PFZ = Preferred Fishing Zone.

The mix of oceanic activities varies in each country, depending on its unique national circumstances and the national vision adopted to show its own conception of a blue economy. In order to qualify as components of a blue economy, activities need to: (a) provide social and economic benefits for current and future generations, (b) restore, protect, and keep up the diversity, productivity, resilience, core functions, and intrinsic value of marine ecosystems, or (c) offer clean technologies to generate renewable energy, and circular material flows that will cut waste and promote recycling of materials (World Bank, 2017).

Interestingly, the blue economy aims to consider economic development and ocean health as compatible propositions. It is generally understood to be a long-term strategy for supporting sustainable and fair economic growth through ocean-related sectors and activities. This type of economy is relevant to all countries and can be applied on various scales, from local to global. In order to become do-able, the blue economy concept must be supported by a trusted and diversified knowledge base, and complemented with management and development resources that help inspire and support innovation.

The Indian Ocean hosts major international shipping lanes (>100,000 ship-transit annually) that transport 66% of crude oil, 55% of container shipment, and 33% of bulk cargo (Fig. 1.2). Several minor ports in the IOR are being developed into major ports with investments in billions of dollars. The coastal areas and islands also offer the sun, sand, sea global tourism that annually runs into billions of dollars. Additionally, about 32% of global supplies of hydrocarbons come from the ocean. Further, the ocean offers vast potential for renewable energy production from wind, wave, tides, thermal, and biomass sources. The exploration of beach placer minerals, phosphorites, polymetallic manganese nodules, hydrothermal sulfide minerals at the tectonic plate boundaries, and above all the extraction of metals chemically from the water column make the ocean hugely significant with regard to resources. Advancement in technology would further open up new frontiers of marine resource development from bio-prospecting to the mining of seabed mineral resources.

The potential of the blue economy could be gauged from the fact that the world-wide ocean economy is valued at around US$1.5 trillion per year, with about 80% of global trade by volume being carried by sea (World Bank, 2017). Highlighting the close connections among the oceans, climate change, and the well-being of the people of the IOR, the blue economy is seen to inspire best use of "blue" resources. Healthy oceans and seas can greatly contribute to inclusiveness and poverty reduction, and are essential for a more sustainable future especially for SIDS and coastal LDC alike. The blue economy provides these regions with a basis to pursue a low-carbon and resource-efficient path to economic growth and development designed to enhance livelihoods for the poor, create employment opportunities, and reduce poverty. However, SIDS and coastal LDC often lack the capacity, skills, and financial support to better develop their blue economy.

1.3 THE INDIAN OCEAN

As the focus in this book will be on the blue economy of the Indian Ocean, it is imperative to introduce a brief account of this ocean, in terms of geography, geology, biology, climate, and historical activities. The physiographic details of this ocean are given (Table 1.4, Fig. 1.3).

1.3.1 GEOGRAPHY

The Indian Ocean is an important entity of the Asian continent and extends from longitudes 20°E to 147°E and latitudes 30°N to 40°S. The ocean has an area of 73.5 million km^2 and covers approximately a fifth of the Earth's surface. Its size makes the Indian Ocean the third-largest body of water in the world. The average depth of this ocean is 3,890 m, but its deepest point at 7,450 m is located in the Java Trench (Eakins and Sharman, 2010). The Indian Ocean is bounded by Asia in the north, Africa in the west, Australia in the east, and on the south by the Southern Ocean (also known as Antarctic/Austral Ocean; Fig. 1.3). The 90°E meridian passes

TABLE 1.4
Physiographic Details of the Indian Ocean

01	Area (km²)	70,560,000	02	Shore to Shelf-Slope Border	19 ± 0.6
03	World Ocean %	19.5	04	Shore to Shelf-Slope Border	47.6 ± 0.8
05	Volume (km³)	264,000,000	06	Depth of Shelf-Slope Border	130–175
07	World Ocean %	19.8	08	Continental Shelf Area	15%
09	Deepest Point (meter)	7,258	10	Drainage Basin Area (km²)	21,100,000
11	Average Depth (meter)	3.741	12	No. of Individual Basins	800
13	No. of Hotspots/ Mantle Plumes	06	14	Distribution % of Basins	AS: 50, AF: 30, AT: 20
15	Maximum Width (km)	7,600	16	Shore to Slope-Rise Border	205.3–255.2
17	Av. Length of Rivers Falling in IO	764	18	Sea Surface Temperature, Sea Surface Salinity	Pre-M: 26-31C, 30–35 Post-M: 27-31C, 21–32

19 **Eight Biodiversity Hotspots in the Indian Ocean Region**

Maputaland-Pondoland-Albany: 8,100 (1,900 endemic) species of plants; 541 (0) birds; 205 (36) reptiles; 73 (20) freshwater fishes; 73 (11) amphibians; and 197 (3) mammals

Coastal Forests of East Africa: 4,000 (1,750 endemic) species of plants; 636 (12) birds; 250 (54) reptiles; 219 (32) freshwater fishes; 95 (10) amphibians; and 236 (7) mammals

Horn of Africa: 5,000 (2,750 endemic) species of plants; 704 (25) birds; 284 (93) reptiles; 100 (10) freshwater fishes; 30 (6) amphibians; and 189 (18) mammals

Western Ghats–Sri Lanka: 5,916 (3,049 endemic) species of plants; 457 (35) birds; 265 (176) reptiles; 191 (139) freshwater fishes; 204 (156) amphibians; and 143 (27) mammals

India-Burma: 13,500 (7,000 endemic) species of plants; 1,277 (73) birds; 518 (204) reptiles; 1,262 (553) freshwater fishes; 328 (193) amphibians; and 401 (100) mammals

Sundaland: 25,000 (15,000 endemic) species of plants; 771 (146) birds; 449 (244) reptiles; 950 (350) freshwater fishes; 258 (210) amphibians; and 397 (219) mammals

Wallacea: 10,000 (1,500 endemic) species of plants; 650 (265) birds; 222 (99) reptiles; 250 (50) freshwater fishes; 49 (33) amphibians; and 244 (144) mammals

Southwest Australia: 5,571 (2,948 endemic) species of plants; 285 (10) birds; 177 (27) reptiles; 20 (10) freshwater fishes; 32 (22) amphibians; and 55 (13) mammals

Notes: 02 = Active margin, km; 03 and 07 = percentages of world oceans in area and volume; 04 = passive margin, km; 06 = meter; 08 = percentage of world shelf area, Shelf = Continental Shelf, Slope = Continental Slope; 14 = percentage of basins in Asia, Africa, Australia, AS = Asia, AF = Africa, AT = Australasia; 16 = km, Rise means Continental Rise; 18 = Sea-surface temperature in centigrade, Sea-surface salinity (psu = practical salinity unit = ~parts per thousand), M = monsoon; 19 = Vimal Kumar et al. (2008); Mittermeier et al. (2011).

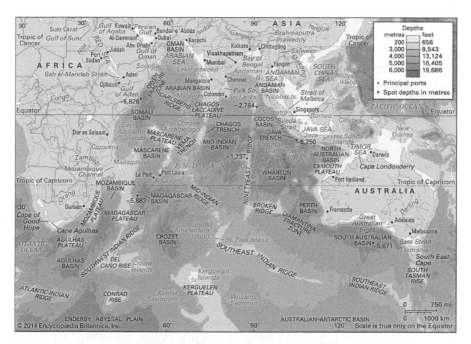

FIGURE 1.3   The Indian Ocean with all its Expansion

through the Ninety East Ridge, which is one of the longest north-south running geological features in the world oceans.

The IOR countries are spread over three continents of Asia, Africa, and Oceania, and also encompass several islands. This ocean touches the coasts of 32 countries and 18 islands. These include 11 countries and 6 islands in Africa, 20 countries and 4 islands in Asia, 1 country and 2 islands in Oceania, and 6 islands in the Southern Indian Ocean (Table 1.1). Many of these countries are in fact archipelagos of hundreds of islands. For example, Indonesia has an official listing of 15,708 tiny islands, while Maldives has 1,200 islands under her belt. The Indian Ocean includes several seas (Arabian, Andaman, Laccadive, Red, and Zanj), Gulfs and Bays (Bengal, Mannar, Aden, Persian/Arabian, Aqaba, Tadjoura, Great Australian Bight, Carpentaria, Bahrain, Kutch, Khambhat, and Oman), and seaways, straits, or channels (Malacca, Sunda, Hormuz, Mozambique, Torres, Palk, Guardafui, and Bab-el-Mandeb). These water bodies contribute their waters and sediments to the Indian Ocean.

While the Suez Canal connects the Indian Ocean with the Mediterranean Sea through the Red Sea, the Indian Ocean is characterized by a number of choke points that are very significant from a strategic point of view. Some of these Choke Points are Bab el Mandeb, Strait of Hormuz, Lombok Strait, Strait of Malacca, and Palk Strait. More of these are discussed in Chapter 5.

India, located at the apex of this ocean, with the Arabian Sea to her west, the Bay of Bengal to the east, and the Indian Ocean to the south, is unique in several aspects. India is in a strategic position with her westward connectivity to the Arabian nations, Pakistan, South Africa, Seychelles, eastwards with Bangladesh, Myanmar, and Indonesia, and southwards with Sri Lanka and Maldives.

In contrast to the Atlantic and Pacific, the Indian Ocean is not open from pole to pole and can be likened to an embayed ocean. This is unique for this ocean, which remains a great influencing factor to shape its climate, development of extreme events, and atypical ocean current pattern. Crowned by the Indian Peninsula, which played a major role in its history, the Indian Ocean has predominantly been a cosmopolitan stage interlinking diverse regions by innovations, trade, and religion since early in human history (Prange, 2008).

With a coastline of 114,172 km, the Indian Ocean is about eight times the size of the USA. This ocean is nearly 10,000 km wide at the southern tips of Africa and Australia and has a volume of 292.13 million km³. Melting of polar ice-caps makes the Indian Ocean actually grow wider by about 20 cm every year. This ocean is geologically significant as three large tectonic plates—Indo-Australian, African, and Antarctic—converge in it, their borders being tracked by the extension of mid-ocean ridges (McKenzie and Sclater, 1971; Mukhopadhyay et al., 2018a). The Indo-Australian plate shoves under the Asian continent at an average speed of 5 cm/yr. A lost continent by the name of Mauritia was recently found submerged in the western Indian Ocean. The physiography and biodiversity potential of the Indian Ocean is briefly touched on in Table 1.4.

The continental shelves of this ocean are largely narrow, averaging 200 km, with an exception off Australia's western coast, where the shelf width exceeds 1,000 km. The continental shelf makes up 15% of the Indian Ocean. More than 2.5 billion people live in countries bordering the Indian Ocean, compared to 1.7 billion for the Atlantic and 2.7 billion for the Pacific (some countries border more than one ocean). Supplied by some large rivers debouching into it, such as Zambezi, Indus, Ganges, Shatt-al-Arab, Brahmaputra, and Irrawaddy to name a few, river runoff of 6,000 km brings sediment to the Indian Ocean. The rivers emptying into the Indian Ocean are, however, shorter in average length (740 km) than those of the other major oceans. The drainage basin of the Indian Ocean covers 20.98 million km², almost identical to that of the Pacific Ocean, or 30% of its ocean surface (compared to 15% for the Pacific). The Indian Ocean is divided into roughly 800 small individual basins, half that of the Pacific, of which 50% are located in Asia, 30% in Africa, and 20% in Australasia (Keesing and Irvine, 2005). Pelagic sediment is predominant in the Indian Ocean, covering about 86% of the seafloor, while the other 14% is floored by terrigenous sediment.

The Indian Ocean has been named after India (*Oceanus Orientalis Indicus*) since at least the year 1515 (Douglass and Miller, 2018). The ancient Sanskrit texts named this body of water as Ratnakara (mine of gems). This ocean is however poorly studied scientifically, only through a few expeditions. The first major investigation was conducted between 1872 and 1876 by the *HMS Challenger*, followed by the Valdivia expedition in 1898–9. In the 1930s, the John Murray Expedition mainly studied shallow-water habitats. The Swedish Albatross Expedition in 1947–8 also sampled the Indian Ocean on its global tour, and between 1950 and 1952 the Danish research vessel *Galathea* scooped up deepwater fauna along transects from Sri Lanka to South Africa. Later the Soviet research vessel *Vitiyaz* visited the Indian Ocean. However, the most comprehensive survey of this ocean was conducted between 1959 and 1965 during the International Indian Ocean Expedition (IIOE;

Demopoulos et al., 2003). IIOE was a large-scale multinational investigation project involving 45 research vessels from 14 countries, including India. An enormous amount of good quality data on the biology, chemistry, and physics of this ocean was collected. The project was sponsored by SCOR (Scientific Committee on Ocean Research) and later by the IOC (Intergovernmental Oceanographic Commission). On the completion of 50 years of IIOE, the second string of investigation (IIOE-2) was initiated from 2015 by a club of global scientific institutions again under the umbrella of SCOR and IOC.

Of the countries in the IOR, Australia and India have taken the lead in marine research. The Commonwealth Scientific Industrial Research Organization (CSIRO), since its formation in 1916 in Canberra (Australia), has advanced Australia with a range of inventions, innovations, and discoveries in marine science. Since the 1960s, a number of institutions of repute have come up in the Indian Ocean. Many leading universities in this region started special schools or streams in Oceanography/ Marine Science/Environmental Science. The National Institute of Oceanography in Goa (India) emerged in 1966 and has been leading the R&D aspects with regard to this ocean. With a mission to understand the seas around the ocean and making the knowledge useful for the society, this multidisciplinary institute has, among others, demarcated a first-generation mine site for polymetallic (manganese) nodules in this ocean.

Named as the Indian Ocean Nodule Field, the area holds the second richest manganese nodule deposit in the world (Mukhopadhyay et al., 2008, 2018a), which is exploitable even at the present market price under certain conditions (Mukhopadhyay et al., 2019). Established in 2006, the Marine Research Institute of the University of Cape Town is spearheading research on behalf of South Africa in the western Indian Ocean. The Mauritius Oceanography Institute, the Kenya Marine Fisheries Research Institute, and the National University of Singapore are some of the other R&D centers in this ocean. Almost all IOR countries, however, now have maritime research institutes in one form or another.

1.3.2 GEOLOGY

The Indian Ocean has the most complicated origin among the three largest world oceans. The geological history of the Indian Ocean starts from the time India formed a major part in nature's jigsaw puzzle of landmass called Gondwanaland. This supercontinent that existed between 550 Ma (Neoproterozoic) and 180 Ma (Jurassic, a time when dinosaurs roamed the Earth), was made up of Australia, South America, Antarctica, Africa, and the Arabian Peninsula as the pieces of the puzzle. This gigantic tectonic plate was subjected to seismic and volcanic activities that resulted in rifting and drifting of the landmasses. The opening of the Indian Ocean began at around 156 Ma when Africa separated from East Gondwanaland. The Indian Subcontinent began to separate from Australia-Antarctica by 135–125 Ma, and as the Tethys Ocean north of India began to close during 118–84 Ma the Indian Ocean opened behind it (Chatterjee et al., 2013; Fig. 1.4).

Because of this break-up over a prolonged geological time, India slowly moved northwards and slammed into the Eurasian Plate at around 50 Ma. As a result, the

	Stage	Age	Events
	Late Stage	10–00 Ma	Arabia leaves **A2** (Africa + Somalia + Madagascar), and Australia separates from **B1** (India + Sri Lanka)
Phase 3	Early Stage	43–10 Ma	Australia joins **B1** (India + Sri Lanka),
	Late Stage	65–43 Ma	Seychelles leaves **B1** (India + Sri Lanka) and joins **A2** (Africa + Somalia + Arabia + Madagascar)
Phase 2	Early Stage	95–65 Ma	Madagascar leaves **B1** (India + Seychelles + Sri Lanka) to join **A2** **B2**, meanwhile, breaks as Australia and East Antarctic separates
Phase 1	Late Stage	135–95 Ma	Both the groups divide further into two sub-groups each. **Group A1** = South America, **Group A2** = Africa, Somalia and Arabia; **Group B1** = India, Madagascar, Seychelles, Sri Lanka, **Group B2** = Australia, East Antarctica.
	Early Stage	160–130 Ma	Gondwanaland splits into two major landmasses: **Group A** = South America, Africa, Somalia, and Arabia; and **Group B** = India, Australia, East Antarctica, Madagascar, Seychelles, Sri Lanka.

FIGURE 1.4 Stages of Dismemberment of the Gondwanaland (Modified after Royer et al. (1992); Minshull et al. (2008); Chatterjee et al. (2013), Mukhopadhyay et al. (1997, 2018a)).

Tethys Sea that existed between Tibet and India was gradually closed. The suturing between Europe and the northern part of the Gondwanaland was complete with the formation of the Himalayan Range that consists largely of distorted, folded, and fractured sedimentary rocks that were formed from the Tethys Sea sediments (Fig. 1.4). The Indian Ocean took its present configuration around 36 Ma. Though the geological activities that pre-empted the formation of this ocean initiated nearly 140 Ma ago, the majority of the Indian Ocean basin dates back only fewer than 80 Ma ago.

The active spreading centers in the Indian Ocean, the youngest of all oceans, are represented by mid-ocean ridges that are a part of a worldwide system of underwater mountain chains. The most prominent landmark in this ocean is the inverted Y junction (also known as the Indian Ocean triple junction, IOTJ), where Indo-Australian, African, and Antarctic plates meet. The Central Indian Ridge (including the Carlsberg Ridge) separates the Indo-Australian Plate from the African Plate, while the Southwest Indian Ridge separates the African Plate from the Antarctic Plate.

The Southeast Indian Ridge, on the other hand, separates the Indo-Australian Plate from the Antarctic Plate. The Indo-Australian Plate, in fact, comprises three plates: Indian Plate, Capricorn Plate, and Australian Plate. All these three meet to form a diffuse plate boundary south of Sri Lanka. The enormous African Plate has been divided since 20 Ma by the East African Rift into the Nubian and Somali Plates (Chatterjee et al., 2013).

In addition to spreading ridges as above, the Indian Ocean is also characterized by prominent zones of plate subduction, of which two are important: the 6,000 km-long Andaman-Java-Sumatra Trench in the Bay of Bengal, and the 900 km-long Makran Trench located south of Iran and Pakistan in the Arabian Sea. Numerous fracture zones, oceanic ridges, and seamount chains produced by mantle plumes/hotspots ornament the Indian Ocean floor (Fig. 1.5). For example, the Reunion hotspot, active

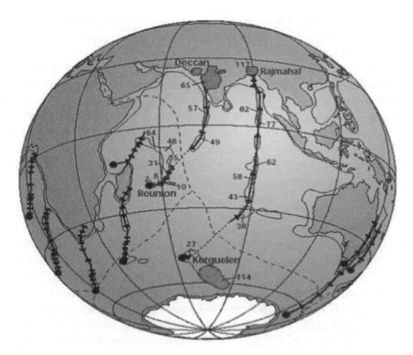

FIGURE 1.5   Mantle Plumes/Hotspots in the Indian Ocean (After www.mantleplumes.org).
Notes: Traces of ten mantle plumes/hotspots in the Indian Ocean manifested on the seafloor as seamount or ridges, shown in solid lines with age in million years. Most prominent are the Deccan-Reunion plumes and the Rajmahal-Kerguelen hotspots/mantle plumes. Ages are increasing toward north, suggesting movement of the Indian Plate toward north after separation from Gondwanaland.

since 72 Ma, formed the Deccan Trap between 68 and 62 Ma at the central west coast of India. This active source of eruptive mantle plume thereafter formed toward the south, in order, the Laccadive group of islands, Maldives, Chagos Archipelago, Mascarene Plateau, Mauritius Island, and Reunion Island in the western Indian Ocean, as the Indo-Australian Plate moved over this hotspot to the north since the mid-Cretaceous. The other major hotspot track in the Indian Ocean was formed by the movement of the Indo-Australian Plate over the Kerguelen hotspot, active since 100 Ma. This resultant submerged aseismic ridge connects Kerguelen Island (of present age) in the Southern Ocean to the Rajmahal Trap (age 116 Ma) located on the eastern Indian landmass, through Kerguelen Plateau and Ninety East Ridge. The third major hotspot is the Marion hotspot (active since 100 Ma) whose erupted material formed Prince Edward Island and the submerged Eighty-five East Ridge. All three hotspots are still active (Muller et al., 1993; Chatterjee et al., 2013; Parthasarat hi and Reillo, 2014, and references therein).

Because of its highly complex ridge topography, the Indian Ocean became the site of many basins that range from a smallish size of about 350 km up to large ones that measure about 10,000 km across. There is a large igneous province (LIP) in the southern Indian Ocean called the Kerguelen Plateau. This plateau is thought to be volcanic in origin and is the largest of its kind in the world.

Compared to other oceans in the world, the concentration of seamounts in the Indian Ocean has been far and few. Summits of these seamounts are located at deeper than 3,000 m water depth and located mostly north of 55°S and west of 80°E. Most of these seamounts were originated at spreading ridges and drifted away along with the plate to a place far away from these ridges. Many of the large seamounts in the Indian Ocean, however, show two episodes of formation: primary formation at the spreading ridge crest, while secondary contribution came from in-situ volcanism at the basin areas (Mukhopadhyay, 1998; Iyer et al., 2018).

1.3.3 BIOLOGY

The Indian Ocean is the warmest among all oceans and may not be very conducive to marine life compared to the other world oceans. Moreover, it has the lowest oxygen content in the world. This can be attributed to the fast evaporation rate in this ocean. The world's lowest and highest water salinity levels were both recorded in the Indian Ocean. Despite this constraint, a huge concentration of phytoplankton can be found in the western part of the Indian Ocean, due to the yearly monsoonal winds (especially around summer). Its water has a high concentration of hydrocarbons (both dissolved and floating).

As mentioned above, among the tropical oceans, the western Indian Ocean hosts one of the largest concentration of phytoplankton blooms in summer, due to the strong monsoon winds. This wind leads to an increased coastal and open-ocean upwelling, which brings nutrients on the sea surface from deep, where enough light is available for photosynthesis and phytoplankton production. This phytoplankton forms the base of the marine food web (phytoplankton → zooplankton → large fishes) and supports the marine ecosystem (Fig. 1.6). The Indian Ocean accounts for the second-largest share of the most economically valuable tuna and shrimp catch. Any variation in sea surface temperature impacts phytoplankton growth, so also the fish production. In fact, marine plankton production has declined by 20% in the Indian Ocean during the past six decades. The tuna catch rates have also declined by about 50–90% during the past half-century, with the ocean warming adding further stress to the fish species (FAO, 2018).

As more species are fast getting included in the endangered list (dugong, seals, turtles, and whales are already on the list), nine large marine ecosystems (LMEs) have been formed in the Indian Ocean to campaign for sustainable fishing. These LMEs are Agulhas Current, Somali Coastal Current, Red Sea, Arabian Sea, Bay of Bengal, Gulf of Thailand, West Central Australian Shelf, Northwest Australian Shelf, and Southeast Australian Shelf. This ocean shelters as many as 246 large estuaries. The coasts, including tidal flats, of the Indian Ocean, cover about 3,000 km². The coastal areas produce fish of 20 tons/km². However, coastal areas are being urbanized with populations often exceeding several thousand people per square kilometer. This often proves destructive beyond sustainable levels.

FIGURE 1.6 Life in the Ocean

Top left: Dolphin off Western Australia. Top right: Swarm of Surgeon fish off Maldives. Bottom left: King Penguins on a beach in Crozet Archipelago. Bottom right: Coelacanth. (Wikipedia).

Coral reefs cover about 0.2 million km², while mangrove covers 80,984 km² in the Indian Ocean region (almost half of the world's mangrove habitat), of which 42,500 km² is in Indonesia. Both coral and mangrove shelter fish; however, of late both the ecosystems have been suffering from habitat loss. Interestingly, the Devonian time (about 410 Ma) animal species Coelacanth found once mostly in Comoros Island became extinct at around 66 Ma, but was rediscovered recently from Indonesian waters (Fig. 1.6). These coelacanths, however, have evolved to inhabit a different environment now—they are brown, and their lungs adapted for shallow, brackish waters evolved into gills adapted for deep marine waters.

1.3.4 CLIMATE

While surface current pattern normally controls the sea-surface temperature (SST), the inflows from the Atlantic Ocean, Red Sea, and Antarctic Intermediate Water influence the deepwater circulation in the Indian Ocean. For example, the minimum surface temperature north of 20°S latitude in the Indian Ocean has been

22°C, increasing to 28°C to the east. Southward of 40°S latitude, temperatures drop quickly. Similarly, annual mean surface water salinity ranges from 32 to 37 parts per thousand, the highest occurring in the Arabian Sea and in a belt between southern Africa and southwestern Australia.

The Indian Ocean is unique in many ways. It is a large-scale Tropical Warm Pool (a high-temperature water mass in western Pacific and eastern Indian Oceans whose intensity oscillates in decades), which on interacting with the atmosphere, affects the climate both regionally and globally. The highlands of Asia (Himalayas) block heat export and prevent the ventilation of the Indian Ocean thermocline (a layer in the ocean that divides the upper mixed layer from the calm deep water below). The Himalayan landmass also influenced the evolution of the Indian Ocean monsoon, the strongest on Earth, which causes large-scale seasonal variations in ocean currents, including the reversal of the Somali Current and the Indian Monsoon Current. The Somali Current is a cold ocean boundary current that runs a major upwelling system along the Somali coast and is influenced by the SW (Somali upwelling occurs only during SW monsoon) monsoon in the Indian peninsula (Fig. 1.7). The unique phenomenon of upwelling occurs near the Horn of Africa and the Arabian Peninsula in the Northern Hemisphere and north of the trade winds in the Southern Hemisphere. The Indonesian Through-Flow is a unique Equatorial connection to the Pacific (Schott et al., 2009). During summer, warm continental masses draw moist air from the Indian Ocean, hence producing heavy rainfall. The process is reversed during winter, resulting in dry conditions.

Dominant currents (flow pattern) in this ocean are mainly controlled by the monsoon and include two large circular currents. One of these occurs in the northern hemisphere flowing clockwise and the other to the south of the equator moving counterclockwise. In the North Indian Ocean, there is a complete reversal of the direction of currents between summer and winter, due to the changes of SW and NE monsoon winds. In summer from June to September the dominant wind is the SW monsoon when the currents are blown from a southwesterly direction as the SW Monsoon Drift. The SW monsoon brings rain from the Northern Indian Ocean (that includes both the Arabian Sea and Bay of Bengal) to the Indian subcontinent. When the monsoon winds change, cyclones sometimes strike the shores of the Arabian Sea and the Bay of Bengal. This SW direction of drift reverses in winter, beginning from November, when the NE monsoon blows the currents from the northeast as the NE Monsoon Drift. Currents in the north will then be reversed to anti-clockwise (Wyrtki, 1973; Schott et al., 2009).

The currents of the North Indian Ocean exhibit the dominant effects of winds on the circulation of ocean currents. The Equatorial Current, however, turns southwards past Madagascar as the Agulhas or Mozambique Current and flows eastwards with the West Wind Drift along the equator as the West Australian Current. Tropical cyclones occur during the months May–June and October–November in the northern Indian Ocean and January–February in the southern Indian Ocean. In the southern hemisphere, the winds generally are milder, but summer storms can be devastating at times in the southwest Indian Ocean (Mauritius and Reunion). Fossil fuel and biomass burning in South and Southeast Asia causes much air pollution (also known as the Asian Brown Cloud) that reaches to as far as 60°S. This pollution has implications on both at a local and global scale (Ramanathan et al., 2001).

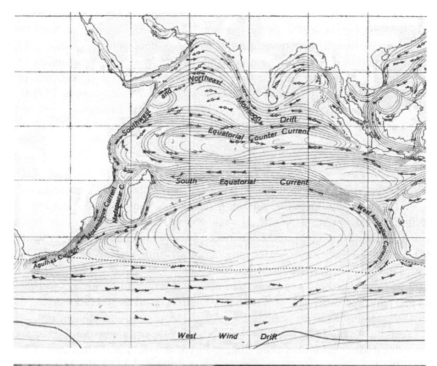

FIGURE 1.7   Currents in the Indian Ocean

Top: The Somali Current off the Somali coast in the context of the Indian Ocean Gyre during (northern) summer. The circular current east of the Horn of Africa is known as the Great Whirl. Bottom: Southward Somali Current during winter Northeast Indian Monsoon (after Mann and Lazier, 2006).

The SW monsoon is extremely important for the entire South Asian economy, which accounts for about 80% of the total annual rainfall in this region with strong bearing on socioeconomic development in these developing countries. Many civilizations perished when the monsoon failed in the past. Huge variability in the Indian Summer Monsoon has occurred pre-historically, with a strong, wet phase during 33,500–32,500 BP (before present); a weak, dry phase between 26,000 and 23,500 BP; and a very weak phase from 17,000 to 15,000 BP, corresponding to a series of dramatic global events (Naidu and Govil, 2010; Sinha et al., 2011).

1.3.5 HISTORICAL ACTIVITIES

The Indian Ocean is credited as being the storehouse of the world's earliest civilization, starting with Mesopotamia, Sumer, Egypt, Indus Valley, Tigris-Euphrates, and the Nile. The earliest known maritime trade is also recorded in this ocean between Egypt and Somalia (c. 3,000 BC), and between Mesopotamia and the Indus Valley (c. 2,500 BC). One of the reasons for the ease of doing sailing and business in the Indian Ocean is that this ocean is far calmer than the Atlantic or Pacific oceans. Sailors used to take advantage of both southwest and northeast powerful monsoons to travel to India and return, respectively. This probably facilitated Indian and Indonesian people to cross the Indian Ocean to settle in East Africa, Mauritius, and Madagascar.

It seems the Arabia-India sea route was opened by the first and second centuries BC, as intensive trade relations between Roman-Egypt and the Tamil kingdoms (Cheras, Cholas, Pandyas, etc.) in India started. Between 1405 and 1433, the Ming Dynasty of China set out for a voyage to East Africa through the Indian Ocean. In the summer of 1498, the Portuguese sailor Vasco da Gama reached Calicut, India, after successfully sailing around the southern tip of Africa (Cape of Good Hope), bypassing the Muslim nations who were controlling the overland spice trade. Thereafter, Portugal, Holland, and France, armed with heavy cannons, dominated the Indian Ocean for trade and commerce. But by 1815, the British started controlling the trade, and security scenes, and with the opening of the Suez Canal in 1869 Europe increased the trade load to Asia several times. However, since World War II, the United Kingdom largely withdrew from the area, as India, the erstwhile USSR, and the USA took control. Meanwhile, developing countries bordering this ocean insisted to support this ocean as a "zone of peace for trade only," however not before the UK and USA set up a military base in Diego Garcia.

The Indian Ocean, together with the Mediterranean Sea and the Persian Gulf, has connected people since ancient times. The history of the concept of an Indian Ocean World (IOW; Campbell, 2017) can be subdivided into Portuguese, Dutch, and British periods. This IOW is also referred to as the "first global economy" wherein monsoon-linked trade connected Asia, China, India, and Mesopotamia. The IOW developed independently until European colonial dominance in the nineteenth century. The Indian Ocean, hence, remains a cradle of the diverse history of a unique mix of cultures, ethnical groups, natural resources, and shipping routes of about seven thousand years (Boivin et al., 2014).

Modern humans spread from Africa along the northern rim of the Indian Ocean. This migration from the coast began in East Africa about 75,000 years ago and occurred intermittently from estuary to estuary along the northern perimeter of the Indian Ocean at a rate of 0.7–4.0 km per year (McPherson, 1984). The fallout of the extremely violent Toba volcanic eruption about 74,000 years ago in Sumatra (Indonesia) that spewed thousands of tons of ash into the atmosphere, and created a decade-long winter, led to massive die-offs of vegetation. It was possible for a catastrophe of this scale to prune a few branches off the early human family tree and cause human migration. It may have eventually resulted in modern humans migrating from Sunda to Sahul (Southeast Asia) to Australia. Rapid migration probably happened during the Late Pleistocene (11,000 years ago) as supported by the DNA results. Waves of migration continued afterwards leading to the Indian Ocean coastal areas getting inhabited long before the first civilizations emerged. In fact, after about 5,000–6,000 years, six distinct cultural centers had evolved around the Indian Ocean: East Africa, the Middle East, the Indian Subcontinent, Southeast Asia, the Malay World, and Australia (Boivin et al., 2014; Campbell, 2017).

Grains, pottery, copper, stone, timber, tin, dates, onions, and pearls were traded from Egypt and Persia by the Sumerians to the Indian subcontinent of the Indus Valley civilization (2600–1900 BC). The archipelagos in the central Indian Ocean, the Laccadive and Maldives islands were probably populated during the second century BC from the Indian mainland. The Maldives became a major center of trade and commerce (Campbell, 2017).

About 4,000 years ago, food globalization occurred in the Indian Ocean. Taking advantage of the cultural centers five African crops—pearl millet, finger millet, hyacinth bean, sorghum, and cowpea—were taken to Gujarat in India during the Late Harappan time (2000–1700 BC). Gujarati merchants became the first to trade African goods such as ivory, tortoise shells, and slaves. Broom-corn millet found its way from Central Asia to Africa, together with chicken and zebu cattle, although the exact timing is disputed. Around 2,000 BC, black pepper and sesame, both native to Asia, were grown in Egypt, albeit in small quantities. Around the same time, the black rat and house mouse emigrated from Asia to Egypt. Banana reached Africa around 3,000 years ago (Boivin et al., 2014).

Between 7,400 and 2,900 years ago, as many as 11 tsunamis struck the Indian Ocean. Geological evidence suggests they may have been caused by large extra-terrestrial meteoritic impacts (Burckle Crater in the southern Indian Ocean in 2,800 BC and the Kanmare and Tabban craters in the Gulf of Carpentaria in northern Australia in 536 BC) resulting in waves extending to 45 km inland. The flooding must have disrupted human settlements and migration pattern (Gusiakov et al., 2009; Rubin et al., 2017).

As was seen above, the Indian Ocean provides major sea routes connecting the Middle East, Africa, and East Asia with Europe and the Americas, and draws its strategic and economic significance from such an enviable place. Petroleum and petroleum products are transported mostly from the oilfields of the Persian Gulf and Indonesia. In addition, offshore oil production of Saudi Arabia, India, Iran, and Australia has been significant. An estimated 40% of the world's offshore oil production comes from the Indian Ocean. Moreover, heavy minerals (rutile, zircon, titanium, diamond, gold, garnet, iron, monazite, thorium, ruby, sapphire, tin, zirconium, uranium, and rare earth elements), sand, and gravel from the beaches and offshore placers extracted by India, South Africa, Indonesia, Sri Lanka, and Thailand are transported through this ocean. Fishing fleets from several nations from outside the Indian Ocean rim transcend to this ocean to catch fish (mostly shrimp and tuna) as relatively warm Indian Ocean surface water keeps phytoplankton production low. Giving communication a fillip, submarine cables from India to the United Arab Emirates and Malaysia, and from Sri Lanka to Djibouti and Indonesia were in place by the start of the last century.

The establishment of the Dutch East India Company in the early seventeenth century led to a quick increase in trade volume, as more slaves from the Indian Ocean countries (about 0.5 million) were transported to work in the Dutch colonies. For example, some 4,000 African slaves were used to build the Colombo Fortress in Sri Lanka. Indian and Chinese traders supplied Dutch Indonesia with thousands of laborers during the seventeenth and eighteenth centuries (Allen, 2017).

The British East India Company was established during the same period. The British mostly brought slaves from Africa and islands in the Indian Ocean to India and Indonesia but also exported slaves from India. In fact, in 1622 its ship first carried slaves from South India to Indonesia. The French colonized Réunion and Mauritius in 1721, and by the year 1807 the slave population in the Mascarene Islands reached 0.13 million. The British captured the islands in 1810, yet the slave trade did not end there despite a ban by the British Parliament to this effect a few years earlier. In all, Europeans traded about 0.57–0.73 million slaves within the Indian Ocean between 1500 and 1850, and a comparable number of slaves were exported from the Indian Ocean to the Americas during the same period. The slave trade in the Indian Ocean was, nevertheless, very limited compared to about 12 million slaves exported across the Atlantic (Allen, 2017).

With the opening of the Suez Canal in 1869, a new era in shipping dawned. Because the majority of the world's goods are transported via sea, the Suez Canal greatly reduced the time and cost of transporting goods between Europe and countries on the Indian and Pacific Oceans. When it opened, the canal was only 25 feet deep, 72 feet wide at the bottom, and 200 to 300 feet wide at the surface; at present it is around 24 meters deep. Additionally, the Industrial Revolution dramatically changed global shipping with the introduction of new food items, industrial items, gemstones, jewelry, and tourism. Throughout the colonial era, islands such as Maldives and Mauritius were important shipping nodes for the Dutch, French, and British. Mauritius, an inhabited island, became populated by slaves from Africa and indentured labor from India. The end of World War II marked the end of the colonial era. The British left Mauritius in 1974 and, with 70% of the population of Indian descent, Mauritius became a close ally of India.

In the twenty-first century China has taken up the cudgel to lead trade and commerce by reviving the centuries-old Silk route (Fig. 1.8; details in Chapter 5) and investing enormous financial and technological resources to actually build the necessary infrastructure to carry out trade in Europe, Asia, and in the Indian Ocean through rail, road, and sea networks. The economic and strategic significance of the Indian Ocean is listed and discussed in the following chapters.

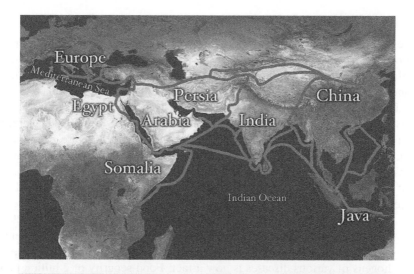

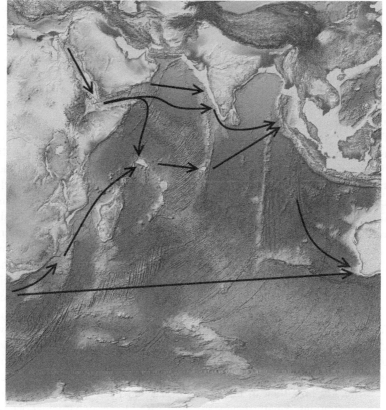

FIGURE 1.8   The Silk Road and New Shipping Routes

Note: The economically important Silk Road (top) was blocked from Europe by the Ottoman Empire in 1453. This spurred exploration, and a new sea route around Africa was found (bottom).

1.4   DESIGNING A BLUE ECONOMY VISION FOR THE IOR

As the world prioritized its focus, after almost two and half decades of procrastination on poverty alleviation on an urgent basis, the philosophy of a green economy in a blue world set the ball rolling. At present about 40% of the world's population and about 50% of the IOR's population live within 100 kilometers of the coast, landward. Pressure on coastal ecosystems would increase with the upsurge in population density and economic activity in the coastal zone. These pressures include increasing loads of pollution, unsustainable development of coastal infrastructure, ocean acidification, disruption in the process of carbon sequestration, and introduction of invasive species through ballast water released by ships. The above factors would impact on loss of biodiversity and sustainability of oceanic environment, resulting in land cover change, human migration, and habitat conversion.

For example, about 20% coverage of corals, 30–50% of mangroves, and 29% of seagrass have disappeared from world oceans since the last quarter of the twentieth century (IPCC, 2014). A threat to human health through toxins in fish and shellfish and pathogens of various diseases is now a fact. Food security of 2 billion people living in the coastal zone will be of paramount challenge for governments worldwide. The high population concentration in the low elevation coastal zone (defined as less than 10 meters elevation) increases a country's vulnerability to sea-level rise and other coastal hazards such as cyclones, storm surges, and any military invasion.

All these threats and consequent opportunities would need a strategic vision (UNCSD, 2017). Some of the important global threats for the blue economy are anomalies in carbon sequestration by seawater causing ocean acidification and formation of hypoxic (low-oxygen) zones, dramatic variations in extreme climatic events (storms, cyclones), rise in sea level, low production of fish, increase in pollution (especially waste material, micro-plastics, ballast water bio-invasion), and the host of problems associated with shipping, commerce, and geopolitics. Impact of climate change, unsustainable extraction of resources, increased marine pollution, insufficient investment in human capital and skill development, inadequate valuation of marine resources and services, poor technical capacities of human/institution, and safety and security issues are a few of the challenges posed to the blue economy.

A blue economy approach must fully anticipate and incorporate the impacts of climate change on marine and coastal ecosystems—both observed and anticipated (Table 1.4). Among these factors, the impacts of acidification on carbon-dependent ecological processes are likely to be the most severe and most widespread. Likewise, the effects of sea-level change will be felt differently in different parts of the world, depending on the ecosystems around which it occurs, and sub-surface geological structure. These impacts will fundamentally affect the management of blue economy resources. In this regard, participation and cooperation among the IOR countries will be extremely important to manage, protect, and keep oceans for the present and future generations. Small island states in the IOR have, in particular, vast ocean resources, compared to their landmass, at their disposal. This presents a huge opportunity to boost their economic growth and to tackle unemployment, food security, and poverty issues (Table 1.5).

TABLE 1.5
An Example of Socioeconomic Returns from the Blue Economy

Country	Blue Economy Magnitude			Employment	
	Year	Output	GDP (%)	Year	No.
Australia	2004	17.00	3.6	-	-
Canada	2004	15.98	1.5	2006	171,365
China	2010	239.09	4.0	2010	9,253,000
France	2006	16.69	1.4	2009	459,358
Ireland	2007	1.9	1.0	2007	17,000
New Zealand	2006	2.14	2.0	-	-
United Kingdom	2008	84.27	4.2	2006	548,674
United States	2009	138.0	1.2	2010	2,770,000

Notes: Output = in US$ billion; GDP (%) = percentage of GDP; Employment = only indicative.
Source: Modified after Report and Information System for Developing Countries (2017, RIS: New Delhi).

The IOR is alive to the great prospect of the blue economy. The First IORA Blue Economy Dialogue on Prospects of Blue Economy in the Indian Ocean was held in Goa, India on August 17–18, 2015. The dialogue deliberated on key aspects of the blue economy that included an accounting framework; fisheries & aquaculture; renewable ocean energy; ports, shipping, and manufacturing services; and seabed explorations and minerals. Immediately after this the First IORA (Indian Ocean Rim Association) Ministerial Blue Economy Conference (BEC) was held in Mauritius on September 2–3, 2015 where the Blue Economy Declaration was adopted (see IORA portal, www.iora.int). Reflecting on the global trends, this Declaration seeks to harness oceans and maritime resources to drive economic growth, job creation, and innovation, while safeguarding sustainability and environmental protection.

Indonesia hosted the Second Ministerial Blue Economy Conference on "Financing the Blue Economy" on May 8–10, 2017 in Jakarta. The Jakarta Declaration aimed at optimizing the use of existing financial instruments in the IORA region to enhance blue economy development in IOR member states. The need for new and creative financing mechanisms and for strengthening collaboration between the public and private sectors, as well as Dialogue Partners, was also highlighted.

The IORA Secretariat has identified the following six priority pillars in the blue economy: Fisheries and Aquaculture; Renewable Ocean Energy; Seaports and Shipping; Minerals and Hydrocarbons; Tourism; and Marine Biotechnology.

However, the essential philosophy is to consider the blue economy beyond a mechanism for simple economic growth. The major environmental problems seen in the open are only symptoms of a far more major disease that is living and growing within human beings. Selfishness, greed, and apathy are the real diseases that need a

quick remedy. A substantial degree of educational, cultural, and spiritual transformation is required to address this lacuna.

The present book discusses all these issues in the light of increasing populations
and their aspirations in the IOR, and the threat and security perspectives in the
region. The book also sketches a revised and enriched version of the blue economy
of what was envisaged by UNCSD in 2012. Such a version based on the BTVG doctrine (Buddha-Tagore-Vivekananda-Gandhi) of ethical governance is proposed to
make business under the blue economy in the IOR region not only sustainable but
also responsible. This book travels through these uncertainties and helps to lay a
plausible roadmap for IOR countries in their quest toward a successful blue economy.

2 The Indian Scenario

Like its atypical geography, geology, climate, and human history, the Indian Ocean is also significant due to its rich economic potential in terms of living, non-living, and energy resources, and scope for expansive maritime activities (Fig. 2.1, Tables 2.1, 1.2; see Mohanty et al., 2015 and references therein). All these entail promising economic and employment opportunities to a country like India. Innovation is considered foremost in making the blue economy successful, with regard to the techno-commercial feasibility of tapping the coastal and marine resources. In this chapter, the present status of the various sectors under the blue economy paradigm in India is discussed. While the status and prospect of the blue economy in ten IOR countries are presented in Chapter 3, the same for another ten countries located outside the Indian Ocean are assessed in Chapter 4.

2.1 LIVING RESOURCES

The oceans are a rich repository of living resources and these encompass the phytoplankton (minute plants that are at the bottom of the marine food chain) to gigantic mammals such as the blue whale. The majority of the population, however, are more interested in fish and fisheries, the most important member of this food chain. In 2016, about 59.6 million people were engaged globally in the primary sector of capture fisheries and aquaculture, with 40.3 million people in fisheries, and the remaining 19.3 million people engaged in aquaculture. And of this, 36% were involved full time, 23% part-time, and the rest were either occasional fisher-folks or stay unspecified (FAO, 2018; Fig. 2.2).

2.1.1 CAPTURED FISHERY

The global coastal inhabitants are largely dependent on fishing and this leads to the spawning of associated activities such as fabrication of nets, trawlers, boats, catamarans, manufacturing of ice for preservation of the catch and its storage, and transport of the catch far and wide. These activities lead to direct and indirect employment and livelihood not only for the local populace but also for migrant workers. Fish, being a major source of protein, is much consumed and sometimes the demand exceeds supply and results in overexploitation. With the

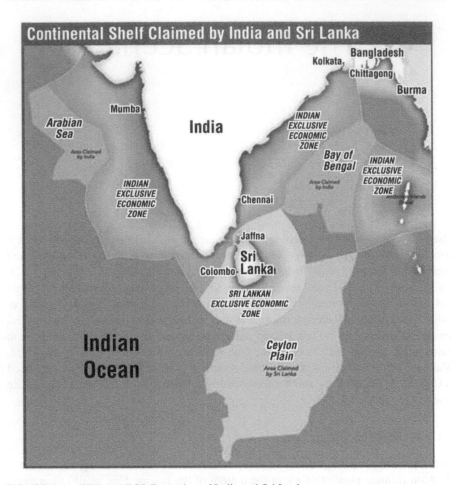

FIGURE 2.1   EEZ and ECS Extension of India and Sri Lanka

Notes: The exclusive economic zone (EEZ) extension of India around its peninsular coast and that of Andaman & Nicobar Islands (bordered by deep pink) is shown. The extended continental shelf (ECS) as claimed by India in the Arabian Sea and Bay of Bengal are shown in light greyish pink outside the EEZ. The ECS area in the Bay of Bengal could be a zone of dispute with neighboring Bangladesh and Myanmar. The EEZ and ECS as claimed by Sri Lanka are also shown.

advent of hi-tech fishing equipment, the catch goes up several notches and sadly even juvenile fish is not spared. In the Indian Ocean, India, China, Indonesia, Thailand, Bangladesh, and Japan dominate fish production and trade. In 2014, of the global population engaged in fisheries and aquaculture, 84% was from Asia, followed by Africa (10%) and Latin America and the Caribbean (4% each). In Asia alone, 94% of the 18 million people were occupied in fish farming (FAO, 2016; Table 2.2).

TABLE 2.1
Scope and Activities of the Different Sectors of the Blue Economy for India

Sectors	Scope and Activities
Ancillary	Offshore financial services (banking, commerce, insurance), consultancy services (hydrographic surveys, environmental, turnkey projects, legal), information and communication technology (ICT), ships, boats and submarines construction, salvage and ship breaking, weather forecast including ocean-observing satellites.
Aquaculture/Mariculture	Increase in production of fish and fish products for consumption within a country and for export.
Research and Education	Education, training, research, development, innovation, patents.
Fishing	Capture fishery, processing of fish and products for local use and exports.
Marine: Bio-products, Bio-technology, Pharmaceuticals	Pharma products and chemicals from marine organisms (fish, corals, bacteria, etc), seagrass, seaweed.
Marine Construction	Construction of ports, harbors, breakwaters, jetties, rigs, pipelines, etc.
Marine Energy	Wave, ocean thermal energy conversion, offshore wind, offshore solar, tidal, salinity power.
Marine Minerals (coast to deep-sea)	Coastal placers (ilmenite, diamond, gold, chromite, tin, gemstones), deep-sea minerals (hydrothermal deposits, cobalt, polymetallic nodules), fossil fuels (oil, gas, hydrocarbons), gas hydrates (methane gas).
Marine Tourism & Recreation	Boating, fishing, sailing, surfing, skiing, scuba diving, swimming, bird watching, trips to islands.
Maritime Culture and Heritage	Archaeological (land and marine), history.
Maritime Safety and Security	Avoidance of sea piracy, intra-country collaboration.
Maritime Trade and Commerce	Imports, exports, roll-on roll-off services.

Globally, pelagic fish contribute 52%, demersal fish 29%, crustaceans 12%, and mollusk 7% in marine fish production. In India, 82% of fish are captured by mechanized trawlers, 12% by motorized boats, and the remaining 1% by man-powered canoe that ply around the coasts (Fig. 2.2; CMFRI, 2016). Demersal fish live and feed on or near the bottom of seas or lakes, while pelagic fish live and feed away from the bottom in the water column. Fisheries in India cater to the livelihood of about 3.79 million fisher population, meet nutritional requirements of a significant part of her population, result in large export earnings, and provide direct employment to about 1 million people.

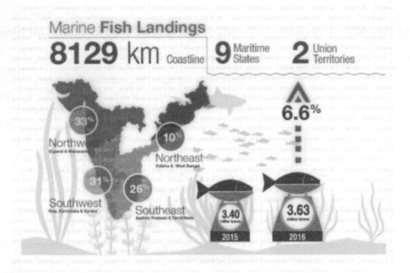

FIGURE 2.2 Marine Fish Landings in India (www.cmfri.org.in/infographics/2)

Note: Landings increased by 0.23 million tons in 2016 from the earlier year.

TABLE 2.2
Fish Catch from Indian Ocean in Comparison to Other Oceans

	Arctic	Pacific	Atlantic	Indian	Southern	Total
Area	14.1	155.6	76.8	68.6	2,1470.3	335.3
%	4.2	46.4	22.9	20.4	6.10	100
Volume	17.00	679.6	313.4	269.3	91.5	1,370.8
%	1.2	49.7	22.6	19.7	6.8	100
Avg. water depth (km)	1.21	4.37	4.08	3.93	4.51	3.62
Max. water depth (m)	4,665	10,924	8,605	7,258	7,235	10,924
Coastline (km)	45,389	135,663	111,866	66,526	17,698	356,000
Fish Capture	–	84,234	24,045	10,197	0.147	118,623
%	–	71	20.3	8.6	0.1	99.00

Notes: Area in million km², volume in million km², fish capture in million tons.

In 2016–17, inland fisheries in India accounted for 68.1% while 31.9% of the resources came from the ocean. At the primary level, the fishery sector employs 11.41 million people and several more along the value chain. The number of landing centers was 1,537 and the number of fishing villages was 3,432. Marine fishery potential in Indian waters was estimated at 4.41 million metric tons (MMT) of which 47% is demersal, 48% pelagic, and 5% oceanic groups (Annual Government Report [AGR], 2016–17).

During 2016–17, India's fishery export was US$37.88 billion, which is about 0.92% of the National Gross Value Added (GVA). India is the second-largest producer of marine fish and also freshwater fish in the world. Fish production increased from 0.415 MMT (0.244 MMT for marine and 0.171 for inland fisheries) in 1991–2 to 11.41 MMT (3.64 MMT for marine and 7.77 MMT for inland fisheries). During the first two quarters of 2017–18 fish production was 5.80 million tons.

India's rank as the second-largest fish producer in the world is nothing to cheer about since it is only a tenth of the production by China, the world's no. 1. In 2012, China had produced 57 million tons of fish and marine algae, which was 63% of its total aquaculture production. Excluding marine algae, China's production of food fish was 41 million tons, while India produced nearly 4.2 million tons, just around 10% of Chinese production. Consequently, India has been showing only a small increase every year from 4.16 million tons in 1991–2 which increased to 9.04 million tons in 2012–13. However, while the marine fish capture has increased by only about 36%, there was a 234% rise in inland fish production. This indicates the potential for the development of the fishery sector in India—a sleeping giant (The Telegraph, 2019).

In 2017, the estimated marine fish landings for peninsular India increased up to 3.83 million tons, which was the second highest ever, slightly below 3.92 million tons in 2012, but was 5.6% more as compared to 2016. In the same year, Gujarat with 0.786 million tons of marine fish landings ranked first among the maritime states in India for the fifth consecutive year (FRAD-CMFRI, 2018; Fig. 2.2). According to the Marine Products Export Development Authority (MPEDA, Department of Commerce, Ministry of Commerce and Industry, Government of India) the quantity of fish caught during 2017–18 was to the tune of 1.38 million tons. The catch included shrimps, squid, finfish, cattle fish, dried items, live items, chilled items, and others and these were exported from 38 points to Japan, the USA, EU, China, Southeast Asia, the Middle East, and other countries. The monetary value of export was US$7,081 million (www.mpeda.gov.in).

The National Fisheries Development Board (NFBD, www.nfbd.gov.in) was set up in 2006 to tap the inland and marine fisheries sectors, and fish capture, culture, processing, and marketing of fish. The overall growth of the sector was enhanced with the application of modern tools including biotechnology (AGR, 2016–17). NFBD prepared a detailed integrated national Fisheries Action Plan 2016 to achieve 15 million tons of fish production by 2019–20.

The fishing ban in India spans about 45–61 days, i.e., between June 15 and July 31 on the west coast of India, and from April 15 until about May 31 on the east coast. The ban helps to curb large-scale fishing that affects the replenishing of the fish population during their breeding season. The ban however does not apply to traditional methods of fishing in territorial waters, where catamarans and non-motorized vessels are used that cause minimal damage to the marine environment. The ban assists to revive the marine lives and restricts fishermen from venturing too far into the rough sea during monsoons. This ban, combined with other management measures such as an ecosystem-based approach, marine protected areas, no-take zones, certification, mesh-size regulation, and minimum legal size at capture, can help effectively arrest the decline in fish resources (www.indiawaterportal.org). The moratorium on fishing, although it affects the country's economy and food availability for the people for a short duration, is necessary to augment the health of the seas and oceans for the remaining months of the year. Hence, the blue economy approach of sustainability to the fisheries activities is vital. To achieve short-term gains one may need to sacrifice the long-term benefits.

Any coastal country has a right to fish within its EEZ (exclusive economic zone) but since the demarcation of EEZs in sea is difficult, the fisher folks could inadvertently cross the EEZ or are tempted to do so in the hope of netting a bigger catch. This is often the case between the Indian states, and also between India and Sri Lanka, and India and Pakistan. The crossovers result in legal problems and high-level diplomacy is needed to free the arrested fisher folks. To avoid such infringements and encourage the blue economy, the IOR countries need to have mutual cooperation by not only sharing technology, knowledge of fish types and catch, and work toward a combined capital investment. All fishing vessels must have satellite-navigated global positioning systems to avoid crossing over to a neighbor's EEZ.

India is associated with several regional and international bodies such as FAO of the UN and Indian Ocean Tuna Commission. India hosts the regional initiative of the Bay of Bengal Program–Inter-governmental Organization (BOBP-IGO). Its mandate is to increase cooperation among member countries, other countries, and organizations in the region and offer technical and management advisories for sustainable development and management of fisheries in the Bay of Bengal. India is also an active member in the Bay of Bengal Large Marine Ecosystem (BOBLME) program. Fisheries issues are debated in the South Asian Association for Regional Cooperation (SAARC), Bay of Bengal Initiative for Multi-Sectoral Technical and Economic Cooperation (BIMSTEC), and Indian Ocean Rim Association (IORA).

2.1.2 MARICULTURE

To even out increasing demand and fall in supply (due to fishing restriction) of fishes in the market, there has been ample scope to boost marine aquaculture or mariculture. This sector not only creates direct and indirect employments but also augments fish catch and fish products. The Indian Ocean being an area of rich biodiversity could better cater to markets of the USA and Europe through sharing and the exchange of modern technology about mariculture. In 2005, the Government of India established the Coastal Aquaculture Authority (CAA) and defined coastal aquaculture as

> culturing, under controlled conditions in ponds, pens, enclosures or otherwise, in coastal areas, of shrimp, prawn, fish or any other aquatic life in saline or brackish water; but does not include freshwater aquaculture. While aquaculture is related to freshwater culture, mariculture is identified strictly with seawater.
>
> (De Silva, 1998)

Mariculture over the years was mainly restricted to mussels, edible oysters and pearl oyster, and to some extent seaweeds. In the last decade, there has been a significant thrust on the development of cage farming (AGR, 2016–17). Since its inception, the CAA registered a total of 34,784 farms.

During 2015–16, the nine maritime states of India (Gujarat, Maharashtra, Goa, Karnataka, Kerala, Tamil Nadu, Andhra Pradesh, Odisha, and West Bengal) produced 81,452 metric tons (MT) of tiger shrimp. The state of Gujarat recorded maximum productivity of 3.12 tons/0.01 km^2/year, followed by Tamil Nadu and Odisha with 2.70 and 2.02 tons/0.01 km^2/year, respectively. Andhra Pradesh came first in production of shrimp (*Litopenaeus vannamei*) during the same period. The maritime states in India produced a total of 406,018 MT of shrimp. Again, the cumulative scampi (freshwater prawn) production from nine states was 10,152 tons. West Bengal was the main producer followed by Odisha and Andhra Pradesh (www.mpeda.gov.in).

Mariculture has shown exponential growth due to constraints on capture fishing and to meet the demand and supply market for fish. Mariculture produced 44.1% of total fish production in the world in 2014, up from 42.1% in 2012 and 31.1% in 2004. All continents have shown a general trend of an increasing share of mariculture production except in Oceania. In 2014, of the top 25 producers and main groups of farmed species (includes fish and aquatic animals and plants), India stood third with 4,884,000 tons after China and Indonesia (FAO, 2016). It is because while India cultures fewer than ten species of fish, China cultures over a hundred species on a commercial scale.

Global mariculture production (including aquatic plants) in 2016 was 110.2 million tons, with the first-sale value estimated at US$243.5 billion. The total production included 80 million tons of food fish (US$231.6 billion) and 30.1 million tons of aquatic plants (US$11.7 billion) as well as 37,900 tons of non-food products (US$214.6 million; FAO, 2018). Recently India has initiated the Blue Revolution which has multidimensional activities with a focus to increase inland and marine fisheries production (www.nfbd.gov.in).

2.1.3 BIO-PRODUCTS, PHARMACEUTICALS, AND BIOTECHNOLOGY

The marine ecosystem with its unrestrained biological and chemical diversities is a rich source of exotic chemicals that could be potentially tapped for use in several industries, such as agrochemicals, food additives, pesticides, fine chemicals, cosmetics, pharmaceuticals, nutritional supplements (nutraceuticals), and insecticides. In addition, biopolymers find a range of application in the dressing of wounds, tissue regeneration, bio-adhesives and biodegradable plastics (Libes, 2009). Pharmaceutical and dietary products from red algae, seaweeds, jellyfish, microalgae, sponges, etc., could be developed for curing ailments such as diabetes, cancer, viral and bacterial infections, allergy, inflammation, etc. (Sarvanan and Debnath, 2013). But tapping all these would involve large investments in bio-prospecting and biosynthesis.

Marine organisms have long been used in the production of traditional medicines in India, Near East, China, and Europe in various forms such as infusions, decoctions, juices, tablets, pills, powders, oils, syrups, fermented liquids, bhasmas (ash) among others (www.indianmedicine.nic.in). For instance, Hippocrates reported that juices from various species of mollusks were commonly used as a laxative, extracts from the sea hare (Aplasia) were used as a depilatory, and essences from gastropod opercula have been used in perfume and incense, while Hebrews and Romans obtained blue and purple dyes from certain marine snails.

The discovery of the sponge-derived nucleosides spongothymidine and spongouridine was made by Bergmann and Feeney (1951). Subsequently thousands of marine natural products (MNP) from marine organisms have been reported, extracted, and patented. This has led to the emerging field of marine biotechnology that has found favor with the pharmaceutical industry (Libes, 2009). For example, diatomaceous earth is used as an insecticide, as a filtration medium, and as an abrasive. Varieties of seaweeds are used to treat dropsy, menstrual difficulties, gastrointestinal disorders, abscesses, and cancer; while sponges have been used to treat tumors, goiter, dysentery, diarrhea, stanching of blood flow, and as contraceptives.

Marine biotechnology has a large role to play and concerns five aspects that are of societal importance, viz. food, energy, health, environment, and industrial products and processes. In 2015 the global market of marine biotechnology was estimated at US$4.1 billion and was projected to reach US$4.8 billion by 2020 and US$6.4 billion by 2025. Some stumbling blocks for marine biotechnology are availability of and access to marine organisms, the dearth of technology for bio-prospecting and bio-screening, de-replication, duplication of work, and protection of intellectual property (Smithers Group, 2015). Governments of various countries are looking for opportunities to have a sustainable economic growth, and in this endeavor, they are also aware of the threat to Earth's ecosystems. Several promising prospects are found in what is termed the "bio-economy," where the application of biotechnology to biomass is resulting in new products and services for population health, sustainable industries, and primary production (Hurst et al., 2016).

It is estimated that marine products make up 32% of the nutraceuticals market (BioMarine Organization, 2012). These products would include food or nutritional supplements, special dietary foods, sports drinks, functional foods, and medically formulated foods. By 2020 the market for Omega 3 poly-unsaturated fatty acid is projected to grow at nearly 14% per annum and would reach nearly $19 billion (KPMG International, 2015). Marine compounds have a long history of use in the cosmetics sector (Balboa et al., 2015). The market for cosmeceuticals is a fast-growing segment of the worldwide cosmetics industry. In 2010 the global market was US$30.5 billion and it was projected to grow at a rate of around 9% per annum between 2015 and 2020 (Global Information, 2015).

Due to the ease of collection and cost, the present-day MNP is from shallow and tropical marine organisms but there is evidence that MNP could also be sourced from deep-sea organisms including microbes. Several MNP have been identified but due to technological and commercial problems, only a few are available in the market and some are in clinical trial stages especially for cancer treatments. Approximately 150 of the MNPs are known to be cytotoxic and about 35 act against tumor cells (Libes, 2009). Some of the commercially available MNP are Ara-A (acyclovir) from marine sponges Tethys crypta as an antiviral drug used in herpes infections and Ara-C (cytosar-U, cytarabine) as anticancer drug used in leukemia and non-Hodgkin's lymphoma (Gopal et al., 2008; Murti and Agarwal, 2010). Even the protective mucus that coats young fish may sometimes contain bacteria that could offer potential new antibiotics. The bacterial strains could be helpful against fungus pathogenic to humans and in inhibition of methicillin-resistant S Aureus (MRSA, www.acs.org).

It was believed that globally there are only 35,000 marine microbial species and 80 bacterial phyla. Recently, Hong Kong University of Science and Technology discovered 7,300 new biofilm-forming species and 10 new bacterial phyla from the water samples of the Pacific, Atlantic, and Indian oceans. The new species also included acido-bacteria, which was known to be present in terrestrial soils that are used to develop new antibiotics and anti-tumor drugs. Acido-bacteria are the first ocean species to contain the gene-editing systems that could help develop new drugs (Zhang et al., 2019). Such an unearthing of new species raises the hope that in future several species of potential use to humans could be found in the oceans.

Hence, in the above aspects, the IORA countries could devise a strategy for joint work with India and China taking a lead to source and tap the marine organisms from the Indian Ocean that are of potential interest in marine biotechnology. Such a measure would help to reduce costs and also avoid duplication of efforts by different countries. Additionally, Small Islands Developing States (SIDS) could be used as incubation and manufacturing centers of drugs from these marine organisms.

2.2 NON-LIVING RESOURCES

The visible and edible living resources are well known to the public, but what is less familiar is a rich repository of several minerals and metals along the coasts and in nearby seas and oceans of India (Fig. 2.3). These resources have day-to-day usage, and could be mined on a commercial scale and profits made. Such mining would reduce the pressure to locate land-based deposits, some of which have either disappeared or vanishing rapidly. Moreover, it would also lessen anxiety about impacts of land-mining on the environment, concerns regarding catalyzing climate change, and most importantly help to further research to look for alternate minerals and metals to replace commonly used ones. To tap marine resources and energies, there is a need to develop and innovate ocean technology and, in this endeavor, India has good prospects.

A majority of the countries concurred with the UNCLOS document (United Nations Convention on the Law of the Sea—the UN regulatory body on ocean management) that was signed on December 10, 1982 in Jamaica and came into force on November 16, 1994. UNCLOS proposed a detailed legal framework for rights and obligations of countries to access, use, and reclaim marine resources in territorial waters and open oceans. As per Article 76 of UNCLOS III, the ocean space under the jurisdiction of a country was classified into several maritime zones (Table 1.2). A coastal nation has the complete rights over resources that can be derived from air, water column, seabed, and sub-surface from her respective Coastal Waters (5.55 km into the sea from the coast), Territorial Sea (5.55 to 22.2 km), and Contiguous Zone (22.2 to 44.4 km). Respective nations can have resources from the water column, seabed, and sub-surface that occur within their Exclusive Economic Zone (EEZ, 44.4 to 370 km), while resources only from the seabed and sub-surface strata can be exploited from the Extended Continental Shelf (ECS/CS, 370 to 647.5 km).

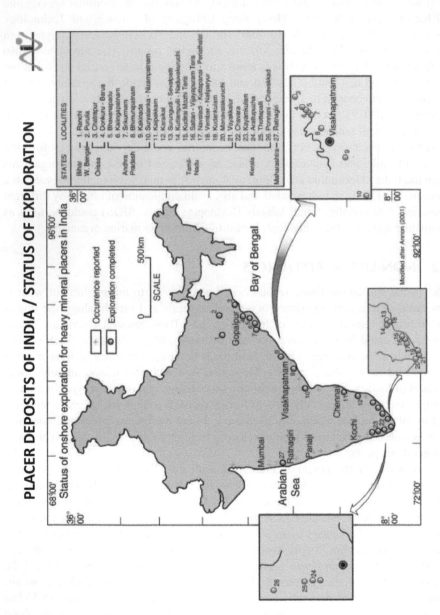

FIGURE 2.3 India, Its Coastline and Placer Mineral Resources (Geological Survey of India).

Resources available beyond this distance from the coast are kept reserved for the common heritage of mankind, and cannot be mined by any country till specifically granted permission by UNCLOS (Article 76, UNCLOS 1982–94). A blue economy involves harnessing of all the resources spread across these five zones in an efficient, best, responsible, and sustainable way.

2.2.1 COASTAL PLACERS AND OFFSHORE MINERALS

Various mineral deposits occur on beaches, continental shelves and slopes, within and outside the EEZ, and in deep-sea (Fig. 2.3). The commonly found beach materials are silica sands, shells, and placer mineral deposits. The minerals found on the beach (and in near shore) are eroded and transported from the hinterland rocks. These minerals in course of time (in millions of years) get concentrated to form placer deposits that could be economically exploited for commercial and industrial purposes. Some examples of placers that are mined are tin by Indonesia and Myanmar, gold by Alaska, tungsten by Myanmar, and precious and semi-precious gems by Sri Lanka. Certain regions of the west and east coasts of India hold significant quantities of magnetite (iron mineral) and ilmenite (titanium) in Maharashtra and Odisha. In addition, chromite in Maharashtra, zircon (zirconium containing mineral used as artificial diamond), monazite with rare earth elements, and thorium (a radioactive element) are found in southern regions of Tamil Nadu, Kerala, and Odisha. Furthermore, garnet, zircon, and ilmenite are found in Tamil Nadu, and also in the offshore areas of Maharashtra (e.g., Ratnagiri; Gujar et al., 2010a, 2010b; Fig. 2.4).

For many years, some coastal areas of Tamil Nadu, Kerala, and Odisha have been mined for strategically important thorium (used in nuclear reactors). Recently there has been a move to allocate mining blocks to private players, both onshore and offshore, by the Ministry of Mines through the Indian Bureau of Mines. Mining the beach placers is quite simple as it requires either manual and/or machine scooping in nearshore areas, while offshore mining involves scooping and/or dredging the deposits with large machines. The sands are transported to beneficiation plants for further processing and extraction. Obviously during mining, refining, and discharge of the tailing, there would be environmental concerns, such as obliteration of sand dunes, removal of vegetation, shift in the high tide line, changes in the coastal morphology, changes in the rate of accretion and erosion of sands, disposal of effluents, misuse and contamination of ground water, and destruction of marine life, etc.

Sand and gravel are important for construction, cement production, glass industries, and also for making asphalt. These are abundant along the coast. The global reserves of sand and gravel along the continental margins are two orders of magnitude more than that of terrestrial ones (Cruikshank, 1974). As mentioned earlier, the value addition of mining offshore silica sands would be for the extraction of placer minerals. In the early 1990s, the Geological Survey of India initiated studies for marine sands in the territorial waters and EEZ off the state of Kerala. It was estimated that areas that are 25–40 km off Kollam have sand reserves of 184 million tons that cover about 180 km^2 and the sand beds occur to a depth of 1.5 meters below the seafloor. Similarly, sand deposits also occur off Kannur (267 million tons covering an area of 500 km²) and Chavakkad (1,336 million tons over an area of 543 km²). Off the Cochin-Kollam stretch, the marine sediments had more than 85% silica sand. Additionally, 8 km offshore of the Beypore-Ponnani coast, good quality quartz and sand occurs in two large basins close to the coast line, where the sand bed is about 3.5 meters thick. River-derived sands that extend as offshore channels (8–20 km away from the coast) are thick and are estimated to be 936 million tons spread over an area of 450 km² (Sukumaran et al., 2010).

The application of the blue economy to explore and exploit placer and sand deposits would need a judicious approach, and has to be carried out in a sustainable manner. For example, once the component of interest has been extracted from the mined sands, the waste sands could be used to replenish the beaches to render it somewhat a pre-mining morphology. Perhaps it is also possible albeit challenging to use these sands, after removal of marine salts and shells, for construction. Offshore mining would result in environmental degradation and formation of sediment plume that in turn would affect the biota in the mined and nearby areas. Hence, enough time should lapse before mining the same areas to allow the beach and nearshore areas to attain a state of equilibrium. These aspects are possible since the marine environment is dynamic, and there is a regular supply of sands that helps rebuild the eroded areas.

2.2.2 DEEP-SEA MINERALS

The deep-sea economic deposits are hydrothermal minerals, cobalt-rich ferromanganese crusts, and polymetallic nodules (Fig. 2.4). These are found respectively along the mid-ocean ridges (70,000 km-long global underwater mountain chains at a water depth of 2,500 m), at seamounts (underwater volcanoes located at variable water depths), and in the deep basins away from mid ocean ridges (water depth more than 5,000 m).

Hydrothermal minerals form due to interaction of cold seawater and volcanic rocks and leaching of the latter leads to formation and deposition of metallic minerals and ores of iron, copper, lead, manganese, gold, silver, platinum, etc., and high-tech minerals/elements. These deposits could be of economic potential depending on the area covered, type and contents of metals, usage and monetary value of the deposits. Such deposits are well surveyed along the Mid-Atlantic Ridge and Pacific Ocean and in recent years have been discovered in the Indian Ocean such as along the Carlsberg, Central Indian, and Southwest Indian ridges. The presence of hydrothermal deposits has been recorded by India, China, Germany, and Korea in the Indian Ocean, and Poland, France, and Russia along the Mid-Atlantic Ridge. These countries are registered Contractors with the International Seabed Authority (ISA, Jamaica), the erstwhile UNCLOS that permits open ocean exploration and exploitation of marine minerals. Similarly, seamounts that have cobalt-rich ferromanganese oxides over rock outcrops have been targeted for in-depth investigations by Korea, Japan, and Russia (Pacific Ocean) and Brazil (South Atlantic Ocean). In contrast to ridges, seamounts are easier to sample since these may be sited at depths as shallow as 500 meters (Mukhopadhyay et al., 2018a).

In the ocean basins at deeper water depths (>5,000 m) polymetallic or manganese nodules occur that are 2-D deposits, unlike the 3-D hydrothermal ones. Since

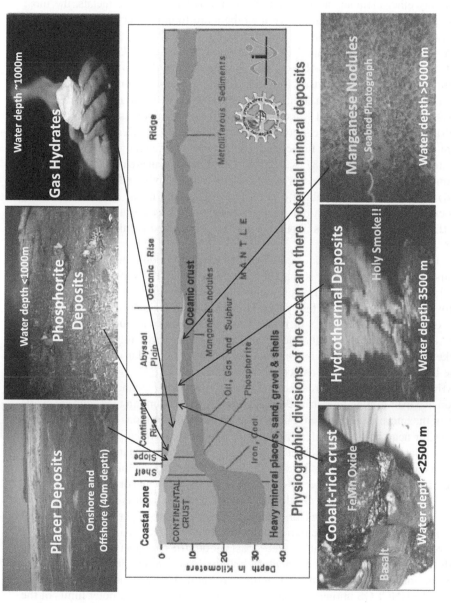

FIGURE 2.4 Mineral Resources in the Indian Ocean (Courtesy National Institute of Oceanography, Goa.)

Note: File Photo of Beach Placers, Manganese Nodules, Hydrothermal Sulfides, Cobalt Crust.

the 1970s extensive research has been carried out on nodules in all the three major oceans. The black color nodules of variable size (mostly 2–6 cm diameter) generally have a nucleus of weathered rock piece and sometimes sharks' tooth, older nodules, and sediment clast, and over the nucleus iron, manganese, copper, cobalt, nickel, zinc and many other elements accrete mainly from seawater. Of these metals, the most important are cobalt and nickel since the others are found in varying abundance as land deposits in several countries.

In the Pacific Ocean, the ISA has allotted an area of 75,000 km² to the Contractors: the Cook Islands, UK, and Northern Ireland (two sites), Singapore, Germany, and China (two sites), Belgium, Kiribati, Tonga, Nauru, France, Russia, Korea, and to the Interocean Metal Joint Organization (Bulgaria, Czech Republic, Cuba, Poland, Slovakia, and Russia) for exploitation of nodules. India is the sole Contractor for the Indian Ocean to mine the nodules from the Central Indian Ocean Basin (CIOB). India was the first country in the world to be recognized as a Pioneer Investor on December 18, 1987, with UNCLOS allocating an area of 1,50,000 km² as a Pioneer Area. The nodule resource estimated for the IONF (Indian Ocean Nodule Field) within the CIOB is about 670 million tons while metal contents (nickel + copper + cobalt) with a para-marginal grade of 2–2.4% constitute 11 million tons and the nodule abundance is about 5 kg/m² on the seabed (Mukhopadhyay et al., 2008, 2018a).

A few preliminary studies of cost-benefit factors of nodule mining, recovery of metals, and sales in the global market have been made. The reports indicate commercial nodule mining could be profitable but one needs to be patient as the return on investment could take a few years. However, once exploitation starts it would be economically viable unless there are unforeseen conditions such as discovery of new sources of metals on land, finding alternatives to the presently used metals, and market values of metals (Mukhopadhyay et al., 2019). Similar challenges exist to mine hydrothermal minerals and cobalt-rich crusts.

Mining of deep-sea resources not only needs specialized technologies but is more expensive than mining terrestrial deposits. Presently some mining systems are available in the international market, but so far commercial operations are yet to take off, except for nearshore placer deposits. Besides the scarce availability of technology, certain related factors limit seabed mining, viz. availability of land ore and mineral deposits, obtaining a license from the ISA to explore and exploit ocean resources, huge capital investments either by consortia or by governments or through a partnership. Once technologies are developed, environmental concerns are addressed and taken care of, and mechanisms for mining are established then perhaps the day is not far off when countries will be retrieving metals from the deep sea.

UNCLOS set the mandate to the coastal countries to document the process of delineating their outer continental shelf and submit the information to the Commission of the Limits of the Continental Shelf (CLCS). India being a signatory to the UNCLOS initiated a national program to delineate her outer limits of the continental shelf beyond 370 km. In this endeavor, India launched a multi-pronged, multi-institutional program that involved mapping, interpretation, and documentation of over 1.2 million km² of offshore data. The areas encompassed the EEZ in the Arabian Sea, Bay of Bengal, and Western part of the Andaman-Nicobar islands.

Under the provisions of Article 76, India handed over the first partial submission to the UN on May 11, 2009 for extending the continental shelf beyond 370 km. The submission is awaiting examination and consideration by the CLCS. The second partial submission for another part of the extended continental shelf is at an advanced stage. Presently the EEZ of India is about 2.37 million km^2 and if the UN approves India's first partial claim then approximately 0.6 million km^2 of seabed beyond 370 km would be added (Fig. 2.1). In that case, India would have sovereign rights of the seabed and sub-seabed resources of the extended area but not of the overlying waters (Rajan, 2018).

2.3 ENERGY RESOURCES

Every nation requires energy for its people, industries, and development. Considering the demand and limited supply of fossil fuels, environmental concerns in tapping these resources, and effect on climate change, it is imperative that we look now for alternative sources of energy. In contrast to the aforementioned non-renewable energies, oceans are the best sources for renewable and clean energies (Table 2.3). Such energies can be generated from offshore solar and wind, tides, waves, salinity difference, currents, and ocean thermal energy conversion (OTEC). The energy needs of several rising economies in Asia (India, China, Japan, Korea, Indonesia, etc.) have led them to look toward the oceans. According to the International Renewable Energy Agency (IRENA), in the next 15 years global demand for primary energy would grow by 40% and of this, a significant need would be from Asia. And by 2050, solar power, with 8,500 Gigawatts (GW) installed capacity, and wind, with 6,000 GW, would account for three-fifths of global electricity generation (IRENA, 2019). Furthermore, the present share of renewable energy in power generation would be expected to rise from 25% to 86% in 2050, and that about 60% of this power would come from solar and wind, with total annual renewable power generation increasing from 7,000 TWh to 47,000 TWh (Terawatts/hour, 1 TW is equal to one trillion (10^{12}) watts).

TABLE 2.3
India's Prospectus for Renewable Energy from Ocean

Energy Type	Potential
Offshore wind	1
Tidal	12.4
Wave	40
Salinity gradient	54.8
OTEC	180

Sources: Das and Ramaraju (1986), AFD and IREDA (2014).
Note: Potentials in Gigawatt.

2.3.1 OFFSHORE RENEWABLE ENERGY

As evident from the Indian Ocean Renewable Energy Ministerial Forum held on January 21, 2014 in the United Arab Emirates, the IORA countries have initiated steps to explore and exploit the potential renewable energy resources (Table 2.3). The Indian government has encouraged tapping of renewable energy by providing investors with tax holidays, permitting 100% foreign investment, generation-based incentives and other inducements. Further, in the Strategic Plan (2011–17) of India there is a stress on renewable energy to help increase India's power generation capacity from 16% to 18% by 2022 (Ministry of New and Renewable Energy, 2011). Under the Sustainable Development Goals (SDG-7) emphasis has been laid that by 2030 India should have less expensive, clean, sustainable, and dependable energy sources. Further, the goal of SDG-14 is to exploit in a sustained way the oceans and seas for their energy and marine resources, especially for SIDS and Least Developed Countries (LDC).

The Government of India (GOI) has earmarked four key sectors of renewable energy: solar, wind, biomass, and small-scale hydropower (GOI, 2011). We discuss some of these aspects of renewable energy from oceans and provide examples of work done in these areas by India.

2.3.1.1 Wave Energy

As the wind blows across the surface of seas and oceans, the waters act as carriers of wind energy, and this can be tapped by using suitable wave energy converters such as attenuators and point absorbers. The quantum of energy sourced from waves would depend on factors such as location, wave height, and wave frequency among others. Depending on favorable situations and installation sites wave energy can be tapped at least to supply to the coastal population.

2.3.1.2 Ocean Thermal Energy Conversion (OTEC)

The sun's rays heat the upper surface of seas and oceans faster in contrast to that at depth. Based on this fundamental understanding the temperature between the warm upper layer and cold seawater up to a depth of 1,000 m has been exploited to produce electricity through OTEC. To operate an OTEC, the variation in temperature between surface and depth should be at least 22°C and this difference in temperature is converted by turbines to generate electricity. Such large thermal changes are common in the tropics and hence OTEC can be effectively implemented between the equator and latitudes 30°N and 30°S. The efficiency of an OTEC is 90–95%, which is among the highest of all power generation technologies in both renewable and non-renewable energy sectors (cf. Mohanty et al., 2015).

An OTEC has a wide application such as for air conditioning, cooling, and for aquaculture (Nihous, 2007). Coupled with desalination technologies, an OTEC plant can help produce fresh water to the tune of about 2.28 million liters each day per megawatt of power generated by an OTEC system (Magesh, 2010). In the case of India, an OTEC is a boon to her inhabited islands such as Andaman and Nicobar and Lakshadweep. Plans are afoot to start an OTEC facility at Kavaratti Island to generate 200 KW of electricity to power, among others, a desalination plant. According to reports, the 2,000 km-long southern coastline of India has a temperature difference of above 20°C that is available throughout the year. This works out to about 1.5 million km^2 of tropical water in the EEZ of India with a power density of 0.2 MW/km^2. The total OTEC potential around India is estimated to be 180,000 MW, even after considering 40% loss of gross power in transportation. Presently the cost-benefit ratio of an OTEC vis-à-vis conventional energy needs to be looked into (www.business-standard.com/, https://niot.res.in).

2.3.1.3 Offshore Wind Energy

Unequal heating of various parts of Earth leads to lighter and warmer air replacing relatively cooler and heavier air. This initiates a substantial movement and circulation of wind over the sea surface. Globally the total annual installation of wind turbines/mills during 2014 was 8,759 MW, with 91% found in European waters. The United Kingdom has the largest offshore wind capacity in Europe and accounts for more than 55% of the total installations. Denmark had set a target of achieving 1,500 MW of energy from offshore wind by 2020. The USA and China have also started to look into the feasibility of offshore wind energy (Mohanty et al., 2015). India is also looking into the prospect of generating power from offshore wind mills (Table 2.4).

The top onshore wind-energy generators have been China with 21,200 MW, USA 7,588 MW, Germany 2,402 MW, India 2,191 MW, Brazil 1,939 MW, France 1,563 MW, Mexico 929 MW, Sweden 717 MW, United Kingdom 589 MW, and Canada 566 MW. Land-based wind power projects added 46.8 GW of new capacity. On the other hand, the total offshore wind capacity grew by 20% to reach 23 GW in 2018 (Global World Energy Council [GWEC], 2018). The top offshore markets were China with 1,800 MW, United Kingdom 1,312 MW, Germany 969 MW, Belgium 309 MW, and Denmark 61 MW. The new installations of offshore projects grew by 0.5% to reach 4.49 GW. GWEC further reported that since 2014, the global wind industry (on land and offshore together) has added more than 50 GW of new capacity each year and it is expected that every year 55 GW or more would be added until 2023.

TABLE 2.4
Theoretical and Technical Potential of Salinity Gradient Energy

Continent	Theoretical Potential		Technical Potential	
	[GW$_{gross}$]	[TWh$_{gross}$/year]	[GW$_{electricity}$]	[TWh$_{electricity}$/year]
Europe	241	2,109	49	395
Africa	307	2,690	63	503
Asia	1,015	8,890	2,081	663
N. America	479	4,195	98	785
S. America	969	8,492	199	1,589
Australia*	147	1,291	30	242
World	3,158	27,667	647	5,177

Source: Stenzel (2012).
Notes: GW = Gigawatts, TWh = Terawatt-hours.
*Incl. Oceania.

2.3.1.4 Offshore Solar Energy

Onshore production of solar energy is commonly implemented in many countries. The constraints of onshore solar panels, however, are that these tend to get dirty resulting in reduced efficiency, and require labor and large quantities of water to keep the panels clean. In contrast, there is a huge potential to tap offshore solar energy by using photovoltaic cells or concentrating solar power. Although installing offshore solar panels is more expensive than onshore ones the disadvantages are minimal and in the long run make up for the cost. India is competent enough to generate 5,000 GW of solar energy (www.ireda.gov.in) and in recent years hundreds of onshore hotels, houses, and offices have installed solar panels to generate and use solar energy. As an important sector of the blue economy, India and many of the IOR countries now need to turn to offshore solar energy to supplement their existing shortfall in power generation.

2.3.1.5 Tidal Energy

Tidal energy, the oldest form of energies, can be dated back to 787 AD in Spain, France, and the British coasts (GWEC, 2018). The gravitational attraction between Earth and moon results in high and low tides, and this causes a tidal cycle that occurs every 12 hours. During high tide, seawater enters through mouths of rivers and into estuaries, and during low tide the flow of water is reversed. During this process of inflow and outflow, the water carries energy and this tidal energy can be tapped depending on area, speed of water flow, and energy differences.

Early tidal power plants used naturally occurring tidal basins by constructing a barrage or dam across the opening of the basin. During high tide, the basin got filled and during low tide impounded water was released through a water-wheel, paddle-wheel, or similar energy-conversion devices. Power was available for about two to three hours, usually twice a day. In the 1960s the first commercial-scale modern-era tidal power plant was built, near St. Malo, France (GWEC, 2018).

Because tides are highly predictable vis-à-vis solar, wind, and wave energies, it is easier to forecast quite accurately the tidal range. The general tidal variation is from 4.5 m to 12.4 m but to obtain significant tidal energy economically at least 7 m high tide is required to start the turbine. Energy derived from tides is commercially more viable since technologies exist to tap energy from tidal lagoons, tidal reefs, tidal fences, and low-head tidal barrages. The countries that have tidal technology are Canada, China, France, Japan, South Korea, Russia, Scotland, Spain, The Netherlands, and the UK. Japan planned to install the first turbine in Gato Island in 2019 in Nagasaki Prefecture. India (in Sundarbans and Gulf of Kutch) and Philippines are planning to install tidal barrages (GWEC, 2018).

2.3.1.6 Salinity Gradient

Saline power is the energy created from the difference in salt concentrations between fresh and saltwater at a river mouth. The energy density from saline gradient can be measured as osmotic pressure between two saline solutions. The global estimation of saline gradient energy is 3.1 TW with Asia having the highest potential, followed by South and North America (Table 2.4; GWEC, 2018).

In the long run renewable energies from the oceans will be cleaner and economical, although variable. For instance, to tap wave energy a detailed study would be required to decipher direction and period of waves and both these quite often vary depending on locations. Small-scale wave energy power plants could have a lower environmental impact on the oceans compared to large-scale plants. Tidal energy plants could change the salinity and hydrology of estuaries where they are sited while offshore wind energy plants could hinder fisheries. But these effects could be much lower than those caused by similar energy generators on land.

Certain requirements to tap renewable energy from the oceans are much-needed technology, cooperation of the people (within a country and among the IOR countries), and the cost involved in setting up, generating, and transmission of electricity produced. In most cases, some of these limitations could be overcome in the long run depending on technologies and scale of operation (Pelc and Fujita, 2002).

India has made some significant progress in exploring renewable energy resources but certain factors placed restrictions on tapping these resources especially for offshore wind energy (Mani and Dhingra, 2013). But more emphasis is to be placed on R&D and support from the government to help reduce the cost of energy production and improved technology for better reliability.

2.3.2 POTENTIAL OF OFFSHORE RENEWABLE ENERGY FOR INDIA

India is endowed with a long coastline, 2.02 million km^2 of EEZ and 0.53 million km^2 of extended continental shelf (AGR, 2016–17). The Ministry of New and Renewable Energy (MNRE, https://mnre.gov.in) is leading India's campaign to accelerate generation of electricity from renewable sources: wind, wave, tide, solar, and salinity gradient.

These regions are potential sources of offshore wind and need to be taken on priority (Tables 2.3, 2.4). In the generation of onshore wind energy India has a total installed capacity of 32 GW (as of March 31, 2017) and is the fifth largest in the world. The National Institute of Wind Energy (NIWE) has measured near-shore wind data at 78 locations along the Indian coasts using meteorological masts. Eight zones each in Tamil Nadu and Gujarat have been identified for offshore wind energy and a total area of 17,706 km^2 has been identified. A preliminary assessment indicated the potential to establish wind farms, each of around 1 GW capacity, along the coastline of Rameshwaram (Dhanushkodi area) and Kanyakumari in Tamil Nadu, and in the Gulf of Khambhat, Gulf of Kutchch, and Saurashtra open coast, all in Gujarat. Presently India does not have offshore wind energy farms but there are plans to increase the generating capacity manyfold by 2022.

The Solar Energy Corporation of India (SECI) signed an agreement with the Gujarat government to set up India's first offshore 1 GW wind energy project. SECI will be in charge of the auction process, and to procure power from the resulting project. It is estimated that the project would result in $2.1 billion investment in Gujarat. Although the offshore wind farms could provide 30% higher energy yield than onshore installations, the areas of concern would remain the higher financial investment at the initial stages and environmental issues.

Being located in the tropics and near to the equator, India has the vast potential to develop the capacity for solar energy. Approximately 5,000 trillion kWh per year energy falls over India and most areas receive 4–7 kWh/m²/day. In 2010, the Jawaharlal Nehru National Solar Mission (JLNNSM) was launched and set a target to develop 20,000 MW of solar capacity by the year 2022. This would help to develop the economy, create jobs, and enhance the domestic production of critical raw materials, components, and products. In July 2015, the GOI increased the target from 20,000 MW to 100,000 MW by the year 2022 and this was to be achieved through deployment of 40,000 MW of rooftop solar projects and 60,000 MW large and medium-scale solar energy missions.

The Indian Renewable Energy Development Agency (IREDA, a government enterprise under the administrative control of MNRE) has identified three locations that could potentially produce tidal energy. The locations are in the Gulf of Khambhat and Gulf of Kutchch (Gujarat, accumulated capacity 12.4 GW) and a plant of 4.65 MW capacity at Durga Duani creek in Sundarbans (West Bengal).

The power potential from the salinity gradient along the Indian coast is 54.8 GW (Das and Ramaraju, 1986). According to MNRE, the total OTEC potential installed capacity in India is 180 GW, and potential wave energy is about 40 GW with the western coastline having higher potential than the eastern one. The maximum power is noted at the southern tip of India, i.e., between Kanyakumari and Koondakulam.

The Chennai-based Indian Institute of Technology set up a 150 kW pilot wave energy plant back in 1991 at Vizhinjam in Kerala on the southwest coast of the country. This was probably the world's first wave power plant working on oscillating water column technology that utilizes change in levels of water inside a caisson as waves approach. As the water level increases in the caisson, the air inside is compressed, and this helps to drive a turbine. Maximum power was generated only during the monsoon months. This project was however decommissioned in 2001. The National Institute of Ocean Technology (NIOT, Chennai, India) is working on projects such as the generation of 0.2–0.33 million liters per day of freshwater using the Low-Temperature Thermal Desalination (LTTD) technology, and to install a floating wave-powered device so as to meet the lighting requirements of small islands (www.niot.res.in).

2.4 MARITIME ACTIVITIES

Other than the tapping of living and non-living resources from the oceans there are several marine sectors that contribute to the success of the blue economy of a country. Several maritime activities or services that form a part of the blue economy are: shipping (ports and harbors), traffic and transport (marine commerce), safety and security (including piracy), tourism and recreation, culture and heritage, and ancillary activities. These sectors either directly or indirectly lead to job creation, innovations, increase in GDP, investments, joint ventures, skill development, cooperation, training, etc. within and among the IOR countries. Some of these maritime services that catalyze the blue economy are expounded below.

2.4.1 COASTAL TOURISM AND RECREATION

Globally, coastal areas are the most preferred holiday destinations because of the salubrious climate that prevails along coasts and respite from heat, dust in the tropics, or icy cold climate common in most temperate countries. Coastal tourism covers a gamut of activities such as boating, surfing, angling, hunt for sunken ships and treasures, swimming, scuba diving in coral-infested areas to look for exotic tropical fish, bird watching, water sports, and so forth. In addition, cruise liners are promising sectors of the blue economy. All such activities led to the development of coastal areas, improvement in infrastructure, increase in employment, and a better life-style for those involved in tourism and recreational sectors. These can be better seen in countries such as Maldives, Sri Lanka, Seychelles, Singapore, and Thailand and many other SIDC. Similar is the case for the Indian states like Gujarat, Maharashtra, Goa, Karnataka, Kerala, Tamil Nadu, Andhra Pradesh, Odisha, and West Bengal, and two Union Territories—Andaman & Nicobar Islands and Laccadive Islands—that cater dominantly to domestic and significantly to foreign tourists.

To ease travel and to increase the footfalls in coastal tourism, India extended the e-visa facility for citizens of 165 countries at 25 airports and five seaports. This pro-gressive step would help enhance tourism in India and a reciprocal visa facility from the IOR countries would be of mutual benefit. Key highlights for the year 2018 and projections for the coming years by the World Travel and Tourism Council (WTTC) revealed that global GDP contribution from the travel and tourism sector was 10.4%. While the global economy grew by 3.2%, the travel and tourism sector was signifi-cantly higher at 3.9%. In 2018, in India, travel and tourism contributed 9.2% of the total economy that translated to US$247.3 billion. Travel and tourism also resulted in generating 42 million jobs (i.e., 8.1% of total employment) and this is expected to rise to 52 million by 2029.

2.4.2 SEABORNE TRADE AND COMMERCE

Significant global trade occurs through maritime routes because transportation by sea is less expensive, reduces carbon footprints, and can carry stupendous bulk volume at one go. Thousands of merchant vessels, such as oil tankers, bulk carriers (coal, grains, pulses, mineral ores, etc.) and container ships crisscross the oceans for trading and of this a majority ply through the Indian Ocean. United Nations Conference on Trade and Development (UNCTAD) opined that over 80% of global trade by volume, which accounts for more than 70% of its value, is done through sea. During 2017 there was a 4% increase in seaborne trade and this was the fastest growth in five years. Total volume of trade went up to 10.7 billion tons of which 50% was dry bulk commodities, containerized trade increased by 6.4% while growth in crude oil shipments fell to 2.4% due to a fall in exports from the Organization of Petroleum Exporting Countries (OPEC; Review of Maritime Transport, 2018).

UNCTAD further predicted that ship-borne trade would grow by 3.8% between 2018 and 2023. Although volumes across all maritime material would increase, containerized and dry bulk commodities are expected to record the fastest growth.

However, the above projection could be affected by increased inward-looking policies and the rise of trade protectionism. A worrisome issue is the trade tension between China and the USA, as well as those between Canada, Mexico, the USA, and the European Union. The other factors that could impede ship-borne trade are ongoing global energy transition, structural changes in economies, and shifts in global value chain development patterns.

An important aspect of maritime transport is to reduce greenhouse gas emissions, which include the Paris Agreement under the UN Framework Convention on Climate Change (UNFCCC) and the 2030 Agenda for Sustainable Development, in particular SDG-13 to take urgent action to combat climate change and its impacts. In 2018, a policy on the reduction of such emissions from ships was adopted according to which, compared to 2008, the total annual greenhouse gas emissions would be reduced by at least 50% in 2050. There are short-, medium-, and long-term strategies, possible timelines, and impacts on States, with particular attention to needs of developing countries, especially SIDS and LDC.

There would be amendments to the 1973/8 International Convention for the Prevention of Pollution from Ships (ICPPS) to make mandatory data collection system for fuel oil consumption of ships to 5,000 gross tons and above, and the data are being collected from January 1, 2019. The global limit of 0.5% on sulfur in fuel oil used on board ships, effective from January 1, 2020, would help to reduce ship-source air pollution and have a positive effect on human health and the environment (Review of Maritime Transport, 2018).

2.4.3 Ports and Harbors

Ports and harbors are essential infrastructures for movement of materials within and outside a country, berthing of cruise ships, and for marine recreational activities. The related activities are towing, usage of tugs, pilotage, and anchorage at berths, repairs of vessels, immigration and customs services, and handling and warehousing of cargoes. The period between 2003 and 2007 was a boom time for shipping when the tonnage handled/transported peaked to 6.5% from 3% in 1990s (Mitroussi, 2013). Container port service increased in India, UAE, Iran, Malaysia, Indonesia, Singapore, Thailand, Oman, South Africa, and Australia. These IOR countries witnessed a remarkable growth in maritime services that included the transport of freight and passengers.

The increase in maritime activities and trade through the Indian Ocean is a boon to IOR countries and an opportunity for SIDS to develop ports and harbors. These could be shared facilities to help reduce capital and recurring costs for the countries involved and to speed up the transport of goods and serve as transit sites for bunkering, marine services, and tourism. It is reported that the Indian Ocean region accounts for 11.2% of shipping in DWT terms (Dead Weight Tonnage), and 9.9% in terms of real nationality, with Singapore, India, and UAE having 1% share in the shipping services (Mohanty et al., 2015). In India, Jawaharlal Nehru Port terminals (Mumbai) attracted 4.8% more business in 2017 (Review of Maritime Transport, 2018).

To harness the full potential of the ocean-based resources, India is trying to strengthen its cooperation with the regional partners showcasing her domestic mega-modernization projects (Maini and Budhraja, 2018). For example, the Indian

Ministry of Shipping has initiated a US$42.857 billion ambitious SAGARMALA project, wherein about 200 major and minor ports will be developed offering IT-based modern services. The project aims also to develop 27 industrial clusters and connect hinterlands through rail and road facilities. This will increase employment opportunities, enhance the maritime ecosystem, and develop the coastal economy. The shipbuilding and repairing industries can develop and accelerate industrial growth along with the large number of associated industries. India's fleet strength of just over 1,200 ships in 2014 is expected to augment to over 1,600 by 2025. This enormity could be of benefit also to the entire SAARC community (Fig. 2.5).

2.4.4 CRUISE LINERS AND LEISURE

Although airplanes are preferred as rapid modes of transport, there are thousands of passengers who go on luxury cruises. As compared to certain sectors of the blue economy that of the global passenger cruise sector is slow but nevertheless steadily growing. This is due to improved economic conditions in several developing countries, better salaries, and availability of loans from banks for travels or payment to tourist operators through equated monthly installments. In this respect, the IOR could be likely markets for both domestic and international travelers. Besides onboard facilities, recreation, and food, the cruise sector involves onshore activities and visits to places of interest (museums, zoos, religious and historical places, restaurants) all of which augment a country's economy.

The Caribbean and Mediterranean Seas are two of the most popular cruise routes that annually cater to thousands of passengers in addition to Northern Europe, Alaska, and South America. Now the cruise liners sector is eyeing the Asian market and with Australia and Africa being looked as potential destinations (International Transport Forum, 2015). The Asian cruise sector forms just 4.4% of the global market yet many cruise companies have offices in China, Hong Kong, Korea, Singapore, and Taiwan. It is estimated that the number of Asian passengers could increase to almost 4 million in 2020 from 1.3 million in 2012. The growth is mostly due to the number of Chinese cruise passengers and with a figure of 697,000 (an increase of 70% from 2012 to 2014) is nearly equal to the combined markets of all other Asian countries (701,000 passengers; Cruise Lines International Association [CLIA[, 2015). A tentative estimation of passengers who would be traveling through the Indian Ocean is given in Table 2.5.

Insisting that type of vessels deployed and duration of cruises offer a hint at the peculiarities of the Asian cruise market, the CLIA found that as compared to Caribbean and European cruises, large and mega cruise ships are few in Asia, while seasonally operated mid- and small-size ships are more common. And these cruises are typically 2–3 days in duration and rarely more than a week. The differences in languages, food habits, customs, and cultures that exist amongst the Asian countries could also support shorter-duration cruises. But given the large population, especially of China and India, opportunities exist for cruise liners to make profit (www.cruisemarketwatch. com). Presently ships ply between Chennai, Kolkata, Vishakhapatnam, and Andaman islands, and between Mumbai and Goa, and between Cochin and Lakshadweep islands. Plans are afoot to start a Cochin-Male-Cochin cruise.

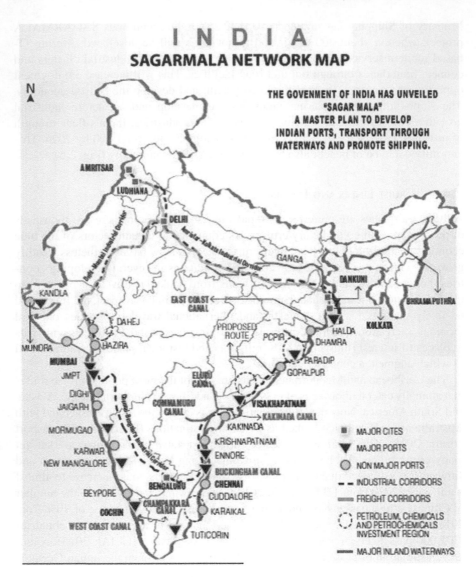

FIGURE 2.5   India's Ambitious Sagarmala Project

Notes: India has undertaken infrastructure upgradation for 200 major and minor ports. A total of INR 1.2 trillion ($16,900 million) will be spent over the next ten years to develop 27 industrial clusters. Hinterlands would be connected through rail and road projects. This will increase employment opportunities, enhance the maritime ecosystem, and develop the coastal economy.

TABLE 2.5
Asia Cruise Traffic Projections for 2020

Countries	Passengers	Increase over Previous Year
China	1,617	242.6%
Japan	766	253.0%
Malaysia	288	100.0%
India	242	82.0%
Singapore	181	98.9%
South Korea	219	200.0%
Indonesia	238	303.4%
Hong Kong	124	134.0%
Taiwan	163	246.8%
Other	102	96.2%
Asia (all)	3,940	193.8%

Source: Chart Management (2014).
Note: Passengers in thousands.

2.4.5 SUPPLEMENTARY SERVICES

India represents diversity of probably the highest nature in terms of races, ethnicities, languages, and religions (Hinduism, Islam, Sikhism, Buddhism, Christianity, Jainism, etc.). Such a multiplicity of factors has resulted in a rich cultural heritage that has influenced the entire Indian Ocean region (Iyer, 2016). Since humans started to navigate the high seas, looked for new lands to conquer, trade, loot, or settle down the signatures of these have been left behind in the coastal areas. In India several sites of antiquities are of interest to people. For example, the drowned city of Dwarka, in Gujarat where evidence shows that a city had existed during historical times. There are several areas along the east and west coasts of India where shipwrecks have been located. Although, unlike in other countries, exploration of shipwrecks for treasure and artifacts are not in the public domain in India, this could be encouraged to boost tourism. The signatures of maritime heritage and trade are found off Goa, Tamil Nadu, and Odisha. Hence, as components of the blue economy these areas could be developed as places of interest to researchers and tourists.

The blue economy has also given rise to a host of ancillary services, such as the development of better ocean-observing satellites, offshore commerce, insurance, banking, construction and leasing of boats and ships, improvement in marine information and communication, the build-up of a database of biodiversity and ecosystems, pollution types and possible remedies, forecast of weather and ocean conditions, prediction of storms, cyclones, typhoons, and tsunamis, and the real-time monitoring of land-ocean-atmosphere conditions. In addition, hotels and resorts, ship chandlers, yards for ship repairing, ship salvaging and ship breaking are important ancillary services. Over 90% of shipbuilding activity occurs in China, Japan, and the Republic of Korea, while 79% of ship demolitions took place in South Asia, notably in Bangladesh, India, and Pakistan (Review of Maritime Transport, 2018). In India the town of Alang in Gujarat is (in)famous for ship breaking and ship salvaging. All these and related activities give an impetus to the blue economy of a country and result in enhanced GDP.

Mariners and the maritime industry always fear for their safety from the wrath of nature. Piracy is an age-old high sea crime yet it continues to haunt the shipping industry and the moment is always dreaded when pirates capture a vessel and/or the sailors are taken captive and held for ransom either for money or for some political reason. Until a few years back Somali pirates were very active in the Indian Ocean and between 2007 and 2012 more than 4,000 sailors were held as hostages.

Besides monitoring of the high seas by naval ships, remote sensing techniques using satellites are also in vogue to find the possible presence of pirates. Recently the Indian Space Research Organization (ISRO) and the French space agency Centre Nat ional d'Etudes Spatiales (CNES, National Center for Space Studies) decided to join forces for maritime surveillance of the Indian Ocean. The network of satellites would aid to detect, identify, and track ships, plying in the Indian Ocean, to safeguard them from pirates and hostile ships (TOI, 2019b). Since piracy could be a major deterrent for a blue economy, it needs to be resolved with the cooperation of the affected countries.

2.5   BLUE CARBON

For a long time, terrestrial forests have been considered as major sinks for natural carbon but subsequently, it has been found that vegetated coastal ecosystems are also highly efficient carbon sinks. This led to the coining of the term "blue carbon" (Nellemann et al., 2009), which denotes the carbon that is captured by oceans and coastal ecosystems. During the daytime all living beings give off carbon dioxide (CO_2, except plants that emit oxygen during the day and CO_2 during the night) that is believed to lead to the greenhouse effect and climate change.

But nature has her own way of alleviating CO_2 by sequestration (or capturing) of carbon by oceans and seas. This is achieved by mangroves, salt marsh, and seagrasses (together known as vegetated marine habitats) that occur along water bodies and this leads to the formation of a carbon sink. These coastal systems not only protect us from storms and tsunamis and harbor marine lives but also help to sequester carbon rapidly and store it underground. For example, over 95% of carbon in seagrass meadows is stored in the soils. Coastal habitats cover less than 2% of the total ocean area but account for approximately half of the total carbon sequestered in ocean sediments (http://thebluecarboninitiative.org). Mangroves, salt marshes, and seagrass globally cover respectively about 152,000 km^2, 400,000 km^2, and 600,000 km^2. If these ecosystems are either destroyed or degraded then the carbon stored for centuries is emitted into the atmosphere and oceans, and becomes a source of greenhouse gases. It is estimated that about 1.02 billion tons of CO_2 get released annually from degraded coastal ecosystems (http://thebluecarboninitiative.org).

The vegetated marine habitats cover less than 0.5% of the seabed, yet store more than 50%, and potentially up to 70%, of all carbon in sediments, as compared to plant biomass on land. The plant biomass such as leaves, stems, branches, or roots can sequester blue carbon for years to decades, while underlying plants in sediments sequester carbon for thousands to millions of years. A worrisome issue is that blue carbon gets affected due to the high rate of loss (annually 2–7%) of these vital marine ecosystems as even compared to rainforests (Nellemann et al., 2009).

Seagrass meadows occur up to 50 m water depth and globally cover 300,000 to 600,000 km^2, with up to 4,320,000 km^2 (Gattuso, 2006) and the loss is estimated at 7% of their known area per year (www.iucn.org/content/seagrass-habitat-declining-globally). Although seagrass forms only 0.1% of the ocean floor the total carbon burial is 10–18% and storage is as much as 19.9 Pg of organic carbon (1 Pg = petagram is equal to one quadrillion grams, or one trillion kilograms; Fourqrurean, 2012). The average carbon burial rate by seagrass is about 138 g C m^{-2} yr^{-1} (grams of carbon per m^2 per year). Seagrass is affected by coastal eutrophication, increase in seawater temperatures, enhanced sedimentation, and coastal development (Duarte et al., 2011).

Mangroves protect the coasts from erosion and extreme events, extreme events, form breeding grounds for fish, provide forest products, and filter nutrients and sequester carbon (Bouillon et al., 2008). Globally, mangroves stored 4.19 ± 0.62 Pg of carbon in 2012, with Indonesia, Brazil, Malaysia, and Papua New Guinea accounting for more than 50% of the global stock. It is reported that 2.96 ± 0.53 Pg of global carbon stock is contained within soil and 1.23 ± 0.06 Pg in living biomass (Hamilton and Friess, 2018).

Mangroves are found along the coastlines of 123 countries located in the tropics and sub-tropics regions of the world (India State of Forest Report, 2017). Worldwide, mangroves cover varies from 150,000 km^2 (Spalding et al., 2010) to 1,67,000-1,81,000 km^2 (Miththapala, 2008). In 2012, the worldwide coverage of mangrove was between 83,495 km^2 and 167,387 km^2 with Indonesia containing about 30% of the entire global mangrove forest area (Hamilton and Casey, 2016). The most extensive areas of mangroves are in SE Asia followed by South America, North Central America, and West and Central Africa. Mangroves in South Asia cover an area of 10,344 km^2 which is 6.8% of the global mangrove cover. The reported mangrove cover in India in 2017 was 4,921 km^2 i.e., 3.3% of the global mangrove cover. This increase in the coverage from 4,046 km^2 in 1987 is due to plantation, regeneration, and conservation. The mangroves of the Sundarbans delta (West Bengal) account for almost half (2,114 km^2) of the Indian mangrove area (India State of Forest Report, 2017; Chand Basha, 2018).

Mangrove forests account for about 10% of global carbon burial (Duarte, 2005) with an estimated carbon burial rate of 174 g C m^{-2} yr^{-1}. Tropical forests account for 3% of the global carbon sequestration, while 14% carbon is buried in mangroves (Alongi, 2012). Marshes are common along most of the tropical coastlines in the world and in such areas the biomass below the ground could be thick. The worldwide coverage of marshes is about 22,000 to 400,000 km^2 wherein the estimated carbon burial rate is 210 g C m^{-2} yr^{-1} (Chmura and Anisfield, 2003).

Algae are other potential carbon sequesters. Unlike land plants algae do not have lignin and therefore carbon is more rapidly released to the atmosphere by algae than sequestered carbon on land (Chung et al., 2011). Algae serve as a short-term storage pool of carbon and this, in turn, can be used as a feedstock to produce several biogenic fuels such as carbon-neutral biodiesel and biomethane, while microalgae could be used to produce biochar that has contents of nutrients (Bird et al., 2011; Kumar et al., 2011). Collaboration between Washington State University, USA and Dalian Ocean University, China led to the use of algae to sequester carbon (Chi et al., 2013). Macroalgae have been used as part of a climate change mitigation program by South Korea through the setting up of the Coastal CO_2 Removal Belt (CCRB) (Chung et al., 2013).

The importance of Wetlands—another component of blue carbon comprising marshy, fen and peatland—in acting not only as a source of water but as a tool to sequester atmospheric carbon, to provide food, fiber, raw materials, and also to prevent flood, land degradation, and desertification, is underlined by some recent initiative of the Government of India. Following the release of National Wetland Atlas in 2011 by the Indian Space Research Organization (ISRO), the government has prioritized 130 wetlands out of a total of 201,503 wetlands identified in the country. Wetlands in India cover an area of 105,600 km² (4.63% of its geographic area) of which about 41,400 km² are coastal wetlands distributed over eight coastal states and two island territories. The government is encouraging engagement of local communities, NGOs, and self-motivated individuals to conserve the wetlands (TOI, 2019b).

If the ecosystems of seagrass, mangroves, marshes, and algae are disturbed or destroyed then carbon capture would be reduced or altered and lead to the release of stored carbon into the atmosphere, and further catalyze the rate of climate change. Although it is arduous to quantify, estimates indicate that if blue carbon ecosystems continue to decline, then in the next century 30–40% of tidal marshes and seagrass and ~100% of mangroves would be lost (Pendleton et al., 2012). The possible causes are human-induced impacts such as constructions, aquaculture, mining, and overexploitation for shellfish, crustaceans, fish, and timber.

In the open ocean, sedimentation is very low and there are fewer plants and animals that could be buried after their death and decay. Therefore, as compared to the coastal ecosystems, carbon burial rates are relatively slow in the open ocean. But there are exceptions such as in the Nazare Canyon near Portugal, where because of extreme depositional environment, burial rates are 30 times greater than in the nearby continental slope. This submarine canyon (steep-sided V-shaped offshore valleys) accounts for about 0.03% of global terrestrial organic carbon burial in marine sediments. Although this figure may appear to be significant, the Nazare Canyon is just 0.0001% of the world's ocean floor area (Masson et al., 2010). Globally there are 9,477 submarine canyons that occupy an area of 4,393,650 km². Submarine canyons in the Indian Ocean number approximately 1,600 that encompass an area of 760,420 km² (second to the North Pacific Ocean where submarine canyons cover 816,580 km²; Harris et al., 2014). Considering the large numbers of submarine canyons one can imagine the amount of area that is available for carbon burial.

The swatch of no ground (SONG), the largest submarine fan in the world, is a 14 km-wide deep-sea canyon in the Bay of Bengal with a depth of about 1,340 m (Curray et al., 2002). The Government of Bangladesh has declared the SONG has a Marine Protected Area (SONG-MPA) which would cover 1,738 km² of the marine area to conserve several species of whales, dolphins, porpoises, and sharks (Ministry of Environment and Forest, 2014). To safeguard the SONG-MPA, Karim and Uddin (2019) proposed the preparation of a management plan and suggested provision of adequate resources to the relevant agencies so as to ensure inter-agency cooperation.

Therefore, the above-mentioned ecosystems need to be protected by India, and the entire world because if these are destroyed then carbon sequestration would be reduced and the previously trapped carbon may be released back into the atmosphere. Hence, it would be imperative for a country to buy and sell carbon offsets from communities who are encouraged to save the environment through protective and restorative measures (oceanservice.noaa.gov/facts/bluecarbon.html). It has been reported that restored seagrass meadows start to sequester carbon in sediment within about four years (Greiner et al., 2013).

2.6   SYNTHESIS

The blue economy presents India with an unprecedented opportunity to meet its national socioeconomic objectives as well as to strengthen connectivity with its neighbors. This type of economy can help in focusing on livelihood generation, achieving energy security, building ecological resilience, and improving health and living standards of coastal communities. The blue economy can, in fact, reinforce and strengthen the efforts of the Indian government as it strives to achieve the SDGs of hunger and poverty eradication along with sustainable use of marine resources by 2030. With a long coastline covering nine states and two union territories, and with an EEZ of 2.02 million km^2, the marine services sector could serve as the backbone of its blue economy. A development like the Sagarmala project is the strategic initiative for port-led development through the extensive use of IT-enabled services.

Considering the large tracts of areas in India that have marsh or wetlands, seagrass, and mangroves, India should participate in the Global Blue Carbon Initiative (www.thebluecarboninitiative.org). This organization assists and guides research, implements projects, and creates policy priorities. Subsequently, the Indian government could create and fund departments or non-governmental organizations to work on blue carbon. Such a progressive step would go a long way to protect and restore near coastal areas and accelerate the process of sequestration of carbon.

3 The Indian Ocean Region

The Indian Ocean offers an important and strategic space in terms of resources, maritime trade, shipping route, and geopolitics under the blue economy paradigm. As mentioned earlier there are 50 countries and islands distributed over three continents—Asia, Africa, and Australasia—that surround the Indian Ocean (Table 1.1). About 35% of the global population (i.e., about 2.6 billion) live in the Indian Ocean Region (IOR). Of these, 22 countries have established the Indian Ocean Rim Association (IORA) for regional collaboration and inclusive development. These countries are Australia, Bangladesh, Comoros, India, Indonesia, Iran, Kenya, Madagascar, Malaysia, Maldives, Mauritius, Mozambique, Oman, Seychelles, Singapore, Somalia, South Africa, Sri Lanka, Tanzania, Thailand, United Arab Emirates, and Yemen (Kannangara et al., 2018; Fig. 3.1). The coastline and exclusive economic zone (EEZ) areas of the IOR countries are also estimated (Table 3.1).

The EEZ areas of the IOR countries are rich in living, mineral, and energy resources, and have strategic importance in terms of maritime trade, traffic, and security (Fig. 3.1). In fact, in 2016, the fish caught in the Indian Ocean accounted for 28% of the global catch, and since then it has been consistently showing an increase. About 16.8% of the world's oil and 27.9% of natural gas reserves are also reported in the IOR (British Petroleum, 2017).

The total shoreline of all countries in this ocean put-together is about 114,172 km (for Australia and South Africa only the coast length falling in the India Ocean is included). The countries such as Indonesia and Australia have a longer coastline, whereas Singapore, Mauritius, and Brunei have shorter ones. Again, Mauritius, Australia, Indonesia, and Seychelles have the widest EEZ compared to their limited coastlines, and hence the GDP of these countries depends much upon the blue resources (Fig. 3.1).

The Indian Ocean holds key routes for maritime trades between Asia and Africa with Europe and the Americas. More than 50% of oil is transported through this ocean (Cordner, 2010). There are as many as 23 high-priority container ports in the IOR handling 166 million TEUs in 2017 (TEU = Twenty-foot Equivalent Unit is an exact measurement of a cargo-container, often used to describe the capacity of container ships). Noteworthy among these are Singapore (capacity 34 million TEUs), Dubai (15 million TEUs), and Port Klang (Malaysia, 13 million TEUs). Smaller

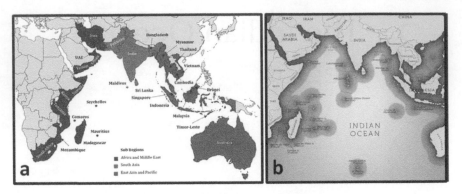

FIGURE 3.1   (a) Indian Ocean Region Countries and (b) Their Exclusive Economic Zones (shaded area; www.en.wikipedia.org)

ports such as Colombo (Sri Lanka) and Mombasa (Kenya) have been growing rapidly in recent times with infrastructural and financial support from China.

This chapter discusses the prevailing blue economy activities in ten countries in the Indian Ocean region. The countries are arranged from the western Indian Ocean to those in the Eastern India Ocean, through the central part. The countries are South Africa, Mauritius, Seychelles, Kenya, Oman, Sri Lanka, Bangladesh, Indonesia, Thailand, and Australia (Table 3.2). The plan, program, preparedness, and implementation of these ten IOR nations vis-à-vis the blue economy are examined. The areas of success and shortcomings are also presented. Also, the contribution of the blue economies in their respective GDPs are assessed. How ready these nations are logistically, economically, and technologically to enact and execute the blue economy concept is presented.

3.1   SOUTH AFRICA

South Africa is encircled by the Indian and Atlantic oceans on its three sides. It has a 3,900 km-long coastline including 1,102 km-long coastlines of two separated islands—Prince Edward and Marion Islands. The country has four coastal provinces. The territorial water covers about 74,000 km^2 and the continental shelf area spreads over 160,000 km^2. The EEZ covers a total area of about 1,535,538 km^2 (the EEZ of the mainland is 1,068,659 km^2, and that of Prince Edward and Marion Islands is 466,879 km^2). The EEZ zone is divided into two regions as they fall in both the Indian and Atlantic oceans (Fig. 3.2).

3.1.1   RESOURCES, STRATEGY, AND ACTIVITIES

The areas targeted for potential economic return under the blue economy architecture encompass aquaculture, marine transport, manufacturing, small harbors and other coastal properties, offshore oil and gas, and marine governance. South Africa initiated the blue economy approach in 2013 under the name "Phakisa" and

TABLE 3.1
Coastline and EEZ Areas of the Indian Ocean Countries

1	2	3	4	5	6	7	8	9	10
Australia	25,760	8,505,348	7,692,116	1.11	Singapore	193	1,067	705	1.51
Bangladesh	580	86,392	143,998	0.60	Somalia	3,025	825,052	637,657	1.29
Comoros	340	163,572	2,415	67.73	South Africa	2,798	1535538	1,221,037	1.26
India	7,200	2,305,143	3,287,263	0.70	Sri Lanka	1,340	532,619	65,610	8.12
Indonesia	54,716	6,159,032	1,904,769	3.23	Tanzania	1,424	241,188	945,787	0.26
Iran	989	168,718	1,628,750	0.10	Thailand	3,219	299,397	513,120	0.58
Kenya	536	116,942	580,367	0.20	UAE	1,318	58,218	83,600	0.70
Madagascar	4,828	1,225,529	586,771	2.09	Mozambique	2470	578,986	786,380	0.74
Malaysia	4,675	334,671	330,803	1.01	Yemen	1,906	552,669	527,970	1.05
Maldives	644	923,322	300	3,077.74	Brunei	161	10,090	5265	1.92
Mauritius	177	1,284,997	2,040	629.90	Cambodia	443	55,600	181,035	0.31
Oman	2,092	533,180	309,500	1.72	Myanmar	1930	532,775	676,575	0.79
Pakistan	1,046	222,255	796,095	0.28	Timor-Leste	706	70,326	14,919	0.47
Seychelles	491	1,336,539	475	2,813.77	Vietnam	3,444	417,663	331,210	1.26
	104,267					**241,84**			

1 & 6 = IOR nations, 2 & 7 = Coastline in km, 3 & 8 = EEZ area km², 4 & 9 = Land area in km², 5 & 10 = EEZ: Land area ratio. If for Australia and South Africa only the coast length falling in the India Ocean are included then the total coastline length of the Indian Ocean will be 114,172 km (after Colgan, 2017)

TABLE 3.2
Economic Background of Ten Selected IOR Nations

	01	02	03	04	05	06	07	08	09	10	11	12	13	14
South Africa	1,221,037	57,725,600	42.4	813.1	29	13,865	89	371.3	6,331	0.699	113	63VH	Christian	Dem
Mauritius	2,040	1,265,577	618	31.7	133	25,029	61	14.8	11,693	0.790	65	35.8M	Hindu	Dem
Seychelles	459	94,228	205	2.92	–	30,486	–	1.56	16,332	0.797	62	46.8H	Christian	Dem
Kenya	580,367	49,364,325	78	191	–	3,867	–	99.2	2,010	0.590	142	42.5M	Christian	Dem
Oman	309,500	4,424,762	15	198	67	46,522	23	81.7	19,170	0.821	48	–	Islam	Mon
Sri Lanka	65,610	21,670,000	327	292.8	61	13,500	91	92.5	4,265	0.770	76	39.8M	Buddhism	Dem
Bangladesh	147,570	162,951,560	1,106	831.7	29	4,992	136	314.6	1,888	0.608	136	32.4M	Islam	Dem
Indonesia	1,904,569	261,115,456	138	3.740	07	14,020	89	1.10	4,120	0.694	116	39.5M	Islam	Dem
Thailand	513,120	68,863,514	132	1.390	–	20,474	–	516.0	7,607	0.755	83	36M	Buddhism	Mon
Australia	7,692,024	25,483,400	3.3	1.313	19	52,191	17	1.500	59,655	0.939	03	33M	Christian	Dem
India	3,287,263	1,324,171,354	403	11.47	03	8,484	119	2.972	2,199	0.640	130	33.9M	Hindu	Dem

01 = area in km², 02 = population, 03 = population density/km², 04 = GDP (PPP) in billion US$, 05 = global ranking in GDP (PPP), 06 = per capita in US$, 07 = world ranking in per capita GDP, 08 = GDP (nominal) in billion US$, 09 = per capita in trillion US $, 10 = Human Development Index, 11 = world ranking, 12 = Gini Coefficient, 13 = major religion, 14 = type of government (Dem = Democracy, Mon = Monarchy)

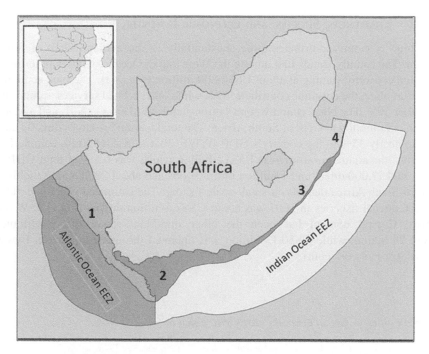

FIGURE 3.2 Exclusive Economic Zone of South Africa with Four Offshore Provinces (https://en.wikipedia.org/wiki/)

Notes: Legend 1: Benguela ecoregion; 2. Agulhas ecoregion; 3. Natal ecoregion; and 4. Delagoa ecoregion.

focused on coastal maritime investment and health (RSA Government Portal, 2018). Stimulating economic growth and developing small harbor and coastal properties were started a little later under the blue economy model. The priority harbors include Port Nolloth, Northern Cape, Port St. Johns, Eastern Cape, Port Edward, KwaZulu-Nata, and fishing harbors at Western Cape. Fishing and fish processing, aquaculture, mariculture, tourism, warehousing, marine transport, ship/boat-building, ship repairs, desalination, and ocean renewable energy are other areas that have recorded fairly good economic returns over the years.

South Africa has about 50 potential harbors that annually cater to about 300 million tons of cargo movement. In addition, nearly 1.2 million tons of liquid fuels along the coast are transported. Nearly 1% of the global market of ship repairing is carried out in South Africa.

The estimated reserves of hydrocarbon oil and natural gas in South Africa's coast, and adjoining areas, is nearly 9 billion and 11 billion barrels, respectively. These resources are enough for consumption for the next 40 years and 75 years, respectively. South Africa is set to drill another 30 wells in the next 10 years. With this initiative, about 370,000 barrels of oil and gas are expected to be produced and 130,000 jobs are likely to be created (RSA Government Portal, 2018).

3.1.2 Challenges, Success, and Economic Potential

The blue economy contributes quite substantially to the national GDP of South Africa. The country stands first among the West Indian Ocean countries in terms of the blue economy valuing at about US$6,800 million (Fig. 3.3). The blue economy activities along the coastline contribute 70%, while other related services account for the other 30%. The blue economy contribution stands third after exports and agriculture in contribution to GDP of South Africa. The total coastal resources directly contribute nearly 35% to the country's GDP (WWF, 2016). In 2016, from commercial fisheries, the annual turnover was US$12 billion which was 0.5% of total GDP. It supported 27,000 direct and 1 million indirect jobs and about 28,388 households in coastal South Africa depend on small-scale fishing. The estimated economic value of recreational fisheries in 2007 was about US$2.6 billion dollars (Fig. 3.3; WWF, 2016). Coastal tourism has been the other important sector that contributed US$1.73 billion, while US$0.13 billion was obtained through eco-tourism (www. globalafricanetwork.com).

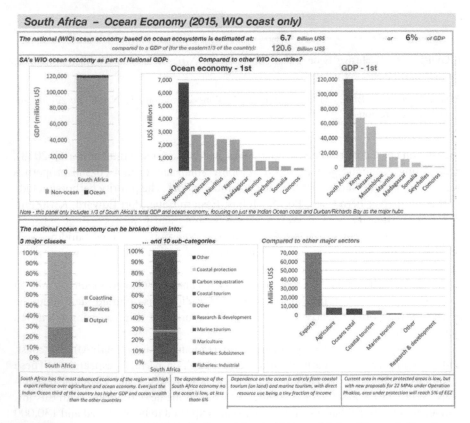

FIGURE 3.3 Blue Economy Contribution of South Africa and Other Western Indian Ocean Countries to GDP (WWF, 2016)

3.2   MAURITIUS

Mauritius, a small developing island country in the Western Indian Ocean, has 2.3 million km^2 of EEZ, which is 1,500 times more than its land size. No wonder its best option is to depend on a sustainable blue economy. In the last few decades the country has wisely shifted its focus from a land-based resource economy to an ocean-based one, while retaining its traditional economy sources from sugarcane farming and textiles at an optimum level of continuance. Marine Protected Areas (MPAs) for mainland Mauritius cover 71.9 km^2, including six fishing reserves (Castro de la Mata, 2012). The lagoon and sea (up to 1 km from the high-water mark line) surrounding the eight islets form the Islet National Park, encompassing an area about 36 km^2 (Fig. 3.4).

3.2.1   Resources, Strategy, and Activities

The main blue economy activities that Mauritius focuses on are seaport, seafood, coastal tourism, and related activities. To ensure sustainable use of seas and oceans, various agencies such as Ministry of Ocean Economy, Marine Resources, Fisheries and Shipping, Mauritius Oceanography Institute, Mauritius Research Council, Ministries of Tourism, Finance and Economic Development, and Environment and Sustainable Development, private sectors, local NGOs, as well as academia had prepared the best practice policy for blue economy activities in Mauritius.

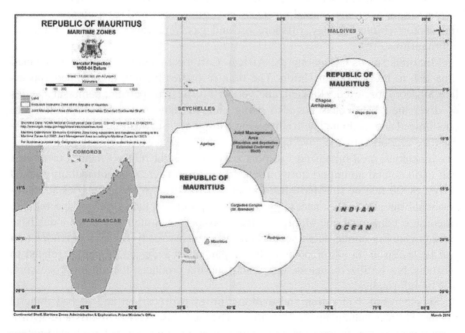

FIGURE 3.4   Republic of Mauritius and Its EEZ (DCSMZAE: Department for Continental Shelf, Maritime Zones Administration and Exploration, Mauritius)

The policy was thoroughly discussed in the United Nations Development Program (UNDP) conference in June 2017. Accordingly, the UNDP has taken up the matter at local and regional levels to support the Mauritius Government to tune the strategic action plan and prepare roadmaps for the development of the blue economy. The implementation of this strategy for Mauritius is in line with the UN's SGD 14 Goal and is monitored by its Ministry of Ocean Economy, Marine Resources, Fisheries and Shipping.

Alongside, with support from the Global Environment Facility (GEF), the bio-diversity of coastal zones and management of MPAs are being executed by the government at the local level. In this regard, the Adaption Fund support is well utilized for coastal protection as well as coral restoration activities on a larger scale. Recording its active role in the blue economy model, Mauritius co-hosted two major international congregations: *Towards COP-22: African Ministerial Conference on Ocean Economics and Climate Change*, and the World Bank-funded *Indian Ocean Rim Association Blue Economy Ministerial Conference*. Equally, Mauritius Oceanography Institute, University of Mauritius, and Mauritius Research Council are also participating in the development of the blue economy through research and innovation.

To become a model country in the blue economy, Mauritius is diversifying into other prime sectors, such as renewable energy, hydrocarbon exploration, and marine ICT (Information and Communication Technology). Mauritius also has an action plan to double its blue economy contribution to GDP with support from the World Bank. The strategy includes: (a) fine-tuning of conventional sectors (coastal tourism, seaport, and seafood-related activities); (b) nurturing promising sectors (aquaculture, maritime-related services, and commerce); and (c) promoting poten-tial sectors (biotechnology, offshore oil and gas, renewable energy, ICT (www. edbmauritius.org).

Mauritius stands top among the African-Caribbean-Pacific Exporters of canned tuna fish to the European Union. The tuna resources include high-valued species like albacore, big-eyed, skipjack, and yellow fin. Besides pelagic, demersal species of diversified fish stocks are also being caught. Industrial fishing operators are encouraged and facilitated with a five-year fishing license. Seafood processing indus-tries are supported by the government. The support services include integrated logis-tics amenities, well-organized Freeport zone, air freight rebate scheme for chilled fish exports, and no import duty on related machinery, etc. For aquaculture related activities, 20 marine sites have been identified for lease to private parties at a con-cessional rate, and other necessary logistic support. As far as the marine biotech research is concerned, the Mauritius Oceanography Institute has mooted a database (moi.govmu.org/online-databases) where all projects ready for commercialization can be accessed. In addition, biotech companies are engaged in studies related to biofuels, bio-fertilizers, marine drugs and pharma products, fish oil bioprocessing, and bioinformatics.

The scope of exploration and exploitation of hydrocarbon and seabed mineral resources are catching up in Mauritius. The new discoveries of offshore hydro-carbon occurrences in the waters of Madagascar and East Africa has created poten-tial interests in the EEZ of Mauritius. Following the discovery of hydrothermal fields off Mauritius jointly with Japanese scientists in 2016, the prospects of sulfide mineral deposits from the nearby mid-ocean ridge are considered for further investigation. Additional seabed surveys are planned jointly with collaborating countries for mapping geological features in unexplored basins in Mauritian EEZ and ECS (Extended Continental Shelf; Fig. 3.4) waters. Besides, exploratory works have started in terms of tapping energy from the ocean, which could be an added booster for blue economy prospects for the country.

In fact, Mauritius is advantageously situated along a key maritime traffic route with an approximate 30,000 vessels crossing its maritime zone. Its government has invested substantially recently to extend the container terminal, installing new cranes, dredging navigation channel to 16.5 meters depth, and toward land reclamation for port-related activities. Additional activities under the Smart Port System are under way for cruise passenger and island terminals. With reference to Mauritius becoming a Petroleum Hub, the present export of 0.36 million tons is expected to rise to 1.17, 1.83, and 2.50 million tons by the years 2020, 2030, and 2040, respectively (www.edbmauritius.org). Besides, petroleum storage facilities, bunkering, ballast water treatment, underwater ship repairing, and cruise line facilities are at various stage of completion.

Activities related to ICT contribute nearly 6% to the GDP of Mauritius, which basically include broadband connectivity, integrated technology platforms, and IT start-ups. These would enhance the prospect of weather modeling, satellite-based surveillance, and ocean sensors developments. Efficient legal frameworks have also been created for better involvement of stakeholder, and people, at large.

3.2.2 Challenges, Success, and Economic Potential

The success of the blue economy in Mauritius depends on the carefully drafted Marine Spatial Plan (MSP) that takes care of different externalities offering best-practice choices for long-term strategies. The MSP includes sustainability and eco-friendly approaches and carrying the concept of valuing ecosystem to future generations, particularly with reference to climate extremes (surges, cyclones, sea-level rise, etc.). In fact, integrating landscapes with oceanscapes for better low-carbon green economy has been the building block of Mauritius's blue economy strategy.

The blue economy of Mauritius represents nearly 18% of GDP with estimated employment of about 20,000 skilled people. The blue economy contributes US$2.4 billion to the overall GDP of US$13.5 billion. Compared to other Western Indian Ocean countries, Mauritius stands fifth, and its blue economy contribution to Mauritian GDP stands third, after exports and agriculture. The fishing, aquaculture, and seafood-related activities alone contribute nearly 1.3% to national GDP and account for about 9.1% of export earnings (www.edbmauritius.org; Cervigni and Scandizzo, 2017).

The three most blue economy activities that contribute substantially to the Mauritian GDP are coastal tourism and marine leisure, seaport-related activities, and seafood-related activities. These three sectors have the momentous capacity in future growth in terms of value-added products and enhanced employment (Fig. 3.5).

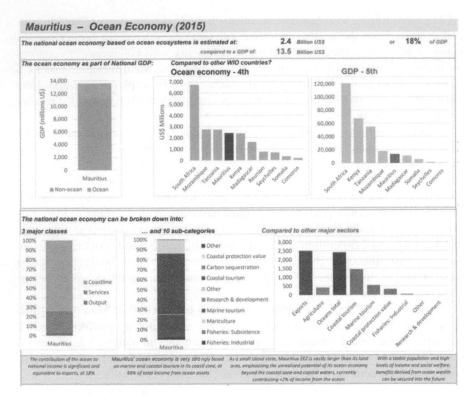

FIGURE 3.5   GDP and Blue Economy of Mauritius (Obura et al., 2017).

3.3   SEYCHELLES

With an EEZ of 1.37 million km^2 in contrast to land area of only 455 km^2, Seychelles, a small island nation, is exceedingly bonded to the ocean resources for its livelihoods and future prosperity (Fig. 3.6). Seychelles comprises 115 islands (41 inner and 74 outer), with 99% of the country's 94,000 population residing in 3 islands alone, viz., Mahe, Praslin, and La Digue. Seychelles covers nearly 6% of the Indian Ocean coral reefs.

3.3.1   RESOURCES, STRATEGY, AND ACTIVITIES

In January 2018, the visionary roadmap of the blue economy was approved by the Government of Seychelles with well-defined strategic plans and investment priorities up to the year 2030. The flagship programs of MSP and marine financing have been highly productive and successful initiatives wherein the country could realize the implementation of such approaches for economic development.

The blue economy vision of Seychelles was built upon the standards of economic competence, innovation and novelty, sustainability, social impartiality and justness, resilience, transparency, accountability and partnerships (Government

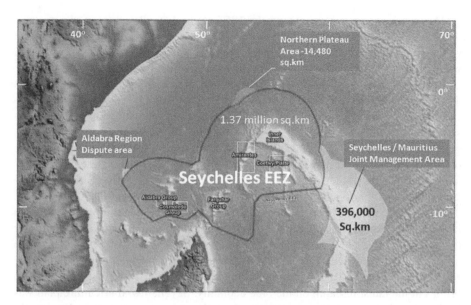

FIGURE 3.6   EEZ Extension of Seychelles (Seychelles News Agency)

Portal, 2018). The most significant blue economy sectors of Seychelles are fishing and tourism. In the absence of land-based food products, coupled with its dependence hugely on imports from other countries, fish remains a prime source of food security for this tiny island nation. Besides, aquaponics (a fusion of aquaculture and hydroponics), a strategic driver for the blue economy, utilizes fish waste to be used as fertilizer for vegetable growth. Micro-, small- and medium enterprises (MSMEs) get prime recognition and opportunity in fisheries, aquaculture, mariculture, blue biotechnology, tourism, and related activities.

Overall, fishing accounts for nearly 31.5% of Seychelles' income (World Bank, 2017). Seychelles contributes 20% of the global tuna catch and this has played a major role in the country's economy for the past four decades, but because of overexploitation, the fish catch has been reduced to 15% (Indian Ocean Tuna Commission, 2018). The traditional fishing sector has been modernized and facilities like handling and storage capacity are included. A recently built central common cold store at Ile du Port has capacity to store nearly 12,600 tons of fish. This facility would be a valuable addition to tuna fishing, and to also control tuna overfishing. Likewise, the multipurpose Port Victoria handled 250,000 tons of tuna in 2016, besides her impressive performance in handling cargo and cruise liners. In October 2018, Seychelles launched the world's first sovereign "blue bond" project to support sustainable marine fisheries projects. Financed largely by the World Bank through several investors, the blue bond is a pioneering financial instrument to harness capital markets for financing the sustainable use of Seychelles' marine resources. Proceeds from this bond will include support for the expansion of MPAs, improved governance of priority fisheries, and the development of Seychelles' blue economy (www.ipsnews.net).

The tourism sector supports nearly 18% of total employment in the country (National Bureau of Statistics, 2017) and encourages and promotes tourism with value-added services and quality. The proposed construction of a maritime museum would attract tourists, besides remaining as an educational asset.

Besides fishing and tourism, the blue economy agenda of Seychelles also includes development of ports and their infrastructure, IT connectivity, sea connectivity, aquaculture, renewable energy, seabed resources mapping, blue biotechnology, and ecosystem conservation. As the blue economy and maritime security are interrelated, security threats such as illegal-unreported-unregulated (IUU) fishing and marine pollution are monitored through coordinated surveillance. Besides these, Seychelles' large EEZ is vulnerable to piracy and human trafficking. Such security risks are scrutinized and controlled by the national maritime security agency.

The blue economy strategy of Seychelles addresses the integration of MSP with land use, research, innovation, maritime safety, and regional cooperation. In order to support Seychelles' blue economy prospects, the University of Seychelles set up an autonomous knowledge center called the James Michel Blue Economy Research Institute (BERI) in 2015. Many international universities lend a hand to BERI to bring out the latest research on the blue economy. Seychelles' mariculture plan was finalized in 2015 to promote culturing of pearls and sea cucumber in Praslin. As a member of the Indian Ocean Commission, Seychelles shares its intelligence and capacity building expertise with neighboring countries to check and control illegal fishing that has emerged as one of the threats to the blue resource of the country.

Since Seychelles entirely depends on imported petroleum for energy generation and consumption, there is an effort to partially switch over to renewable energy (Report on Energy Policy of the Republic of Seychelles 2010–2030, 2015). The small and medium enterprise loan scheme supports entrepreneurship in developing Ocean Thermal Energy Conversion (OTEC) and offshore wind-power generation. Oil and gas exploration is another area where the Seychelles Government has put in a lot of efforts during the last few years. Seychelles and Mauritius signed an agreement in 2012 to jointly manage the seabed resources of a common area of 396,000 km^2 (Joint Management Area, JMA) within their EEZ and ECS.

Since agriculture is not much developed, tourism and fishing are the main pillars of the economy of this island, while shipping and port developments are considered as supporting services. The Seychelles Ports Authority (SPA) extends facilities such as trans-shipping space, cargo handling, bunkering, warehousing, cold storage, and inter-island terminals, etc. Upholding small-scale fishing port amenities for sustainable fishing and high-quality inter-island transport services as communication facility are prime concerns of the government. Under prospects of the blue economy, marine renewable energy and marine biotechnology are however taken up as potential sectors.

3.3.2   CHALLENGES, SUCCESS, AND ECONOMIC POTENTIAL

Since Seychelles waters host a rich marine biodiversity including giant tortoises and corals, the environmental sustainability is of main concern under the blue economy developmental program. To address this issue, marine spatial planning was

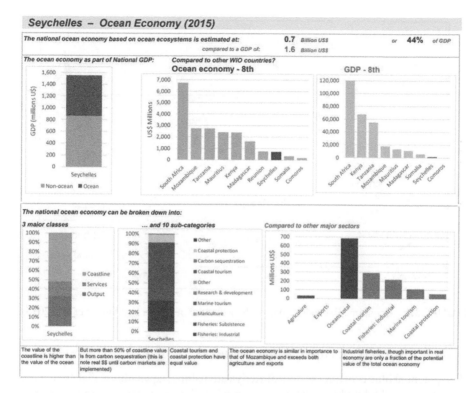

FIGURE 3.7 Blue Economy Contribution to Seychelles GDP (WWF, 2016)

undertaken first, and the offshore coverage of the MPAs has been expanded from 15% of EEZs to 30% (www.seymsp.com). The Seychelles government, UNDP, and Global Environment Facility (GEF) collaborated to start an MPA finance program to the tune of US$6.8 million.

Overexploitation of tuna fishing is a major problem and restrictions are made especially for yellow-fin tuna. For sustainable fishing, development of an artisanal fisheries management plan and a participatory approach by local communities are encouraged. Moreover, there is a plan to use fish products for pharmaceutical and cosmetic industries (Report of Nordic Atlantic Cooperation, 2018). The tourism sector of Seychelles contributed 25% to her GDP in 2015 (equal to more than US$400 million), while the fisheries sector accounted for 8% of GDP (World Bank, 2017; Fig. 3.7).

3.4 KENYA

The Republic of Kenya has a 640 km-long coastline with an EEZ of 230,000 km². In addition, it claims 103,000 km² under ECS (Fig. 3.8). Nearly 4 million people live in the coastal area. The Kenya Marine and Fisheries Research Institute (KMFRI) spearheads the fisheries research in the country.

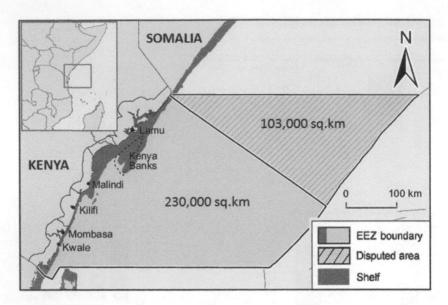

FIGURE 3.8   Economic Exclusive Zone of the Republic of Kenya Le Manach et al., (2015)

3.4.1   Resources, Strategy, and Activities

The potential blue economy sectors of Kenya are fishing, aquaculture, maritime transport, tourism, shipbuilding and repairing, ship handling, port-related services, marine cargo logistics, safety and security, offshore mining, and marine biotechnology and bio-prospecting. Of these, tourism, fisheries, and shipping jointly contribute 4% to national GDP. There is a huge scope for Kenya to develop blue economy strategies following the examples set by nearby island nations such as Seychelles and Mauritius. After new strategies were formed in 2018 by the Kenyan Government, aquaculture received a committed focus. This resulted in construction of nearly 48,000 ponds and creation of four mini-processing and cold storage facilities.

In May 2016, an umbrella ministry was created to synchronize the activities of all sectors related to the blue economy. Further, a special committee was formed by the government in September 2016 with emphasis on fisheries, maritime transport, and marine logistics services. Subsequently, the "Presidential Blue Economy Task Force" was formed in 2017 with a view to use blue resources for the economic development of the country by ensuring stakeholders' participation. Besides addressing the sustained livelihood and creating employments, the task force monitored the implementation of action plans in promising sectors such as living resources (fisheries and aquaculture), non-living resources (mining of hydrocarbon and minerals), tourism, maritime transport, and environmental conservation (www.unenvironment.org).

Following the first global conference on the sustainable blue economy with focus on creating economic growth, ensuring healthy waters, and building safe communities held in Nairobi in 2018, the prime policies on financing blue economy and people-oriented sustainable strategies were evolved. Consequently, nearly US$172 million was committed to the development of blue economy activities by the international community through joint ventures and capacity building (Government Report, 2018). Further to the continuation of the committed activities on blue economy, various measures such as launching new coastguard vessels (to fight piracy and illegal fishing), establishing a new fishing corporation and new fishing ports in S himoni, Kilifi, and Lamu (with a potential of creating 12,000 new jobs), and expansion and modernization of several other ports were initiated.

Vision 2030 has accelerated Kenya's efforts at development of the blue economy. A timely attention is being accorded by Kenya to maritime transport and logistics services by building many port infrastructures along the coast and retuning the National Shipping Line for coastal shipping arrangement. In recent times, capacity-building measures for fish processing, shipbuilding, and ship repairs are highly encouraged. Four special economic zones are being established at Mombasa, Lamu, Kisumu, and Naivasah. Efforts are also being made to reinforce activities to cut Kenya's carbon emission through mangrove reforestation.

The fishery industry that includes fishing, fish processing industries, boat-building and repairing activities, and other downstream industries, employs nearly two million people. Inland fisheries (mainly from Victoria Lake) still accounts for about 90% of fishing in Kenya. The country has about 14,000 km² of suitable land for aquaculture with a potential to harness 14 million tons catch earning nearly US$500 million (FAO, 2014). The inland aquaculture production was 24,096 MT in 2014 but declined to 14,952 MT in 2016 due to the end of the economic stimulus program. But marine aquaculture is limited to 100 MT. Coastal shipping in Kenya is not an active sector. There are only two private shipbuilders, viz., African Marine and Engineering Co. Ltd and Southern Engineering Co. Ltd, who also repair vessels. With refurbishing of Mombasa port, the cargo handling has improved significantly from 14.4 MT in 2006 to 27 MT in 2017 (Marete, 2018).

3.4.2 CHALLENGES, SUCCESS, AND ECONOMIC POTENTIAL

Kenya has many serious challenges in developing the blue economy at its optimal level. Besides the long-time dispute with Somalia over the maritime boundary, the other pressing issues include IUU fishing, maritime terrorism, illicit trade in crude oil, arms, smuggling of drugs, and human trafficking. To append further concern to the existing challenges, the unorganized sand mining, destruction of coral reefs and coastal forests, and the dilapidation of marine ecosystems due to discharge of oil and the dumping of toxic wastes are threatening activities. The offshore area of Kenya has large oil and gas reserves (Wairimu and Khainga, 2017). However, a part of Kenya's ECS (about 100,000 km² beyond the distance of 370 km from the coastline) is under dispute with the neighboring countries. The Fisheries Management and Development Act 2016 has been successful in bringing IUU fishing under control by regulating licensing and movement of fishing vessels. For blue carbon strategies, many developments on mangroves, seagrasses, seaweeds, shrimps, lobsters, and

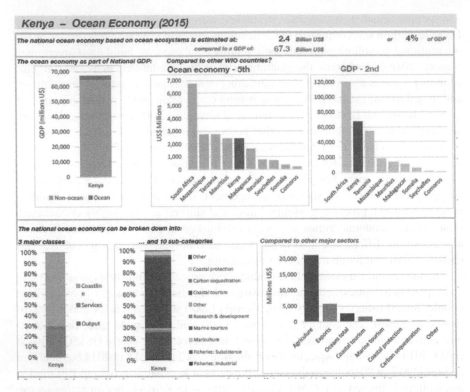

FIGURE 3.9 Blue Economy Contribution to Kenyan GDP (WIO Ocean Economy Report (Obura et al., 2017))

prawns are being effectively implemented in recent times. The Beach Management Units manage all these critical habitats along the coast.

Kenya's GDP in 2015 was US$67.3 billion, which in 2018 increased to US$85.98 billion. The GDP of Kenya has been second among the West Indian Ocean countries, and it is fifth as far as the blue economy contribution to GDP is concerned. In 2015, blue economy sectors contributed 4% to the GDP, which amounts to US$2.4 billion (Fig. 3.9). Although the fisheries sector contributes only 0.5% of GDP, nearly 2 million people are employed in this sector (Obura et al., 2017). Tourism, however, adds substantially to the GDP with a significant share coming from coastal tourism (Benkenstein, 2018). Given the vast ocean resources and when compared with its neighbors, Kenya has potential to augment and excel in blue economy sectors through best practice strategies and action plans.

3.5 OMAN

Oman has a 3,165 km-long coastline, with UAE and Yemen as its maritime neighbors. Out of its 309,500 km² of total land area, 82% is desert, 15% is mountainous terrain, and the remaining 3% lies in the coastal plain (Fig. 3.10). The EEZ of Oman is

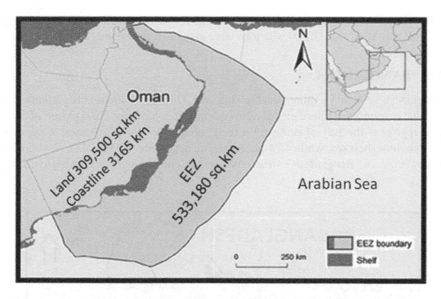

Oman

Land 309,500 sq.km
Coastline 3165 km
EEZ 533,180 sq.km

Arabian Sea

EEZ boundary
Shelf

0   250 km

FIGURE 3.10   Exclusive Economic Zone of the Republic of Oman (www.seaaroundus.org)

533,180 km². The fertile coastal plain at Al-Batina is the most industrialized city and has several ports and fishing villages.

Around the early 2000s, the IORA played a significant role in establishing the Fisheries Support Unit in Oman. In 2011, Oman was selected to set up the Regional Maritime Transport Council with support from the Oman Chamber of Commerce and Industries. By 2013, Oman was accounting for one-third of foreign trade for IORA countries (Wippel, 2013). Oman's vision and perspective on the blue economy were detailed in 2019 in the Ocean Economy and Future Technology conference held in Muscat. The main sectors identified are fisheries, aquaculture, shipping transport, offshore oil and gas, offshore renewable energy, and marine mining (www.omanobserver.com).

Tourism is a potential blue economy sector of Oman due to its long coastline and historic places including four UNESCO heritage sites. In 2013, the tourist inflow was 1.96 million and the government aims to target 12 million by 2020 (NCSI, 2014). As a new project, major upgradation and expansion are under way at Muscat international airport to handle 12 million passengers. Likewise, Salalah international airport and minor airports in Sohar, Duham, Adam, and Ras al-Hadd are been upgraded. The fisheries sector employs nearly 15,000 fishermen and supports 100,000 people's livelihood. To give a sense of resource potential, the fish catch in 2011 from Oman waters was 158,723 tons (NCSI, 2014).

The GDP of Oman in 2013, 2014, and 2015 was US$78.93 billion, US$81.03 billion, and US$69.83 billion, respectively. There was a decline in economic growth in 2016 due to underperformance of the oil sector and lack of fresh investments. But Oman returned to high GDP in 2018 with US$79.23 billion. Since onshore wells are deep, the cost of drilling, production, and developing remains high, while the offshore wells are more readily manageable. In 2014, the contribution from tourism to GDP was 3.3% and 37,000 new jobs were generated (www.mmis.com).

3.6 BANGLADESH

Bangladesh's sea area is more than her land area. Thus, it has focused her attention on the blue economy for future growth. This country encompasses 118,813 km^2 of maritime regime in the Bay of Bengal. Of these, about 20% include coastal waters and 35% shallow shelf sea, while 45% are located in deep waters (Fig. 3.11). A whopping nearly 90% of Bangladesh's trade is carried out through sea. Approximately

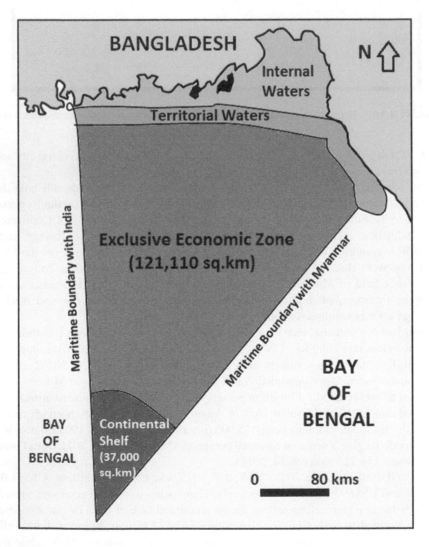

FIGURE 3.11   Maritime Boundaries of Bangladesh (MoFA, 2014).

30 million people are dependent on ocean economic activities. After Bangladesh resolved the maritime boundary problems with Myanmar in 2012 and with India in 2014, the blue economy is fast emerging as an important tool to develop the country, socially and economically.

3.6.1 Resources, Strategy, and Activities

The key blue economy sector of Bangladesh is fishing. Its EEZ has four potential fishing grounds covering a total of 14,600 km^2 in the Bay of Bengal (Shahidullah, 1983; Fig. 3.11). In 2016–17, 16% of the fish catch of Bangladesh came from sea, 28% from captured sources, and the other 56% were the cultured products. Among the fish available in Bangladesh waters, hilsa, rohu, katla, shrimp, mud crabs, and estuarine eels remain the major living resources. In addition to fisheries, the other sectors that can strengthen the blue economy and make it attractive in Bangladesh are mangrove tourism, exploration of minerals (including salts) and hydrocarbon, and probably power generation from tides and wind.

Bangladesh is one of the world's leading fish producing countries with a total production of 4.134 million tons in 2016–17, 0.08 million tons more than what was targeted for the year. Of this, fish from aquaculture contributed 56.44%. In the last ten years the average growth performance of this sector was almost 5.43%. Government is trying to sustain this growth performance, by ensuring a production target of 4.55 million tons by 2020–1. According to FAO statistics from 2016, Bangladesh is ranked fifth in world aquaculture production and has achieved self-sufficiency in fish production (FSB, 2017).

Hilsa (locally called Iilish) is the national fish of Bangladesh and has been declared as a Geographical Indicator. Hilsa accounts for about 12% of the country's total fish production. To achieve an increased target, the government has implemented a unique coordinated management program. This resulted in an enhanced production from 0.199 million tons in 2003–4 to 0.496 million tons in 2016–17. Shrimp is also one of the major export items in Bangladesh. Total shrimp and prawn production including through captured mechanism has improved from 0.160 million tons in 2002–3 to 0.246 million tons in 2016–17 (FSB, 2017).

Nearly 6 million people in Bangladesh are engaged in the sea salt production industry and the shipbreaking industry (Patil et al., 2018). Salt mining and associated downstream industries created 5 million jobs and supported 25 million livelihoods (Al Mamun et al., 2014). The coastal sand consists of heavy minerals such as zircon, ilmenite, rutile, kyanite, garnet, magnetite, and monazite (Hossain et al., 2015) and could be explored for mining. After the resolution of the maritime boundary with Myanmar, the newly added offshore areas showed the potential of oil and gas reserves. Especially, in the southern island district of Bhola, huge reserves of natural gas were discovered to the tune of 700 billion cubic feet (www.theindependentbd.com/post/120371). The shipbuilding sector is also projected to increase by 10–15% in the next decade.

The integrated plan of Bangladesh's blue economy envisages: (a) expanding fish catch through best practices; (b) identifying more suitable places for aquaculture; and (c) controlling anthropogenic pollution. The strategy further includes assessment of the stock, identification of migratory routes, and suitable catching techniques. The 7th Five-Year Plan (2016–20) of Bangladesh highlights the development and management of fisheries including deep-sea fishing, improved aquaculture, renewable energy from the ocean, new maritime industries (like shipbuilding), eco-tourism (marine cruises and inland waterways), shipping infrastructure, and capacity-building in marine science and research. About 1.7 million tons of salt are produced in each year, which is mostly consumed domestically (Al Mamun et al., 2014). In 2012, nearly 231.5 million passengers were transported through waterways. During the 2006–15 period, maritime trade earned US$95 billion as freight charges to shipping (Alam, 2015).

3.6.2   CHALLENGES, SUCCESS, AND ECONOMIC POTENTIAL

The unique conservation regulatory mechanism of Bangladesh with regard to hilsa fish species, a valued commercial food product, has become a learning model to all the South Asian Association for Regional Cooperation (SAARC) countries especially India, Myanmar, and Pakistan. The framework regulates overcatching, and catching at the breeding grounds. There is still a need to: (a) develop MCS (management and conservation security) strategies to monitor the activities of mechanized trawlers as well as semi-mechanized boats; (b) construct more surveillance check posts as there is only one in Chittagong; (c) create sufficient infrastructural facilities such as minor fishing ports, and efficient maritime communication (Nasiruddin, 2015). The application of marine biotechnology has high scope for improvement.

In 2000, the population in the coastal lowlands of Bangladesh was about 64 million and is projected to increase to 85 million in 2030, and further over to 100 million in 2060. Bangladesh is vulnerable to climate change and inundation by a rise in sea level. It is estimated that sea-level flooding could inundate more than 17.5% of the land, unsettling the livelihood of millions of people and damaging substantial economic activities. Judging by all these opportunities and threats, Bangladesh requires "ocean auditing" in an effective manner and a plan for specific stages of transition to initiate blue economy activities (World Bank, 2016; Mukhopadhyay et al., 2018).

Bangladesh exports frozen shrimp and other fish and fishery products to more than 50 countries, including Belgium, the UK, The Netherlands, Germany, the USA, China, France, Russian Federation, Japan, and Saudi Arabia. In 2016–17, the country earned US$~508 million. In 2015, the blue economy contributed US$6.2 billion to the Bangladesh economy (about 3%). Of these, tourism and recreation contributed 25%, marine fisheries and aquaculture 22%, shipping and transport 22%, offshore gas and oil extraction 19%, shipbuilding and breaking 9%, and minerals 3% (Patil et al., 2018).

3.7   SRI LANKA

Sri Lanka is situated in a strategic location of important sea trade route in the Indian Ocean. A population of 20.9 million people resides in a land area of 62,700 km². Its

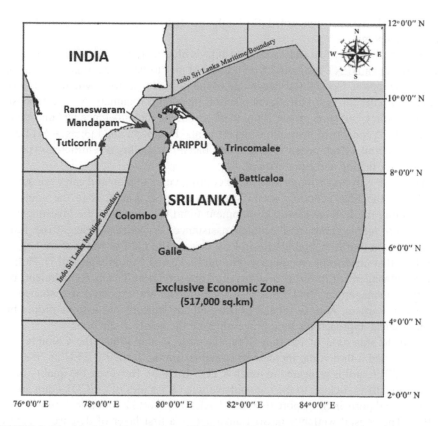

FIGURE 3.12 Maritime Boundaries of Sri Lanka (Flanders Marine Institute 2018 Maritime Boundaries Geodatabase).

EEZ area is more than eight times than its land area (Fig. 3.12). The coastal area is administered by 14 well-defined districts covering 23% of the total land area, and shelters nearly 25% of the total population. This population solely depends on ocean resources and related activities (Azmy, 2013).

3.7.1 Resources, Strategy, and Activities

Captured fishes have been the main source of protein (70%) for the one-quarter of the country's population that lives along the coast (FAO, 2014). In addition, for food security aquaculture and mariculture are encouraged by the authorities. However, the fishing gears used by the fisher-folks, more particularly the explosives used during fishing, have been criticized, warranting a coordinated surveillance by the authorities. Besides damaging the coastal ecosystem, the blasting of explosives threatens the coastal tourism that accounts for three-quarters of the entire tourism infrastructure of the country. Nearly two-thirds of global sea-borne trade are routed through Sri Lanka, which could be a potential resource to develop in future. As far as the ocean-based energy is concerned, Sri Lanka needs to project suitable plans for tidal and wave energy.

In 2016, Sri Lanka had endorsed the country's blue economy initiatives by adopting the "Sri Lanka NEXT program" with a coveted aim to become a regional maritime hub in the region. Coastal tourism is one of the potential sectors, which needs a significant facelift through best practice action plans and roadmap. A comprehensive national policy and strategy got going to weed out the incompetence and unaccountability of the concerned institutions/departments. Developmental plans of Sri Lanka aim to undertake deep ocean research, and incentivize using scientific eco-friendly methods for fishing and aqua/mariculture.

The MSP will play a major role in the coming years to plan and carry out actions to elevate the blue economy sector. Accordingly, Sri Lanka is in the course of establishing the Indian Ocean Development Fund to appreciate the investments in the form of loans, grants, and technical assistance (Wickremesinghe, 2016). Issues with India on maritime boundaries in Palk Bay and in Gulf of Mannar are ongoing conflicts that need solutions quickly (Goonetilleke and Colombage, 2017). Holistic coastal management in terms of ecosystem conservation, beach nourishment, bioresource management, and coastline development are expected to enhance the quality of tourism. Actions related to licensed aquaculture and using a proper biotechnological approach for food processing are the needs of the hour.

With the initiation of the transshipment trade in the 1980s, the Colombo port has witnessed a three-time increase in container traffic, from 1.7 TEU in 2000 to 6 TEU in 2017, and is projected to touch 8 TEU in 2020. However, the quality of the Colombo port is sticking to the regional average (4.5 scales of OECD), compared to the high-end ports in Singapore (6.7) and UAE (6.2, World Economic Forum Report, 2017). The coastal wetlands in Sri Lanka act as a first layer of defense to control floods (IUCN, 2007).

3.7.2 CHALLENGES, SUCCESS, AND ECONOMIC POTENTIAL

Coastal tourism is one of the successful activities especially in the western and southern provinces of Sri Lanka; particularly whale watching has become very popular in recent times. The incredible growth of the Colombo port since 2000 is attributed to its strategic location, and to the recent enhancement of facilities including opening up four new terminals and deepening of the main channel to house larger container ships.

Although progressing well, this island country has failed to live up to expectation during the last few decades. For example, sustainable fisheries management suffered from a lack of reliable data on illegal and unauthorized fishing. There appears no end to the dispute with India on border fishing activity (Goonetilleke and Colombage, 2017), so also the trafficking of drugs and human in Sri Lankan waters (Dissanayake, 2015). Equally damaging issues to the marine ecosystem are the accidental oil leakage and overfishing.

The merchandize export and import of Sri Lanka since 2005 has been dynamic, and in a way reflect her ambition to be the hub of the Indian Ocean trade in near future. The rate of growth of merchandize was 10.2% in 2017. The total export varied from US$6,347 million in 2005, to US$8,602 million in 2010, to US$10,505 million in 2015, and to US$11,360 million in 2017. In contrast the total merchandize import during the same years respectively was US$8,834 million, US$13,512 million, US$18,935 million, and US$20,980 million. The balance of trade has been negative and increasing from US$-2,487 million in 2005 to US$-4,909 million in 2010, US$-8,430 million in 2015, to US$-9,619 million in 2017. Of the exports, food items account for 67%, and manufactured goods 25%. Similarly, the total export in transport services increased from US$673 million in 2005 to US$2376 million in 2017, against an import of US$1985 million in 2005 and US$3576 million in 2017 (UNCTAD Maritime Profile, 2019).

3.8 INDONESIA

With more than 17,500 islands enveloping a total marine area of about 6.32 million km^2, Indonesia is an ocean-based society. The country has a long coastline of 108,000 km, with more than 50% of people of the country being associated with coastal resources. It has 6,159,032 km^2 of EEZ with shelf area of 2,039,381 km^2. The MPAs are 2.2% of territorial waters and coral reef covers about 51,020 km^2 and mangrove spread over 31,894 km^2 (Fig. 3.13; World Bank, 2013).

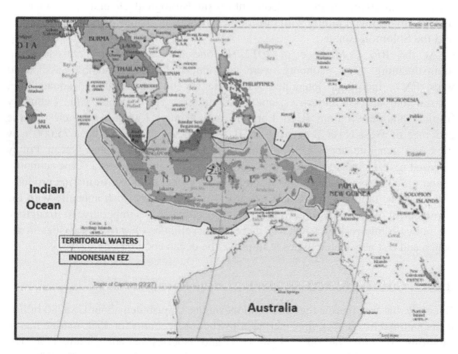

FIGURE 3.13 Territorial Waters and Exclusive Economic Zone of Indonesia (www.quora. com/What-countries-border-Indonesia).

3.8.1 RESOURCES, STRATEGY, AND ACTIVITIES

Indonesia has a good potential for oil and gas in its maritime regime. In fact, about 90% of promising oil and gas basins in Indonesia are found in the marine regime: 14 close to the coast and another 40 in the offshore region. The reserve potential of oil is to the tune of 11.3 billion barrels and gas reserves are about 101.7 trillion cubic feet. Besides, the new discoveries of gas hydrates and biogenic gas off the west coast of Sumatra and southwest of Java add significance to the natural gas potentials (Kuswardhani et al., 2013). The fisheries and aquaculture industry employ about 7 million people. The country has also increased its potential in cargo/shipping trade sector as more than 70% of global sea trades passes Indonesia through the Strait of Malacca and the Strait of Timor. Moreover, about 76% of the coral species of the world lies within the Coral Triangle Area in which Indonesia is a major partner. And coral reef tourism fetches more than US$3.1 billion annually. As far as coastal tourism is concerned, nearly 44% were visitors from abroad.

The blue economy strategy of Indonesia is based on the integration of various marine economic activities with the conservation of the same ecosystem and existing sociocultural regime. The strategy encourages tapping of so-far untapped potential of bio-prospecting, marine mineral resources, desalination, ocean energy, coastal tourism, and ICT. As a research partner, Germany supports Indonesia to dwell into R&D research related to marine biodiversity and blue carbon ecology (coral reef and mangroves). In 2006, Indonesia proposed the "Regional Coral Triangle Initiative," at the 8th Convention on Biological Diversity, to conserve marine environment jointly with Philippines, Malaysia, Timor, Papua New Guinea, and the Solomon Islands.

At the Rio Conference in 2012, Indonesia's zero carbon emission and mainstreaming blue economy were announced. In continuation, associated with FAO, Indonesia integrated its strategies on tuna fishing, aquaculture, coastal tourism, salt industries, and pearl culture. The Lombok Blue Economy Implementation Program (East Lombok and Central Lombok regencies in the province of West Nusa Tenggara—NTB) is expected to create about 75,000 new jobs and generate about US$115 million annually. The National Ocean Policy released in 2017 includes roadmaps and protocols for the Indonesian blue economy (Dinarto, 2017). Indonesia is the second-largest global producer of marine products and its aquaculture sector produces nearly 4 million tons of fish annually with the involvement of about 4% total workforce of the country. It is imperative to strengthen the policy framework to watch and increase surveillance of various activities in the ocean and test new technologies to improve the ocean economy.

3.8.2 CHALLENGES, SUCCESS, AND ECONOMIC POTENTIAL

Recently, the vast marine regime of Indonesia has contributed about US$280 billion to its national income. Of this, marine construction accounts for 35%, marine manufacturing 26%, mineral, oil, and gas 16%, and fisheries and aquaculture 11%. The rest is contributed by marine tourism (10%), and marine transport (1%) (Ebarvia, 2016; Indonesian Ministry of Marine Affairs and Fisheries, 2018).

However, there are many threats to the blue economy pattern in Indonesia. For example, more than 50% of Indonesia's wild fish stocks were overfished in 2017. Marine plastic debris causes an annual revenue loss of US$140 million in the tourism sector and US$31 million in the fisheries sector. Shortcomings in implementing blue economy strategies in Indonesia include lack of infrastructure networks such as ideal electricity grids, processing and storage amenities, besides lack of incentives from the government to aid start-ups in upcoming and promising sectoral activities in the blue economy. Though nearly 70% of sea trades traverse through Indonesia (via the Strait of Malacca and the Strait of Timor), due to under-development of ports, shipping, and inadequate infrastructure, it stands to lose huge revenue. Moreover, the country is struggling with issues such as marine pollution, overfishing, IUU fishing, and degradation of coral reef. The IUU fishing alone inflicts a loss of US$3 billion annually. As a punishment, 81 such captured vessels were made to sink in 2017 alone. Bleaching of corals is predominant and nearly 82% of bleaching has been reported due to ocean acidification (Burke et al., 2002).

More than 1.29 million tons of plastic debris were reported to have been dumped in the sea in 2010 (Jambeck et al., 2015). About 108 maritime piracy cases that occurred close to ports were registered in 2015 by the International Maritime Bureau. Mangroves and associated landforms of sea grasses and tidal marshes are dwindling in coverage due to expansion of shrimp aquaculture and palm oil plantation. In view of all these critical issues, it is suggested in the national plan to restore and conserve the existing marine ecosystem. It is also planned to create an Indonesian Marine Pollution database to aid policymakers and law enforcement officers for implementing action plans and enforcement strategies (www. wri-indonesia.org). Appropriate action to control wastewater pollution (through treatment) in core tourism zones caused by local residents needs to be taken up. The overtourism in some parts of Bali and Lombok is required to be checked, as it exceeds the carrying capacity of the region.

In 2008, the contribution of the ocean economy to the Indonesian total GDP was US$73 billion, which was about 13% of GDP (Indonesian Maritime Council, 2012). This contribution got amplified in 2013 to US$256.5 billion and to US$280 billion in 2018 (Fahrudin, 2015; Indonesian Statistical Council, 2015). In 2015, the total value of Indonesia's maritime eco-system service was US$412 billion. The blue economy contribution to GDP in 2016 was 6.7% and about 5.3 million people were employed in ocean-related activities. The fisheries sector contributed nearly US$26.9 billion, which is about 2.6% of GDP. The potential of marine resources that can actually contribute to the Indonesian GDP has been projected to a much higher estimate of US$1,096.62 billion/year (Indonesian Ministry of Marine Affairs and Fisheries, 2018).

There are about 64,028 fish processing units in Indonesia and nearly 4.83 million tons of fish were processed in 2012. The conventional use of mangrove products as the blue carbon product is valued to US$3,000/0.01 km^2/year (IUCN, 2007). The maritime sector employs 4.12% of the total workforce of Indonesia, of which the aquaculture and fish sub-sector contributes 0.79% and the fish processing sub-sector engages 0.70%.

3.9   THAILAND

The coastline of Thailand is 3,148 km long, while its EEZ covers about 420,280 km² (Fig. 3.14). Of the total coastline, nearly 830 km faces critical coastal erosion. National Marine Park covers about 4,791 km² of land and proclaimed areas of non-hunting ground encompass about 72 km². The environmentally protected area spreads to 4,518 km².

3.9.1   RESOURCES, STRATEGY, AND ACTIVITIES

Both fisheries and aquaculture are major contributors to Thailand's ocean paradigm with nearly 90% of fish being exported. These industries provide employment to about 650,000 (Department of Fisheries, 2015). According to recent statistics, blue carbon resources significantly complement the resource potentials of Thailand as far as the blue economy is concerned. Out of 2,837 km² of coverage by blue carbon resources, the spread of mangroves, sea grass, and coral reefs shares 2,441 (86%), 205 (7.5%), and 191 (6.5%) km², respectively.

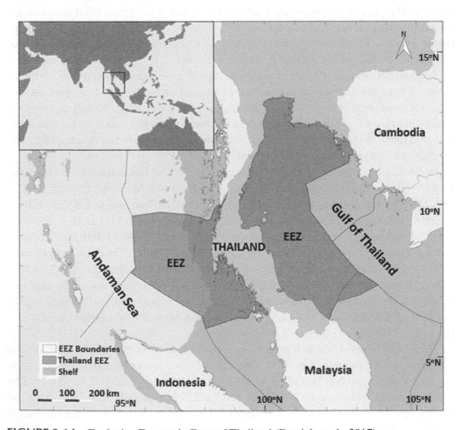

FIGURE 3.14   Exclusive Economic Zone of Thailand (Derrick et al., 2017).

The plan and implementation strategy of Thailand's blue economy was enacted in 2015 in the form of the Marine and Coastal Resources Management Act. The strategies on the resource conservation and management were effectively redrafted in which community participation and ecological governance are to play a significant role. Coastal tourism and marine transport are developed as promising sectors in the blue economy concept. Due to data limitation on ocean economy, the economic evaluation leading to the predictive growth trend in blue economy sectors has its own confinement. It is recommended that its national policy on assessment of ocean economic activities must also include economic accounting (Nabangchang and Krairapanond, 2015). In response to the above recommendation, Thailand has taken up new research projects to determine the economic value of mangrove ecosystem, work out strategy for blue economy development of coastal provinces, and zoning of blue economy resources (Proceedings on Blue Economy Forum, 2017).

Thailand has also recently commenced mangrove rehabilitation plans, fish stock assessment, low-cost wastewater treatment schemes, offshore and coastal wind power, eco-tourism, and low-carbon tourist destination plans. In fact, mangrove products have been contributing more than a quarter of per capita GDP (IUCN, 2007).

3.9.2 CHALLENGES, SUCCESS, AND ECONOMIC POTENTIAL

Thailand is now witnessing a rapid decline in mangrove forest coverage due to accelerated shrimp farming, coastal erosion, and unregulated urbanization. Considerable stretches of beach areas are increasingly being utilized for building hotels and resorts for recreational activities. Another undeniable issue is the bleaching of corals due to ever-increasing ocean acidification, and an increase in sea surface temperature. Moreover, free-for-all coastal tourism and effluent discharge from industries and domestic sources are posing threats to marine mammals (Naban gchang and Krairapanond, 2015).

While both fisheries and aquaculture contributed US$3,473 million in 2013, the contribution from the non-living resources was US$17,703 million in 2014. The marine manufacturing industries, on the other hand, earned US$22,686 million in 2014 (Department of Marine and Coastal Resources, 2014). Living and non-living marine resources in Thailand waters are estimated to be around US$685.70 billion. In 2015, the ocean economy of Thailand was US$118.19 billion (gross value added) and the value of ecosystem services was to the tune of US$36 billion. The GDP of Thailand in 2018 was US$505 billion, which is expected to grow by 3.9% in 2019 and by 3.7% in 2020. The blue economy overall contributes about 28% to the GDP of Thailand.

3.10 AUSTRALIA

Australia has a maritime estate of 13.86 million km^2, which is double the size of its landmass (Fig. 3.15). The estate spreads over the Indian, Southern, and Pacific oceans, and is the third-largest marine jurisdiction in the world. The length of the coastline of Australia is about 60,000 km. Apart from the EEZ associated with its mainland and islands Australia has about 2 million km^2 EEZ area in Antarctic territory. Several

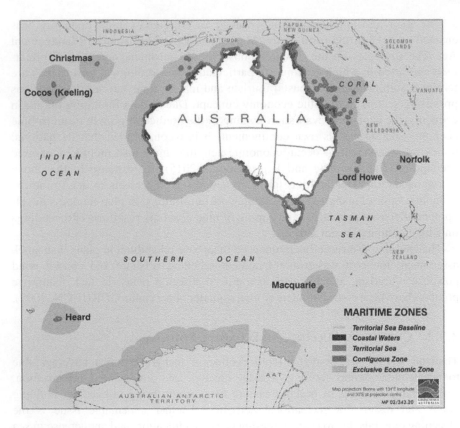

FIGURE 3.15   Maritime Regime of Australia (Courtesy of Geoscience Australia).

global heritage sites such as the Great Barrier Reef, the Ningaloo coast, Shark Bay, etc., are found in Australian waters.

3.10.1 RESOURCES, STRATEGY, AND ACTIVITIES

The major blue economy sectors of Australia include tourism, fisheries, aquaculture, ports, shipbuilding, offshore oil and gas along with emerging sectors such as ocean energy (wind, wave, and tide), biotechnology, and undersea mining. Australia's blue economy roadmap has been drafted as the National MSP (NMSP) for 2015–25. The blue economy value of Australia for the year 2015 was US$100 billion. The plan identified an array of research and development strategies focusing on energy security, food security, conserving marine ecology, coastal urban development, marine security, climate change, and balanced resource allocation (National Marine Science Committee, 2015).

Realizing the potential of the blue economy and being endowed with the world's third-largest EEZ, Australia allotted in April 2019 a considerable amount for research and innovation by establishing the world's first offshore blue economy platform through the Cooperated Research Centre (CRC). While the University of Tasmania will lead the research, 45 partners from national and international research institutions and industries will participate. As a floating marine laboratory, the offshore research platform will host facilities for renewable energy, aquaculture, and marine engineering activities. The CRC is anticipated to generate about US$3 billion for the country's economy (www.maritime-executive.com). Under the ten-year plan on "A Vision in Blue," the marine industries will contribute US$67.84 billion each year to the economy, in addition to US$16.96 billion from ecosystem services (www.aims.gov.au). Australia promotes regional marine security with trade partners China, Japan, and the Republic of Korea to decrease and eradicate IUU fishing through investments in surveillance vessels and submarines (Defence White Paper, 2016).

Among the prevailing activities, the fishing sector showed a high rate of growth in 2008–9 but later it started to decline, though mariculture grew strongly. The marine-based aquaculture contributed US$1 billion in 2012, which almost doubled in 2015. The offshore oil and gas sector impressively recorded a growth from 33.4% in 2002–3 to 53% in 2011–12. The LNG (liquefied natural gas) export earned US$16.3 billion in 2013–14. Marine tourism accounted for US$14 billion in 2012. Nearly 95% of petroleum production was from offshore basins in 2009–10, worth US$21.8 billion, and 10,000 people were employed.

The recent discovery of new oil fields and creative offshore technologies has again put Australia in a stronger position as far as future potentials are concerned. Since appreciable measures are used for ecosystem services such as climate regulation, nutrient cycling, etc., the contribution from this sector could be expected to reach US$25 billion in the future.

3.10.2 Challenges, Success, and Economic Potential

The NMSP (2015–25) places a large emphasis on resource allocation to various blue economy sectors. The challenges and the economic potential of various domains of the Australian blue economy campaign are given separately.

Food: Australia now imports nearly 72% of its seafood as the fish catch and aquaculture sectors are yet to be developed to their best level to take care of the domestic needs. The decisive research challenges in marine aquaculture include developing cost-effective but high-quality aquaculture feeds, designing aquaculture spatial planning frameworks, and producing healthy stocks.

Energy: The export of LNG was to the tune of US$11.5 billion in 2013–14. The production consistently rose to 20 million tons in 2012 and is expected to reach 76.6 million tons by 2020. The success story of innovation in the offshore LNG industry is the floating production facilities. This facility is a floating platform of about 488 m long and 88 m wide installed in the northwest shelf of Australia. This facility serves both as a production platform and an offshore port and does not need any additional land-based operating plants and laying of pipelines, minimizing pollution in the process. Equally promising is the prospect of generating electricity from wave and tide. The most potential locations to generate energy from waves are found in the western coasts, while those from tides are identified along the southern and northwestern coasts.

Coastal environment: The zone of 50 km from the coast is a significant region in Australia as nearly 85% of the population lives in and around this area. In addition, all the coastal and marine-related facilities like port, shipping, marine industries, fishing and associated industries, mineral resources, renewable energy, etc., are located in this region. A range of specific strategies has been put in place to tackle climate extremities (cyclone, tsunami, landfall, uncertainties, and sea inundation).

Conservation: As a world leader in biodiversity conservation, Australia has a forceful legal framework such as Australia's Biodiversity Conservation Strategy 2010–30 and the Commonwealth Marine Bioregional Plan. New discoveries of marine species are continuously being reported from the deep and remote marine areas of Australia, and the existing conservation strategies are getting modified all the time.

Security and safety: In order to keep up peaceful relations with neighboring countries bordered by sea and to protect resources and trade-related activities, the well-defined marine safety strategies are equipped with new surveillance systems. Moreover, safeguarding maritime properties and coastal settlements from natural hazards and climate extremities, marine security is considered as a key blue economy sector (Brewster, 2015).

Climate change: As Australia is susceptible to the impacts of climate change like coral bleaching, tropical cyclone, ocean warming, etc., various real-time and dynamic models are prepared to understand, check, and mitigate the extreme events in best-suited ways. These impacts have already started affecting the population of the high-value fisheries, oysters, lobsters, and large predatory fishes. Moreover, understanding the effects of the El Nino Southern Oscillation (ENSO) phenomenon, the Indian Ocean Dipole (IOD), and the Southern Annular Mode (SAM) help device models related to the weather prediction and climate projections for coastal communities including fishermen. Even a rise in sea level by 1.1 meters as projected for the year 2100 would give an estimated loss of consolidated coastal resources of around US$150 billion.

Economy: In 2009–10, tourism activities earned US$8.1 billion, fisheries and aquaculture produced US$1.5 billion, and the shipbuilding sector contributed US$4.7 billion. The offshore oil and gas industry is growing fast and is expected to contribute US$46 billion to GDP in 2020 (Brewster, 2015).

In 2012, the blue economy of the country contributed US$33 billion to its total GDP. Due to the implementation of NMSP and well-structured strategies and management plan, the growth rate of the blue economy would be more than two to three times the rest of the country's economy, and it is anticipated to contribute US$70 billion by 2025 (Brewster, 2015). Over the next decade, the blue economy is expected to grow by about 7.5% per annum to exceed the proposed 2.5% growth rate in GDP (National Marine Science Committee, 2015).

3.11   SYNTHESIS

In summary, of the 50 countries that surround the Indian Ocean, this chapter has discussed only 10, besides India (see Chapter 2). These countries have shown an inclination to make the blue economy an important contributor to their overall growth (GDP), and a tool for conserving and sustaining the limited resources of this part of the globe (Table 3.2). However, the geographical, political, economic, and cultural multiplicity among the IOR countries may have all these years delayed the development of forming regional collaboration and institutions to maintain sustainability (Roy, 2019). In this direction the formation of IORA is a blessing. A few things that probably pre-empt the establishment of a successful blue economy regime are peaceful coexistence, military security, effective monitoring and law enforcement, policy certainty, ethical governance, and private investment. In the following chapters a new blue economy paradigm will eventually be outlined.

4 Global Status

Activities of the blue economy, which has a mandate to use oceanic resources in a sustainable manner, broadly into two categories—ocean industry and maritime services. The ocean industry generally includes extraction of oil and gas, mining of minerals, salt manufacturing, extraction of chemicals, biomedicine, energy and power generation, shipbuilding, engineering, and infrastructure development. The maritime services, on the other hand, include navigation, communication, shipping and transportation (of goods and hydrocarbons), tourism and leisure, scientific research, education and training, and also the suppliers and customers of ocean industries.

But as nations extract various ocean resources, the sustainability of the oceanic environment has increasingly been challenged. While highlighting the conflict between economy and ecology in the blue economy paradigm, this chapter assesses cases of ten selected ocean-based economies located in the Atlantic and Pacific oceans. These countries are China, the USA, Japan, Canada, New Zealand, Brazil, Pacific Islands, the United Kingdom, Mauritania, and Nigeria. The contribution of various ocean sectors in terms of economic output and generating employment is analyzed. The plan, program, preparedness, and implementation of each of these nations vis-à-vis the blue economy are discussed. Further, the areas of success and shortcomings are identified, together with the government initiatives in protecting the oceans and maintaining the marine ecosystem, besides listing what IOR countries could learn from such global experience.

4.1 CHINA

With its vast coastline extending for about 14,400 km, continental shelf spanning over 1.83 million km^2 (FAO, 2016), and a huge unverified exclusive economic zone (EEZ), China has great potential to develop its blue economy. Consequently, China developed the idea of a blue economy that involves both utilization of ocean resources and protection of the oceanic environment (Wang, 2016) and worked systematically for its implementation.

4.1.1 THE BLUE ECONOMY ACTIVITIES

China did not have a clear vision to develop its blue economy until the early 1970s (Zou, 2012; Conathan and Moore, 2015). The beginning of the blue economy in China can be traced back to the late 1970s, which gained momentum through the five-year plans. Though the 7th Five-Year Plan (1986–90) laid emphasis on building ports and marine-related equipment, the real interest in the blue economy began only during the 11th Five-Year Plan (2006–10). The blue economy registered an impressive growth of 13.5% per annum during this period and created 33 million jobs. Taking a cue from the 11th Five-Year Plan, the 12th Five-Year Plan (2011–15) outlined specific targets for the blue economy. It aimed to increase the share of the blue economy to 10% of national GDP by 2015 and up to 15% by 2035. The 12th Plan announced that developing the ocean economy would be a key national development strategy and introduced the concept of land–sea coordination. The 13th Five-Year Plan continued and pursued the idea of land–sea coordination and tried to enhance it to a higher level. The 13th Five-Year Plan has the stated goals to develop marine resources, improve maritime economy, protect marine environment, and safeguard marine rights and interests.

China's planned strategy to advance the ocean sector and blue economy is a success story and has achieved most of the targets. China was the largest producer and exporter of fish in the world in 2015 with a production of 65.2 million tons of fish. It is to be noted that 73% of this production came from aquaculture and 27% from capture fisheries. For this, approximately 1.04 million fishing vessels were engaged providing employment to 14 million people. Besides, 15.9 million people found employment in allied and associated services with its fish exports worth US$19.7 billion in 2015.

Shipping, transportation, and tourism are the other major components of the blue economy of China and it has made huge strides in these sectors in recent years (www.maritimecyprus.com/2018/10/31). It owns commercial fleets with a capacity of 147.2 million gross tons. China is the third-largest fleet owner in the world, with a global share of 11.5%, the first being Japan (Wartsila, 2018). China's maritime industry is one of the strongest in the world at US$1.2 trillion (nearly half of India's GDP), and it has been growing at an average rate of 7.5% per annum in recent years. It accounted for 67% of Asia's cruise-line passenger volume and about 2.1% of the world's total cruise passengers in 2016 (CLIA, 2016). China is also one of the global leaders in the maritime world with regard to ship owning and financing. China has a strong position in the world of ports. Seven of China's ports figure in the top ten ports listed globally, with Ningbo-Zhoushan port occupying the top position in terms of cargo handling (www.joc.com/port-news/) and Shanghai port ranked first in terms of numbers of containers handled (www.seatrade-maritime.com/ports).

Although China made a foray into cruise ship tourism only in 2013, it declared the same year as the maritime tourism year (www.llanda.com/chinas-marine-tourism-year-presents-great-opportunities/). This declaration was intended to bring more foreign tourists to the coastal cities. The cruise industry is expected to contribute US$8 billion to China's economy by 2020. During the period 2002–17, coastal tourism increased by almost ten times, from approximately US$20.7 billion to US$204.8 billion (Wai Ming To and Lee, 2018).

China does not lag behind in case of offshore mining either. China's seabed mining industry is emerging as one of the strongest in the world. It has invested heavily in the seabed mining industry and technology and is self-sufficient is exploration activities. China has the privilege of being the only country to obtain an exploration license from the International Seabed Authority (ISA) for all types of deep-sea mineral deposits (for manganese nodules and cobalt crust in the Pacific Ocean, and hydrothermal sulfides in both the India and Pacific oceans; Sharma, 2015).

In 2017, the maritime economy of China generated approximately US$1.09 trillion. The size of this economy is expected to reach US$1.4 trillion by 2020 (Xinhua Net, 2018). The maritime economy of China is classified into three sectors, viz. primary, secondary, and tertiary. The primary sector includes marine fishing and aquaculture industries. Offshore oil and gas, mining, shipbuilding, etc. are included in the secondary sector. The tertiary sector encompasses the transport industry, coastal tourism, and associated service industries. The primary, secondary, and tertiary sectors have witnessed an annual growth rate of 11.1%, 12.8%, and 14.2%, respectively. The share of these sectors to the nation's ocean GDP in 2017 was 4.64%, 38.77%, and 56.59% in primary, secondary, and tertiary sectors, respectively. The contribution of these sectors has changed over the years. While the share of the primary and secondary sectors has decreased, the share of the tertiary sector has increased (Fig. 4.1). The shares of the secondary and tertiary sectors remained more or less the same until 2010. Since then, the gap between the share of the tertiary sector and the secondary sector has only widened.

Though China achieved rapid economic growth in its Gross Ocean Product (GOP), this growth has not been uniform throughout the country. The intensity of blue economic activities varied widely among the 11 coastal subregions of the country (Ren et al., 2018). The GOP has been very high in regions like Guangdong and Shandong, but less in areas such as Hainan, Guangxi, Hebei, and Liaoning. Guangdong has

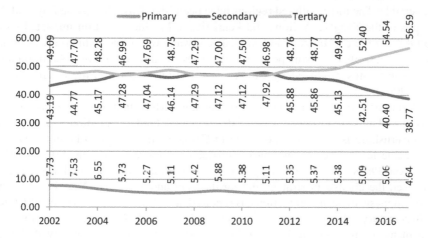

FIGURE 4.1 The % Share of Primary, Secondary, and Tertiary Sectors in China's Ocean GDP (Based on the sectoral data provided by Wai Ming To and Lee (2018)).

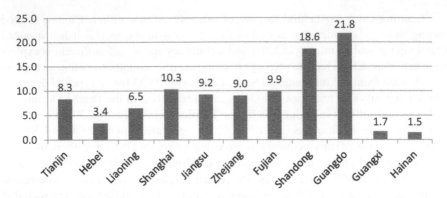

FIGURE 4.2   The % Share of 11 Sub-regions in China's Ocean GDP in 2017 (Based on the data provided by Ren et al. (2018)).

a share of 22% of the total ocean GOP, while Hainan has a share close to just 1% (Fig. 4.2). The uneven development of science and technology, socioeconomic conditions, and non-availability of natural resources are probably the key factors responsible for the regional differences (Sun et al., 2018).

4.1.2   Marine Environment and Conservation

According to State Environmental Protection Administration (SEPA), though China envisaged its blue economy strategy for developmental purposes, the overall quality of China's marine environment remains in dire states (World Watch, 2019, www.worldwatch.org/sepa). The study by SEPA was based on marine pollution, coastal ecosystem deterioration, and marine accidents. The study revealed that in some coastal areas, the damage caused by marine pollution had been severe and irreversible. The most polluted regions have been Bhai Bay, Hangzhou Bay, and the mouths of the Yangtze, Yellow, and Pearl rivers. Marine pollution impacted more than 80% of the coastal zone in China. The sewage water and industrial wastes are the major culprits of marine pollution. It was estimated that out of 8.8 million tons of plastic wastes produced by China, 3.53 million tons reach the ocean (Zhang et al., 2004).

Marine pollution is only one side of the story. More than 60% of eco-sensitive coastal wetlands are lost due to coastal development, land reclamation, sea-level rise, etc. Overfishing is another major issue in China as the fish catch is more than the sustainable level. The overemphasis on ocean economy during the 11th and 12th Five-Year Plans was mainly responsible for marine environmental degradation. However, the change in strategy during the 12th and 13th Five-Year Plans giving more emphasis to ocean protection has renewed hope for the blue economy of China.

In summary, China's blue economy has made tremendous progress under the five-year plans and it is one of the emerging sectors of the Chinese economy. Although the growth of the blue economy is taking place at a fast rate, the share of each region to GOP differs widely. Marine pollution, overfishing, loss of coastal zone, etc. are major environmental concerns for China. Protecting the marine environment lies at the heart of the blue economy of China.

4.2 UNITED STATES OF AMERICA

The United States of America (USA) is truly a maritime nation being surrounded by three oceans with a coastline of 19,924 km, and its EEZ encompassing an area of 11.5 million km^2 (FAO, 2016). The blue economy activities in the USA are carried out in eight regions, namely, Great Lakes, Gulf of Mexico, Mid-Atlantic, North East, North Pacific (Alaska), The Pacific (Hawaii), South East, and West Coast (Table 4.1), and spread over six activity sectors: living resources, offshore mineral extraction, marine construction, ship and boat building, marine transportation, and tourism and recreation (Table 4.2).

4.2.1 BLUE ECONOMY ACTIVITIES

The USA is amongst the top five fish producers in the world, even though commercial fishing, aquaculture, and seafood processing and marketing accounts for only 2.7% of employment and 3.7% of GDP (NOAA, 2018). Of these, seafood processing, the largest subsector, contributes about 48.4% of GDP and 45.7% of employment.

The major components of offshore mineral extraction are oil and gas, while limestone, sand, and gravel are also mined. Though this sector contributes 26.4% of

TABLE 4.1
Regional Profile of the USA Blue Economy

Regions	Activities	Employment	GDP
Great Lakes	Freshwater fishing, recreational fishing	300,000	18
Gulf of Mexico	Offshore mineral extraction (oil and natural gas)	601,000	125
Mid Atlantic	Tourism and recreation sector—provides greatest employment	755,000	54
North East	Fishing, fish harvesting, shipping and shipbuilding	250,000	18
North Pacific (Alaska)	Commercial fishing, seafood harvesting	51,000	11
South East	Tourism and marine transportation	114,000	8
The Pacific (Hawaii)	Economy depends on ocean-related employment	397,000	26
West Coast	Deepwater ports—marine transport sector, tourism and recreation	707,000	60

Source: NOAA Report (2018) on the US Ocean and Great Lakes Economy: Regional and States Profiles.
Note: GDP in billion US dollars.

TABLE 4.2
Contributions of Ocean Economy Sectors in the USA

Sl no.	Sector	Industry	No. of Industries	GDP (%)	Employment (%)
1	Construction	Marine Related Construction	1	1.9	1.4
2	Living Resources	Fish Hatcheries & Aquaculture, Fishing, Seafood Markets, Seafood Processing	4	2.4	2
3	Minerals	Oil & Gas Exploration & Production Sand and Gravel Mining	2	33.4	4.9
4	Ship & Boat Building	Boat Building & Repair Ship Building & Repair	2	5.6	5.1
5	Tourism & Recreation	Amusement & Recreation Services, Boat Dealers, Eating & Drinking Places, Hotels & Lodging Places, Marinas, RV Parks & Campgrounds Scenic Water Tours, Sporting Goods, Zoos, Aquaria	7	36	72.3
6	Transportation	Deep-Sea Freight, Marine Passenger Transportation, Marine Transportation Services, Search & Navigation Equipment Warehousing	5	20.6	14.3

Source: Modified from NOEP (2016) and NOAA (2018).

the ocean economy component to the nation's GDP, its effectiveness in generating employment has been low (only 4.1%). This sector is capital intensive and requires highly skilled workers who have to work in hazardous conditions at sea, and so the average salary of these workers is much higher than the national average. This sector however witnessed a decline both in employment by 15.9% and GDP contribution by 17.6% between 2015 and 2016 (NOAA, 2018).

Marine construction involves dredging of navigation channels, beach nourishment, and dock building. This sector is a tiny component of the ocean economy of the USA as it contributes only 1.4% of the employment and 1.9% of the blue economy component of GDP (NOAA, 2018). Marine construction activities are largely concentrated in four states, namely, Texas, Florida, California, and Louisiana. These four states accounted for 55.9% of the employment and 54% of the GDP in the ocean sector.

The ship- and boat-building sector not only involves construction of commercial ships, commercial boats, recreation boats, commercial fishing vessels, ferries, and other marine vessels, but also caters to their repair and maintenance. This sector has accounted for 5.1% of employment and 5.6% of the blue economy component of the national GDP (NOAA, 2018). The growth of this sector has fluctuated both in terms of employment and contribution to GDP.

The marine transportation sector includes all the businesses related to transportation such as traffic of deep-sea freight, marine passenger services, pipeline transportation, warehousing, manufacture of navigation equipment, etc. It accounts for 14.3% of employment and 20.6% of the blue economy component to GDP (NOAA, 2018). It is an integral part of the ocean economy, with warehousing being the largest component of the marine transport sector in terms of employment.

The tourism and recreation sector is labor-intensive, and employs more people than all other ocean sectors taken together (72.4%), and records high contribution to ocean GDP (36.1%; NOAA, 2018). In general the employees in this sector were paid low annual wages (US$24,000) in 2015. One of the reasons is that employment in this sector is seasonal, and most jobs offered are part-time. Hotels and restaurants near the shore areas are the primary providers of jobs. Hotels and restaurants account for 93.9% of employment in this sector. Approximately, three-fourths of the ocean economy jobs created are in tourism and recreation.

The ocean economy contributes substantially to the national economy of the USA. For example, in 2016, it accounted for 154,000 business establishments, 3.3 million employees, and contributed US$129 billion in wages and US$320 billion in GDP (NOAA, 2018). The ocean economy witnessed a higher growth rate as compared to the USA economy as a whole. In 2016, employment in the ocean economy had increased by 2.7% over 2015, while the national average employment growth was 1.2%. While the US economy as a whole grew by only 2.7%, the ocean economy's contribution to GDP grew by 5.7% (NOEP, 2016; Table 4.2).

The contribution of the ocean economy in each State, which lies along the coast, is more or less proportionate to their size. Alaska and Hawaii have contributed the maximum in terms of employment and GDP. Hawaii contributed 18% to employment and 19% to GDP whereas Alaska contributed 11% and 10% to employment and GDP respectively (NOEP, 2016; Table 4.2).

4.2.2 MARINE ENVIRONMENT CONSERVATION POLICY

As a responsible agency to protect the ocean and to ensure its sustainable use, the government has to reduce the over-use of ocean resources. Nevertheless, the lack of strong initiative from the government has hampered the progress of the implementation of National Ocean Policy (NOP) in the USA. In fact, the country has not implemented an organic NOP sanctioned by the Congress (Torres et al., 2015). The government established a National Ocean Council (NOC) to implement NOP, with the involvement of 27 federal agencies, departments, and offices. However, it remains uncertain whether the federal government would succeed in NOP implementation. Implementing a comprehensive ocean policy assumes a lot of significance in balancing the economical and ecological benefits of the ocean.

The health of oceans, coastal, and Great Lakes ecosystems is declining day by day, creating concern about whether the blue economy can continue to provide goods and services in a sustained manner. The US ocean policy has paid greater attention to reaping economic benefits from the oceans. However, it did not pay adequate attention to the protection of the environment. There is reason to believe that the USA has not done enough to protect its marine resources as part of its ocean policy.

Like many other nations, the USA is also a party to polluting oceans with plastic. About 0.11 million tons of water-borne plastic garbage originates from the USA annually. The USA is 20th in plastic-based ocean pollution (Johnson and Ketker, 2017). In 2013, the US Government revised the Joint Marine Pollution Contingency Plan (JCP) promulgated by the Canadian and United States Coast Guard. The revised JCP is intended to provide non-binding guidance to the US and Canadian Coast Guards in controlling marine pollution.

The USA is also an active member of the International Whaling Commission (IWC), which was established in 1946 for the proper conservation of whale stocks and controlled development of the whaling industry. A number of Marine Protected Areas (MPAs) in the form of marine sanctuaries, estuarine research reserves, ocean parks, and marine wildlife refuges still maintain the reference point. There are 13 national marine sanctuaries and 28 National Estuarine Research Reserves controlled and maintained by States in partnership with the National Ocean Service (NOS). NOAA's National Coastal Zone Management Program tries to address the coastal issues through cooperation between the federal government and coastal and Great Lakes states and territories.

4.3   JAPAN

Japan has enormous potential to develop its blue economy as it is surrounded by oceans. Its coastline extends up to 29,751 km with an EEZ of 4.05 million km^2 and a shelf area of 233,000 km^2. Japan's ocean area is one of the largest in the world and is 12 times larger than its land area. Nevertheless, Japan did not pay much attention toward ocean governance. The blue economy activities, which laid immense emphasis on sustainable development of oceans, began with its inclusion in the Basic Act of Ocean Policy in 2007. The salient features of this policy was to promote scientific study of oceans, healthy development of marine industries, and joint working of government and private sectors in the management of oceans (Terashima, 2012). The Ocean Policy stressed on continued coexistence of oceans and mankind and brought ocean development within the framework of blue economy. The First, Second, and Third Basic Plans on Ocean Policy were implemented in 2008, 2013, and 2018 respectively.

4.3.1   BLUE ECONOMY ACTIVITIES

Japan has classified ocean industries as ocean space activity (primary), marine resource utilization (secondary), and materials and services (tertiary) (Park and

Kildow, 2014). Japan is one of the world's top producers of fishery products that reached a peak level of 12.28 million tons in 1984, though it fell to a mere 5.32 million tons in 2010. One of the reasons for the fall in fish catch was the introduction of the system of total allowable catch (TAC), to control excessive fish catch. The fishery production further decreased from 4.765 million tons in 2014 to 4.389 million tons in 2018. While production of captured fishery decreased from 3.713 million tons in to 3.33 million tons, the production from aquaculture had slightly increased from 0.988 million tons to 1.003 million tons during the same period (Statistics Japan, 2019). Japan devotes 90% of its fishery production for domestic consumption. Ironically, Japan is the third-largest importer of fish products in the world, after the EU and USA as there is huge demand for fish in Japan. Most of the imports of fishery products for Japan are from China (Popescu and Ogushi, 2013).

Ocean mining is one of oldest ocean economy activities in Japan as it had started the exploration of ocean minerals in the 1970s. Japan has huge potential in off-shore mining and is all set to get further access to sea-based mineral ores (www. japanesenews.co.jp/news/2016/business/japan-plan-search). Mining is a major blue economy activity in Japan and it considers development of technology to exploit seafloor minerals as a national priority. Japan is in the process of developing technology to mine manganese nodules and crusts, cobalt, nickel, and rare earths elements.

Shipping plays an important role in Japan's international and domestic trade. Around 40% of transportation of domestic goods rely on coastal shipping. Almost 96% of supplies that enter and leave Japan are transported through sea (Keiji, 2011). However, currently, it is facing stiff competition from China and South Korea. Japan's fleet ownership is 11.7% of the global market and that of shipbuilding remains at 24.88%. The total fleet increased from 16,013 in 2005 to 37,536 in 2018 (UNCTAD, 2019). The total fleet includes oil tankers, bulk carriers, general cargo, container ships, and other types of ships.

Being an island nation, Japan's tourism industry depends more on oceans. In 2016, the number of tourists visiting Japan touched 24 million. Tourism has contributed significantly to the Japanese national economy. In 2010, tourism accounted for 2.4% of gross national product (Yang and Yue, 2018).

4.3.2 MARINE ENVIRONMENT

Japan ranks second in terms of per capita volume of disposable plastic waste. Plastic pollution is causing serious threats to the Japanese seas. The Japanese government disposes, on average, 100,000 tons of seaside trash per annum (Ripton, 2015). Global organizations such as The Ocean Conservancy, The Surfrider Foundation, etc. are engaged in ocean clean-up activities in Japan. Japan's Environment Ministry is also actively participating in clean-up operations throughout the year. Japan and UNEP are also working jointly to develop counter measures against plastic pollution.

There are a lot of initiatives from the non-governmental sectors in dealing with the ocean-related issues. A Study Group on ocean policy assists the government to address ocean and coastal issues and on the conservation of marine environment.

4.4   CANADA

With a continental coastline of 243,042 km, the longest in the world, and a shelf area of 7.1 million km² and EEZ of 2.9 million km² (FAO, 2016), Canada considers the oceans as an integral part of its identity and takes pride in its management (FAO, 2016). Around 20% of Canadians live near the marine coast. Canada's marine coast is divided into three large regions, namely, East Coast, North East, and West Coast (Lemmen et al., 2016). These regions differ widely in terms of endowments of marine resources. All provinces and territories of Canada, except two, Alberta and Saskatchewan, share the coastline.

Canada introduced its first ocean strategy in 1987, even before its Ocean Act in 1997. This Act was lauded as the world's first holistic and ecosystem-based law when it came into force ten years after the strategy was made. The management and protection of ocean resources and environment were the primary goals of the strategy, along with coastal economic development and promotion of ocean industries. Three guiding principles of ocean management, as envisaged in the Oceans Act, are: sustainable development, integrated management, and precautionary approach. However, its implementation was far from satisfactory due to inadequate governance mechanism and commitments (Jessen, 2011; Bailey et al., 2016).

To increase the productivity and efficiency of ocean-related companies, Canada formed several ocean super clusters in an innovative way. These clusters are composed of ports, shipping companies, seafood processing plants, and offshore industries. To make the marine industry more competitive, Canada has used these clusters as a major instrument of policy, where knowledge and technology are created by processes of interactive learning and networking of informal and formal sectors. These networks reach out to organizations such as small and medium enterprises, large companies, entrepreneurs, knowledge-generating and transmitting organizations, and innovation intermediaries (Doloreux and Shearmer, 2018).

4.4.1   BLUE ECONOMY ACTIVITIES

The ocean economy has been contributing quite consistently to the growth of Canada's GDP. The contribution of the ocean economy increased from US$17.7 billion in 2006 to US$26 billion in 2016. It provided jobs to around 300,000 in 2016 as against 171,340 in 2006 (Pinfold, 2009; Bailey et al., 2016).

Between the years 2012 and 2015, contribution of fisheries and other ocean sectors to the national GDP and employment generation had increased substantially. The seafood industry is the largest contributor to the blue economy in terms of its contribution to GDP and employment in 2015. This industry includes commercial fishing, aquaculture, and fish processing. Though Canada started oil and gas exploration way back in 1959, the first commercial production of offshore oil began only in 1992. Being one of the world's largest producers of petroleum, offshore oil and gas drilling contribute substantially to GDP but less in terms of employment and income to the workers. The number of jobs provided by oil and gas exploration was only 15,487 in 2012, and reduced further to 9,850 in 2015.

Marine transportation is a major segment of the blue economy of Canada and contributes significantly to its economy. This sector stands second in terms of

TABLE 4.3
Maritime Sector in Canada: GDP and Employment in Marine Sector
of Canada

Maritime Sectors Gross Domestic Product ($ Millions) and Employment by Industry
(2012–15)

Industry	2012		2015	
	GDP	Employment	GDP	Employment
Seafood	6,300	73,911	8,652	100,549
Oil & Gas Exploration/ Extraction	8,438	15,487	4,640	9,850
Marine Transportation	6,160	59,741	6,650	64,466
Marine Tourism & Recreation	3,694	59,883	3,531	56,681
Total Manufacturing & Construction	1,742	32,683	1,776	35,393
Fisheries & Oceans	1,465	15,274	1,605	17,170
Total Economy	1,822,808	17,492,500	1,994,911	18,000,60

Source: Government of Canada: infostat@dfo-mpo.gc.ca. GDP is in millions of Canadian dollars.

contribution to GDP and employment amongst the marine subsectors (Table 4.3). Canada's marine transportation carries more than C$200 billion of Canada's international trade (infostat@dfo-mpo.gc.ca). The tourism and recreation sectors are labor-intensive and have high potential for employment generation. Their contribution to employment is very high, though their GDP contribution is low due to low wages and low returns against the invested capital. The number of jobs created by marine tourism and recreation witnessed a decline from 59,883 to 56,681 between 2012 and 2015 (Table 4.3). Manufacturing and construction is the sixth-largest sector in terms of both income and employment. The gross domestic product of manufacturing and construction (ship- and boat-building, ports, and harbor construction) has grown over the years marginally (Table 4.3).

The contribution of the ocean sector also differs regionally. The Atlantic coast accounted for 3.5% of regional GDP and 2% of regional employment. The Pacific coast ocean sector (British Columbia) contributed 2.6% to provincial GDP, but 3.5% to employment (infostat@dfo-mpo.gc.ca). The contribution of the Arctic region to GDP and employment was not assessed, due to lack of a developed industrial/service economy and weak linkages among the sectors.

4.4.2 Marine Environment and Conservation Policy

The Government of Canada has established several Marine Protection Areas (MPAs) and has also taken active steps for the prevention of pollution in oceans and coastlines by monitoring ship-traffic movement in the Canadian waters. In 2017, The National Aerial Surveillance Programme detected 2,878 liters of oil spills in Canadian waters. Less than 11% of Canada's plastic gets recycled, while the rest ends up in lakes, parks, and oceans damaging the ecosystem in the long run. In this regard, Canada has joined the UN Clean Sea Campaign to tackle plastic marine debris.

Canada was the 144th country to ratify the United Nations Convention on the Law of the Sea (UNCLOS). Keeping in mind the objectives of sustainable development, Canada took many important and rigorous steps from the 1970s to the mid-1990s to protect its ocean territory (Jessen, 2011). It extended the fisheries jurisdiction to 200 nautical miles in order to restrict overfishing by European nations.

Canada's marine environment is of considerable economic, social, cultural, and political importance. With 27 departments and agencies, it has a well-developed administrative set-up to manage the oceans and to supervise its marine-related activities.

4.5   NEW ZEALAND

New Zealand's continental coastline of 15,134 km is the eighth longest in the world. The country has a shelf area of 1.7 million km^2 and an EEZ of 4.09 million km^2 (FAO, 2016). New Zealand's EEZ and territorial seas area are the fourth largest in the world. New Zealand is an archipelago of over 300 islands scattered in the Pacific Ocean (Peart, 2005). New Zealand is endowed with abundant marine life and rich deposits of offshore oil, gas, and minerals. These resources support various blue economy activities such as seafood, mining industries, marine biotechnology, shipping, and marine tourism.

4.5.1   BLUE ECONOMY ACTIVITIES

The marine economy contributed about 1.9% to New Zealand's GDP in 2013. If we consider both direct and indirect activities of the marine economy, its contribution rises to 3.5% of GDP. Regrettably, the contribution of the marine sector to GDP reduced from 1.9% in 2013 to 1.4% in 2017, although its contribution to total employment remains more or less similar at 2% since 1997. In 2017, this economy generated 32,964 jobs (Statistics New Zealand, 2018).

The commercial fisheries sector of New Zealand is relatively small, unlike that of China and Canada. Yet, the fisheries sector played an important role in meeting domestic demand and is among the leading exporters. In 2017, fishing and aquaculture contributed 28.9% to the marine economy and provided employment to 14,712 individuals, i.e., 44.6% of the marine sector employment (Statistics New Zealand, 2018). In fact, the importance of captured fisheries and aquaculture started declining after 2007, but went up again in 2016 and 2017. For example, in 2013, this sector contributed 22% to the marine economy and 0.4% to the national economy, while in 2017, the contribution was 28.9% and 0.5% respectively (Table 4.4).

Though there are several oil and gas deposits within the marine jurisdiction of New Zealand, commercial exploitation of these deposits did not take place for a long time. Commercial production first started in the Taranaki basin located 35–50

TABLE 4.4

Contribution (% Share) of Various Sectors to the New Zealand Economy

	Shipping			Offshore Minerals			Fisheries & Aquaculture			Marine Service		
Year	1	2	3	1	2	3	1	2	3	1	2	3
2007	0.6	36.1	49.5	0.5	28.7	3.2	0.4	26.3	44.5	0	0	0
2010	0.5	23.4	54	1.1	54	3.6	0.4	17.9	39.1	0.1	3.7	2
2011	0.5	22.8	55.3	1.1	53.7	4	0.4	19.6	38.4	0.1	3.1	1.1
2012	0.5	23.6	49.5	1	51	3.8	0.4	20.1	43.8	0.1	3.9	1.7
2013	0.4	24.3	48.9	0,9	48.9	4.2	0.4	21.2	44	0.1	4.3	1.9
2014	0.4	27.3	49.2	0.7	45.5	3.7	0.4	22.1	44.3	0.1	4.7	1.7
2015	0.4	26.2	50.5	0.7	42	3.8	0.4	23.4	43.8	0.1	6.9	1.8
2016	0.5	33.8	49.9	0.4	26.4	2.5	0.5	32	44.3	0.1	6	1.9
2017	0.5	37.3	48.1	0.4	26.7	2.2	0.5	28.9	4.6	0.1	4.8	1.4

Source: Statistics New Zealand (stats.govt.nz/indicators/marine economy).
Notes: 1 = Percentage share of National GDP; 2 = Percentage share of Marine GDP; 3 = Percentage share of marine employment.

km west of the Lower North Island. The majority of New Zealand's petroleum production comes from this basin. Over 1.8 billion barrels of barrel of oil equivalent (BOE) have been discovered, of which 70% is gas. More than 400 wells have been drilled throughout the basin, in about 20 fields (Kroeger et al., 2013). New Zealand produced, on an average, 35,500 barrels of oil per day in 2013. While oil outputs are mainly exported, natural gas is used mainly to meet domestic demand. During 2007–13, crude oil exports contributed an average of 4.2% to total goods exported and 3% of total exports of goods and services. Offshore minerals, largely oil and gas, occupied the position of the largest contributor to the marine economy. It contributed 48% of marine economy in 2013 (Statistics New Zealand, 2016). In 2017, the contribution of offshore minerals to the marine economy decreased to 27%. During the same period (2013–17), employment contribution of offshore minerals to the marine economy fell from 3.2% to 2.2 % (Statistics New Zealand, 2018).

Shipping is a major component of New Zealand's marine economy. Domestically, 6 million passengers rely on harbor ferries for their day to day activities while approximately 12 million tons of oil are annually transported within the country. A fleet of 1,500 commercial fishing vessels ply on seas. More than one million people use over 900,000 recreational marine craft. Marine manufacturing is one of New Zealand's largest sectors with an annual turnover of around US$1.04 billion (Maritime New Zealand, 2017). The contribution of the shipping industry decreased after 2007 but has shown signs of revival since 2013 when shipping sector contributed 24.3% to the marine economy GDP and 0.4% to the GDP of the national economy.

Shipping contributed 47% of the filled jobs and 47% to the earnings of the total marine economy. In the year 2017, shipping was the largest contributor to the marine economy at 37% and it contributed 48% of employment in the marine economy (Table 4.4).

The marine industry is the largest specialized manufacturing industry in New Zealand and consists of manufacturing and servicing of commercial vessels and recreational vessels and also provides services to the marine defense industry. Together with sea transport, the boat building industry contributes significantly to New Zealand's economy (Statistics New Zealand, 2016). Marine services in New Zealand include marine surveying, mapping, marine business and consultancy, and services to water transport. For example, in 2017, the contribution of marine services to the marine economy was 4.8% of the marine economy and 0.1% of the national economy. The contribution of marine services to total employment in the marine economy was 1.4% in 2017 (Statistics New Zealand, 2018).

Marine tourism and recreation have fared indifferently and contributed only 1% to the total marine economy and 0.02% to the Kiwi economy in 2013. These sectors represented 0.9% of filled jobs and 0.6% of earnings of the total marine economy (Statistics New Zealand, 2016). However, things are changing and the true potentials of these sectors is being realized. Cruise-liner tourism has been expanding and, during 2010–16, cruise tourism increased from 100,000 to 250,000 persons (Yeoman and Akehurst, 2017). In 2017, marine tourism and recreation contributed 2.2% to the marine economy. The employment contribution has been very low, only 3.6% of total marine economy employment (Statistics New Zealand, 2018).

4.5.2 MARINE ENVIRONMENT AND CONSERVATION POLICY

New Zealand did not have legislation to protect its marine biodiversity for a long time. This changed in 2006 when the government launched the MPA Policy and Implementation Plan. New Zealand is internationally renowned as an innovator in natural resource governance (Peart, 2005). New Zealand's marine environment is very eco-sensitive. There are certain marine species found only in New Zealand seas that are not found elsewhere in the world. New Zealand's marine ecosystem is home to a large number of seabirds. However, between 2013 and 2016, 90% of seabirds, 80% of shore birds, 22% marine mammals, and 9% of sharks faced the threat of extinction (www.stats.govt.nz/indicators). Protecting marine species is going to be a major challenge for planners and policy makers.

New Zealand has been very innovative in ocean management and has introduced many models and legislations. The Comprehensive Fisheries Management System based on individual transferable quotas (ITQs), Resource Management Act 1991 (RMA), the Fisheries Act, and Quota Management System (QMS) are considered very important tools in the efficient management of oceans. The Marine Mammals' Protection Act and the Marine Resources Act are other Acts which look into marine environmental issues.

The QMS is intended to set sustainable harvest levels of fish species within the EEZ and territorial area through the system of total allowable catch (TAC). Under QMS approximately 45,000 tons of fish are sustainably harvested every year (Williams et al., 2017).

4.6 BRAZIL

The ocean boundary of Brazil represents 32% of its national boundary and it is almost entirely covered by the South Atlantic Ocean. While the continental coastline extends for a length of 8,400 km, the shelf area is 0.8 million km². The EEZ of Brazil covers 3.69 million km² and it is known as the "Blue Amazon" (FAO, 2016; Seraval and Alves, 2011). In addition to the 3.69 million km² EEZ, the UN Commission has agreed to add another 0.71 million km² under its extended continental shelf for Brazil to harvest living and non-living resources (Abreau, 2015).

4.6.1 Blue Economy Activities

The coastline generates lot of wealth for Brazil, as half of the country's population lives in coastal areas. Marine-based production contributes around 19% of the country's GDP. The key blue economy activities of Brazil include oil, gas, bio-fuels exploration, fisheries, tourism, and housing expansion.

Brazil depends more on captured fisheries as aquaculture fishery constitutes around 38% of the total fish production. The low yield of captured fishery from a comparatively huge maritime area of 4.4 million km² is due to very petite natural productivity, and the use of obsolete fishing fleet. It was estimated that 3.5 million people directly or indirectly depend on fisheries and aquaculture in Brazil (Barone et al., 2017). Fishing in Brazil takes place both on an industrial scale and on a small scale. Marine fisheries fleet is categorized into three types: coastal, long distance, and artisanal.

Oil and gas exploration is a major blue economy activity in Brazil. The country has the second-largest oil reserves in South America (12.6 billion barrels) and offshore oil reserves constitute more than 94% of the total oil available in the country. In 2017, it produced 2.73 million barrels of oil per day on an average (www.theoiland gasyear.com/Brazil/market). In 2017, Brazil's natural gas production amounted to 115 million m³ per day. In fact, about 80% of Brazil's gas production comes from offshore fields. Brazil ranks 37th in the world in case of largest proven gas reserves (369.4 billion m³), of which 303.3 billion m³ is located offshore. Gas production registered a growth of 7.2% per annum between 2008 and 2017. Oil and gas industry accounts for 13% of Brazilian GDP (Natural Agency for Petroleum, Natural Gas, and Biofuels, 2018).

The marine transportation and marine industry in Brazil is not very well developed as compared to its potential. This is despite the fact that Brazil depends mainly on sea routes for its trade and 84% of its exports and 74% of imports are by sea (www.brazil. gov.br/about-brazil/news/2017/11). Although Brazil is a large country, its shipping is still not fully developed and remains a potential area for future growth. Brazil was a leader in shipbuilding in the 1980s but drastically declined in the early 1990s, but later recovered at the end of the 1990s and early 2000s. The government has invested considerable sums in the shipbuilding industry to transform it into one of the major industries of the Brazilian economy. At present Brazil has 34 public ports and around 100 private ports facilities.

Tourism and recreation as a blue economy activity is also not well developed in Brazil. Brazil's participation in world tourism is not very commendable though it has great potential. Accordingly in recent years the country witnessed a massive investment in tourism infrastructure particularly in the coastal regions where a vast majority (90%) of tourism activities concentrate (Santhana, 2003).

4.6.2 MARINE ENVIRONMENT AND CONSERVATION POLICY

Brazil came up with a strategy for biodiversity conservation and it ratified the Convention on Biological Diversity (CBD) in 1994. In 2000, the government approved the law implementing the National System of Protected Areas (Abreu, 2015). One of the major concerns is the threat of extinction of fish stocks. The adverse effects of human activities on the coastal environment include loss of habitats, declining commercial and artisanal fisheries, decrease in living and non-living resources, and pollution of beaches, increased erosion, and coastal flooding. Overall, Brazil has made significant strides in achieving its marine biodiversity targets.

The Government of Brazil designated two new MPAs in 2018 to maintain the ocean ecosystem and to protect species like sharks, turtles, rays, and whales. A new initiative by the name "Brazil Blue Fund" is entrusted with conservation of biodiversity in Brazilian coastal and marine areas. The Fund aims to invest US$140 million by 2022. The Fund's activities include the restoration of endangered species, recovery of fish stocks, and development of sustainable tourism (brazil.gov.br/about-brazil/news/2017).

4.7 PACIFIC ISLANDS

The Pacific Islands comprise the Melanesia, Micronesia, and Polynesia regions (Fig. 4.3). These islands are economically self-sufficient and fishing has been a significant contributor. Fiji is the largest country in the group, and it has a population of 900,000 while Tuvalu and Nauru are the smallest countries and have an

FIGURE 4.3   A Section of Islands Scattered over the Pacific Ocean

estimated population of 11,000 each. As these countries are small in size, the available resources are minimal, and their economic base is not very broad and stable. Further, these countries are more vulnerable as they are prone to natural disasters and suffer from problems related to climatic change.

4.7.1 FIJI

Fiji is an archipelagic nation with about 300 islands and has a land area of approximately 18,700 km². It is one of the economically most developed countries in the South Pacific. Its coastline is about 1,130 km long and has an extended sovereignty over 130,450 km² of territorial waters. The EEZ of Fiji encompasses an area of 1.29 million km², with a shelf area of 43,264 km² (FAO, 2016). A majority of Fijians depend on ecosystem services provided by the marine environment.

4.7.1.1 Blue Economy Activities

The fisheries industry is the third-largest resource-based sector in the Fijian economy. Though its contribution to GDP is only around 1.8%, it contributes 7% to Fiji's total export earnings (Ministry of Fisheries, 2018). Like many other countries, Fiji also classifies her fishery sector into three main areas, namely, offshore fisheries, inshore fisheries, and aquaculture (Commonwealth Foundation, 2015). The government encourages private sector investment in the sector. Overfishing is a major issue in Fiji and so the fundamental economic causes of overfishing have to be first addressed (DeMers and Kahui, 2012).

The deep ocean floor surrounding the Fijian islands contains metals like zinc, gold, copper, and silver. The Fiji Islands group is bounded by the North Fiji Basin (NFB) to the west, the Lau Basin to the east, and the North Fiji Fracture Zone (NFFZ) to the north. Exploration activities started as early as 1987, under the Tripartite II program that was jointly funded by Australia, New Zealand, and the United States. Under this program, the Central North Fiji Basin was explored for evidence of hydrothermal mineralization. The government has issued exploration licenses for large areas of the ocean floor. However, this action has come under severe criticism as exploration (and eventually mining) of mineral resources could potentially impact the marine ecosystem adversely. Hence, the World Bank has urged governments to be cautious while granting licenses to mine the ocean floor in future (Folkersen et al., 2018).

Tourism has played a crucial role in Fiji and has great potential to develop even further. In 2015, it contributed 14.1% to GDP directly and, along with indirect contribution, it was 38% (Folkersen et al., 2018). It is vital to ensure environmental quality to sustain tourism.

4.7.1.2 Marine Environment and Conservation Policy

There is an attempt within the country to restore its marine resources. To ensure sustainable use of marine resources, it started "Fiji Locally Managed Marine Area (FLMMA) Network." This is a network of civil society organizations, government departments, and academic institutions. Over 400 communities work together to promote sustainable use of Fiji's marine resources (Commonwealth Foundation, 2015).

4.7.2 Papua New Guinea (PNG)

PNG is the largest Pacific island country with a total land area of 462,000 km². PNG has a population of 7 million people with a density of around 9 persons per km². This is the lowest population density registered in the South Pacific. The length of the continental coastline is 5,152 km with an EEZ of 2.40 million km² and a shelf area of 186,819 km² (FAO, 2016).

4.7.2.1   Blue Economy Activities

PNG's major commercial activity is tuna and shrimp farming. The tuna fishery sector is further subdivided into purse seine sub-sector, long line sub-sector, and hand line sub-sector. While domestic and foreign vessels are allowed to operate in the purse seine sub-sector, only PNG citizens are permitted to work in the other two sectors. PNG has access agreements with Korea, Taiwan, the Philippines, and China. A multilateral treaty also exists with the USA. These treaties and agreements are important as they decide the number of vessels and the access fee. Around 130 foreign vessels fish in PNG waters every year (National Fishery Authority, PNG, 2019). The total tuna catch in 2010 in PNG waters was 702,969 tons and the foreign vessels accounted for 78.7% of the catch (Asian Development Bank, 2014).

PNG has a good amount of mineral resources in the form of sulfides of cobalt, copper, gold, silver, and nickel. In addition, petroleum and natural gas production is a major contributor to the island economy. Mineral extraction is the key source of foreign exchange earnings for PNG with mining and petroleum making up approximately 75% of its merchandise exports. Of these, mineral exports accounted for 49%.

Though PNG has 17 commercial ports, most of them are quite small and in poor condition. The traffic carried by these ports is insignificant. Moreover, many of the ports are unusable in poor weather. Hence the contribution of marine services to the PNG's economy is low. Tourism however is a major activity in PNG and has great potential to create jobs and earn foreign exchange. Tourism activities in PNG are mainly developed through private initiatives with little government support (Asian Development Bank, 2014).

4.7.2.2   Marine Environment and Conservation Policy

Population growth and modernization has resulted in considerable economic activities that in turn led to the degradation of marine ecosystems. In the Pacific Ocean, acidification is a major issue and may harm marine ecosystems and services. This may happen through a rise in sea temperature and coral bleaching that in turn will affect public amenities, tourism, and economic development. The projections show that the sea level in the Pacific region is likely to rise by 40 to 80 cm by the end of this century (Commonwealth Marine Economies Programme, 2018).

The coastal and marine ecosystems support a variety of critical industries including fisheries and tourism. Hence, any degradation of the marine ecosystem will make these industries unsustainable. Continuing mining of minerals in PNG is causing immense environmental problems, and their impacts are likely to continue even after the mines are closed. The fisheries sector together with processing and canning industries also create environmental problems.

4.8   UNITED KINGDOM

With a long coastline of 18,000 km, the continental shelf area of the United Kingdom (UK) extends to about 10,525 km², while the EEZ encompasses an area of 3.22 million km² (FAO, 2016). The UK has a very strong maritime heritage. The UK's maritime policies are a set of sector-based policies with a strong legislative base, although the integrated policies in coastal management have emerged only in the 1990s. The UK has shown a lot of interest in blue economy activities and is a signatory to all major conventions concerned with shipping, fishery, pollution, law of sea, and environment (Balgos et al., 2005). The UK has also built huge infrastructure for marine research and education.

4.8.1   BLUE ECONOMY ACTIVITIES

In the UK, two sectors, the offshore oil sector and coastal tourism, dominate the blue economy. The tourism sector employed 221,000 persons in 2016. Overall, blue economy Gross Value Added (GVA) increased by 16.4% in 2016 as compared to 2009, reflecting an increase in all blue economy sectors, such as in shipbuilding by about 90%, in transport by 52%, in living resources by 34%, in port-related activities by 30%, in offshore oil exploration by 5%, and in tourism by 3% (European Commission, 2018). On the employment front, the blue economy's share was 1.3% in 2016, which however is a 6% drop compared to 2009 (Table 4.5), indicating a technology-driven growth at the cost of labor-intensive activities.

TABLE 4.5
UK Employment and Income under Blue Economy

Sectors	Persons Employed		Value-Added at Factor Cost	
	2009	2016	2009	2016
Living Resources	42.2	41.4	1,908.3	2,552.2
Marine Extraction of Oil and Gas	30.2	33.4	16,562.1	17,442.8
Ports, Warehousing, and Water Projects	27.9	36.4	3,106.8	4,032.4
Marine Transport	17.2	15.9	2,600.9	3,960.7
Shipbuilding and Repair	42.7	39.4	1,573.7	3,002.7
Coastal Tourism	252.1	220.6	7,662.1	7,887.6
Total	412.3	413.6	33,414	38,878
Share of National Economy (%)	1.52	1.52	2.14	1.82

Source: European Commission (2018).
Note: Employment figures in thousands, cost in million euros.

The UK's marine companies range from small and medium enterprises (SMEs) to highly competitive globally recognized names. In the leisure marine sector, SMEs are the major players and the UK is the world leader in the production of quality powerboat and sailing yachts. It boasts a robust commercial sector under the blue economy, and occupies an important position in merchant shipbuilding (UK Marine Industries Alliance, 2011).

The fisheries fleet in UK is the second largest in the EU, in terms of gross tonnage. However, recent years have observed a decline in fish landings and employment in the fishery sector. The UK has a robust fish processing sector that provided 13,554 jobs in 2015 (UK Government Office for Science, 2018). The UK was a net importer of fish in 2015. China and Iceland are the major fish exporters to the UK, whereas the UK's export partners include France, Netherlands, and Ireland. Though the importance of the fishing sector is declining, it remains an important socioeconomic activity in coastal regions, particularly in remote coastal communities in Wales, Scotland, and South West England (Austen and Malcom, 2011). Aquaculture has gained considerable importance in recent years in the food sector. Atlantic salmon is the principal species for which there is a growing demand from China, Japan, and North America.

The UK has a large potential to extract renewable energy from offshore. The UK has the largest installed offshore wind capacity in the world to generate electricity and it occupies around 36% of the global size. The industry is witnessing rapid growth due to technological innovations in the field. Wind energy is environmentally friendly and cost effective and is well-supported by the government. Deep-sea mining, considered as an alternative to terrestrial mining, is seen as a great potential for the blue economy, as significant amounts of deposits of cobalt, nickel, and manganese exist on the seafloor in the form of manganese nodules. The UK's industry is exploring the possibility of mining deep-sea resources.

The shipping industry is an essential segment of the country's economy, and around 95% of total exports and imports are carried by sea. The UK's maritime sector is the largest in Europe and major trade partners are EU countries. Though the UK's shipping sector has great potential for growth, certain environmental factors may affect shipping and trade adversely. For example, climate change is going to have some significant impact on the shipping industry as it is likely to disrupt routes due to floods, hurricanes, etc. The UK has the world's leading maritime business service sector and UK firms have the largest share of maritime insurance premium (26%) and ship breaking in the world. The UK is second only to the USA in terms of exports of maritime services as a proportion of world exports (Government Office for Science, 2018). The UK's port sector is the second largest in the EU and handles around 5% of the world's total marine freight traffic volume. In 2015, 53 major ports of the UK handled 486 million tons, and other smaller ports dealt with an additional 11 million tons.

4.8.2 MARINE ENVIRONMENT CONSERVATION POLICY

The UK government has taken initiatives in minimizing biodiversity losses and has designated 298 MPAs in its waters. Besides, there are 50 Marine Conservation Zones protecting special species in a 20,000 km² area (UK Marine Industries Alliance, 2011).

While only 3% of global seas are protected, the UK can take credit in protecting 23% of its waters. This is very important as around a third of its population live within a range of 5 km of the coast.

The Marine and Coastal Access Act (2009) has been addressing some issues related to the protection of the marine environment. Britain has immense interest in the protection of the environment and called for earmarking 30% of the world's oceans as MPAs by 2030 to reverse the damage done due to overfishing, plastic pollution, etc. There are a host of legislations in protecting marine environment and most of these originate from the EU that helps the UK in implementing marine protection.

The UK seas are home to a wide range of species, and the maintenance of the health of the marine ecosystem depends on how well these species are protected. Although the future of the seas depends on conserving this biodiversity, over the last 100 years, the increase in human activities in the seas has put tremendous pressure on marine life.

4.9 MAURITANIA

Mauritania is a country with 80% of its land area being desert. It has a coast line of only 754 km and a shelf area of 36,256 km^2 with an EEZ of 173,728 km^2 (FAO, 2016).

4.9.1 BLUE ECONOMY ACTIVITIES

Marine fisheries is one of the main pillars of the national economy of Mauritania and it is estimated that the contribution of fisheries to the GDP varies between 4 and 10%. This sector employed approximately 55,000 persons in 2013–14 and the fish catch was around 800,000 tons (Masti, 2018). Traditionally 95% of the catch from the EEZ is exported, which accounted for about US$380 billion in 2014. Government has launched a National Responsible Management Strategy for Sustainable Development of Fisheries and Marine Economy (NRMSSDFME) to ensure sustainable exploit-ation of fishery resources and to strengthen the integration of the fisheries sector into the national economy. It also aimed to promote the development of inland fisheries and aquaculture.

In order to develop its fishery sector, Mauritania signed bilateral fishery deals with several neighboring countries like Morocco, Algeria, and also with Russia, Japan, and the EU. The EU and Mauritania Fisheries Partnership Agreement, signed in 2012, is an important agreement in fisheries development of Mauritania. As per the agreement, more than 100 European vessels from 11 EU countries are allowed to fish in Mauritanian waters, in return for which the EU pays Euro 70 million a year (Katsarova, 2013).

Despite all these efforts, the performance of the fishing industry was below expectation. There has not been sufficient integration of fisheries with the national economy as envisaged in the National Responsible Management Strategy. More than 95% of fish caught is exported as gross frozen products, and not much effort has been put into processing activities onshore. There is a feeling that European vessels are contributing to overfishing in Mauritania due to their highly specialized instruments.

The productivity of European vessels is 50 times higher or even more, than the traditional vessels.

Mauritania has made major advances in offshore oil and gas exploration, in recent years. It has rich reservoirs of offshore oil and natural gas, and has production facilities too. Recently new gas reserves were discovered in the offshore and this will enhance the production of oil and gas. In 2018, the Government of Mauritania signed two production-sharing contracts with Shell Exploration and Production Company for exploration and future production of offshore hydrocarbons. Exxon Mobil, a Texas-based company, is in negotiations with Mauritania's government for offshore oil and gas exploration contract/license (news.exxonmobil.com/).

4.9.2 MARINE ENVIRONMENT AND CONSERVATION POLICY

A quarter of Mauritania's population depends on natural resources for their subsistence. Two national parks—Banc d'Arguin and Parc National du Diawling—are rich in biodiversity, and marine and coastal resources are well conserved. However, these two parks are facing increasing pressure from human intervention and climatic change. Offshore oil and gas development raises a big question over the sustainability of coastal livelihoods and abundant fishing. The country needs more of NRMSSDFME type framework to encourage sustainable exploitation of maritime resources and to facilitate the integration of maritime economy into the national economy.

4.10 NIGERIA

The coastline of Nigeria extends for over 853 km with Niger Delta portion covering about 80% of the entire coastal length. The continental shelf area is about 216,325 km^2, while the EEZ covers 179,839 km^2. The Niger Delta with an area of 70,000 km^2 is one of the largest wetlands in the world.

4.10.1 BLUE ECONOMY ACTIVITIES

Nigeria has great potential to develop her fishery industry. The contribution of fisheries toward employment, food security, enterprise development, and foreign exchange earnings are considered very important. It is also an important source of livelihood and nutrition for many in the rural areas. The estimated annual per capita fish consumption in Nigeria in 2013 was 13.3 kg. In 2015, total fisheries production was estimated at 1,027,000 tons, 33% of which were from marine catch, 33% from inland waters, and 31% from aquaculture. Despite this, Nigeria is a net importer of fishery products, and this sector contributed only 0.5% of the national GDP in 2015 (FAO, 2016).

Exploration of oil and gas, its production and sales are major components of the Nigerian economy. This sector contributes about 55% to GDP, 95% of export earnings, and about 7% of governmental revenue.

4.10.2   MARINE ENVIRONMENT AND CONSERVATION POLICY

The coastline of Nigeria, especially the Niger Delta area, is one of the most polluted and damaged marine ecosystems in the world due to frequent oil spills from pipelines and drilling platforms (Chinoneyeren et al., 2017). Pipeline bursts account for 70% of the oil spill incidences in Nigeria. According to the National Bureau of Statistics, between 1976 and 2001, as many as 6,817 incidents of oil spillage occurred (69% happened offshore) with a total loss of about 3 million barrels of oil.

4.11   SYNTHESIS

In an effort to understand and evaluate the importance and sustainability of the blue economy paradigm in various countries of the world, this chapter has briefly described the prevailing blue economy activities in nine countries and two island nations. The countries chosen for this study are different in many respects, viz. in terms of total area, length of coastline, extension of EEZ, economic growth, population, etc. The chapter hence provides an idea of how blue economy works in different resource, technological, and cultural environments. The contribution of oceans to these economies is undoubtedly substantial.

The contribution of the blue economy toward GDP of these ten countries varies between 1% and 10%. The major components of the blue economy include fisheries, mining, oil and gas exploration, shipping, shipbuilding, marine transport, recreation and tourism, and other related activities. The significance of these sectors differs from country to country. While fisheries play a significant role in some country's economy, mining and tourism play a vital role in others.

The follow-up discussion reveals that most of the countries have focused on the development of suitable technologies to tap ocean resources that has given a significant impetus to the growth of the blue economy. No two countries have strictly followed the same classification of blue economy activities for assessing its contribution to GDP. The contribution of different states/regions to the blue economy is also not uniform in most states or provinces within the country. Regional differences exist due to differences in marine endowments and technological developments. Although most of the countries have enacted Ocean Policies and have initiated measures to protect the environment, almost all the countries have failed in conserving marine diversity and marine environment.

The economy-ecology conflict is highly prevalent in all these countries. There are various issues of environmental concern. Overfishing is one among them. It is a fact that different measures have been adopted by multiple governments and bodies to control overfishing and protect the marine environment. Most countries have laws banning fishing activity during breeding time. There are also laws restricting the use of certain kinds of nets that would adversely affect the fishery resources in the ocean. There are restrictions imposed on fishing in EEZ areas. It is a big question whether these laws are effectively implemented in all countries and different regions within the country. The practical implementation of these laws involves, among others, colossal expenditure in the form of patrolling, and compensating fishermen during the fish ban period, etc.

Seabed mining wastes are a significant concern in many countries. When additional offshore mining licenses are issued, ecosystem damage will emerge as a major concern in the years to come. The waters around most of these countries suffer from plastic pollution, though the severity differs. Japan is facing a grave issue due to plastic pollution of her seas, and considerable amounts have been spent to remove the pollutants. Some reports reveal that the amount of plastic in the sea will soon exceed the amount of fish available in the ocean. Plastic pollution and loss of biodiversity are going to adversely affect the tourism sub-sector of the blue economy.

5 Threats to the Blue Economy

July 2019 was the Earth's hottest month on record, when the global temperature went up by 1.2°C (Celsius) above preindustrial levels. Though the Paris Climate Agreement in 2015, world leaders had committed to prevent the globe from warming more than 2°C, and have been trying to keep global warming even more limited to 1.5°C, relative to preindustrial levels (UNFCCC, 2015), an immediate impact of such an increase in temperature would be on the ocean and climate, and consequently on the development and prospect of the blue economy.

The Indian Ocean is famous for strings of both natural calamities (cyclone, earthquake, tsunami, drought, flood, tidal surge, sea-level rise, subsidence, shore-line change, ocean acidification) and anthropogenic accidents (pollution, hydrocarbon spillage, ballast water invasion, piracy, oil discharge, fires, leakage of poisonous gases, illegal dumping, mangrove degradation, and unlawful exploitation of marine resources). It is no surprise that the Indian Ocean Region (IOR) is sometimes called the "World's Hazard Belt."

As the impacts of these calamities, caused by the rise in global temperature and change in climate, on poverty, famine, societal imbalance, and other resultant tragedies are enormous, managing the disasters following a strategic roadmap must be a priority of the blue economy paradigm. The mitigation process to such calamities is essentially multidisciplinary in nature, as it encompasses the participation of a host of stakeholders, such as people, governments, non-governmental organizations, regional and international partners, donors, civil society, private sector, scientists, academicians, and professionals.

In addition to climate change, there is a considerable degree of geopolitics and military intimidation that poses threats to the establishment of the blue economy. The interests of the great powers in the Indian Ocean were once appropriately reasoned by the maritime strategist Rear Admiral Alfred Thayer Mahan that whoever attains maritime supremacy in the Indian Ocean would be a prominent player on the international scene, and whoever controls the Indian Ocean would dominate Asia. Further looking into the future, Mahan said that this ocean is the key to the seven seas in the twenty-first century and that the destiny of the world will be decided in these waters. Presently the USA, China, and India are the main players in the Indian Ocean and all are involved in spreading their tentacles in terms of grants, capacity building, infrastructure support, and security services to various IOR countries to increase their footprints in this ocean (Ahmad, 2018; Tasir, 2018).

Change in climate and emerging geopolitics will possibly alter every sphere of human existence, and may modify the cognition, attitude, and behavior of the IOR nations and of their populace. In this background, the ever-rising gap between the aspirations of people to live a high-end life, compared to their skill and availability of resources, is proving to be beyond the carrying capacity of the family/society/nation. Analyzing the social dynamics of this region, it appears that the definition of progress and development has so far been either inadequately or wrongly defined.

However, there is some good news even in the dreadful situations as described above. In fact, all the threats offer countless opportunities. The measures to mitigate such threats have potential to create jobs, innovate new technology, and spur worldwide economic growth. The measures could also bring about a new social order of empathy and responsibility. In a true sense, understanding the social dynamics of the region is essential to recognize the rise in IOR's economy (Ward, 2017).

In this chapter, we therefore evaluate three major threats to the blue economy: climate change, geopolitical intimidations, and human mindset that ill-defines the development. The success of the blue economy at the global level appears to depend on perceptible international cooperation on climate change, geopolitics-cum-maritime security issues, and change in the current materialistic mindset. Our assessment will seamlessly spread over three echelons, i.e., South Asia at the local level, the Indian Ocean on the regional scale, and world oceans on the global platform. A successful blue economy would hence demand collaboration and cooperation at the local, regional, and international levels. At the end, a broad canvas of peace, safety, and progress has been drawn up.

5.1 CLIMATE CHANGE

The IOR feeds one-third of the world's population, and as indicated earlier, with its distinctive location, maritime trade routes, strategic choke points, and abundant natural living-mineral-energy resources leading to significant blue economic growth. In fact, about 80% of transport of hydrocarbon passes through this ocean from the Persian Gulf to various destinations all over the world. Visited by both southwest (June–September) and northeast (November–January) monsoons, this ocean is still the hottest sea on the planet, possibly due to its disconnect from cold Arctic waters. Hence, this ocean demonstrated a consistent warming by 0.7–1.2°C between 1901 and 2012.

In the blue economy conducting all activities in a way that is consistent with the long-term capacity of marine ecosystems remains the core philosophy. However, such an economy is threatened very often these days from various angles, particularly in the IOR. For example, 70% of the world's natural disasters occur in the IOR presenting a high humanitarian risk. The IOR is again a front-line area to receive the brunt of change in climate. The vulnerability of Maldives, Mauritius, Seychelles, and Bangladesh to sea-level rise is of major concern. A recent report states that Jakarta, the capital of Indonesia, is sinking and there is a plan afoot to shift the capital to some other city. The coastal areas of several IOR countries are thickly populated, as these areas shelter megacities, energy plants, and industries. Hence, the susceptibility of coastal infrastructure has emerged as an important issue.

5.1.1 Climate Change Impact

The demand for a review of the effectiveness of the prevailing economies (market economy/capitalism) has acquired more importance and significance over the years owing to their possible contribution to the destruction of the environment. Many view climate change as probably the outcome of blatant malfunction of the global market economy, and the uneven discriminatory growth that the world has witnessed since the 1850s. The global warming and climate change (GWCC) is caused by both natural and anthropogenic activities (Fig. 5.1). The adverse impact of GWCC on resources like food, water, agriculture, human health, ecology, and environment will only increase in years to come (Mukhopadhyay et al., 2018b). By 2050 climate change could alone cause a loss of US$10 billion in annual revenue loss to global fisheries (Ortuno and Dunn, 2016).

Climate is changing globally and impacting beyond national boundaries. For example, the threats from increased frequency of hurricanes and cyclones, melting of glaciers, sea-level rise, and ocean acidification are very real. All these will have boundless consequences on the coastal blue economy, particularly for the poor regions like the IOR. The inadequate infrastructure, high population growth, and extreme poverty in this region further compound the problem.

Therefore, the global carbon politics and multilateral negotiations thereof will be of great significance, even to the IOR. In fact, any new concept of the economy needs to address the gamut and impact of climate change while framing its approach and policies. Such a concept discussed in this book (a revised version of the blue

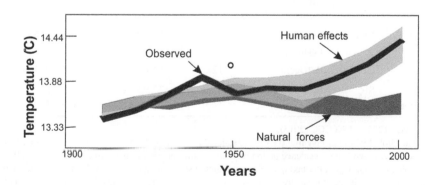

FIGURE 5.1   Global Emission of Carbon Dioxide by Natural and Anthropogenic Activities

Notes: CO_2 contribution from natural activities (blue) gets combined with man-made activities (pink) to make the temperature rise (black) and warm the globe (IPCC, 2014).

economy; see Chapter 8) has to investigate the mutual circular relationship between socioeconomics and climate change.

In fact, the marked surge in concentrations of GHG (greenhouse gases) in the atmosphere since 1850 has become a major concern for the entire human race. Among the GHG, carbon dioxide (CO_2), methane (CH_4), nitrous oxide (N_2O), and chlorofluorocarbons (CFC) are the most prominent (Table 5.1). GHG create a suffocating effect and cause an increase in temperature by several degrees. The balance between energy received by the Earth from the Sun and that released back to the atmosphere is primarily responsible for global warming (Fig. 5.2; Intergovernmental

TABLE 5.1
Characteristics of Major Greenhouse Gases

Greenhouse Gases	Chemical Formula	Source	Residence Period	GWP	Pre-1750	Current Concentration
Carbon Dioxide	CO_2	Fossil-fuel combustion, Land-use conversion, Cement Production	~100	1	280,000	388,500
Methane	CH_4	Fossil fuels, Rice paddies, Waste dumps	12	25	700	1,870/1,748
Nitrous Oxide	N_2O	Fertilizer, Industrial processes, Combustion	114	298	270	323/322
Tropospheric Ozone	O_3	Fossil fuel combustion, Industrial emissions, Chemical solvents	hours-days	N.A.	25	34
CFC-12	CCL_2F_2	Liquid coolants, Foams	100	10,900	0	0.534/0.532
HCFC-22	CCl_2F_2	Refrigerants	12	1,810	0	0.218/0.194
Sulfur Hexafluoride	SF_6	Dielectric fluid	3,200	22,800	0	0.00712 / 0.00673

Source: IPCC (2014); UNFCCC (2015).
Notes: The Global Warming Potential (GWP) indicates the warming effect of a greenhouse gas in a 100-year time horizon, while the Residence period in years is atmospheric lifetime expressing the total effect of a specific greenhouse gas after taking into account global sink availability. The residence indicates how long the gas remains in the atmosphere and increased radiative forcing quantifies the contribution to additional heating over an area. Concentrations of GHGs in Troposphere (in ppb) before 1750 and Current (present time) are given. The vast majority of emissions are carbon dioxide followed by methane and nitrous oxide. Lesser amounts of CFC-12, HCFC-22, and Sulfur Hexafluoride are also emitted and their contribution to global warming is magnified by their high GWP, although their total contribution is small compared to the other gases.

Panel on Climate Change [IPCC], 2014). The low-lying mega coastal cities of the Indian Ocean will receive the maximum brunt of such impacts (MEC, 2005; Mukhopadhyay, 2019).

As can be seen from Fig. 5.2, of the 341.3 Wm^{-2} (watts per square meter) shortwavelength solar radiation received annually from the Sun, 79 Wm^{-2} is reflected back by clouds, aerosols, atmospheric gases, and another 23 Wm^{-2} is reflected by the Earth's surface, while 161 Wm^{-2} is absorbed by the Earth's surface. The other 78 Wm^{-2} of energy is absorbed into the atmosphere by warming the air in contact with the surface, and by evapo-transpiration. Solar radiation is the radiant energy (particularly electromagnetic energy) emitted by the Sun. About half of the radiation is in the visible short-wave part of the electromagnetic spectrum. The other half is mostly in the near-infrared part, with some remaining in the ultraviolet part of the spectrum. Evapo-transpiration, on the other hand, is the sum of the process by which water is transferred from the land to the atmosphere by evaporation from the land-soil, ocean, river, other water bodies, and also by transpiration from plants.

The outgoing infrared long-wavelength radiation from the Earth involves a complex mechanism. For example, atmosphere emits 169 Wm^{-2} of such radiations, which is added further by 30 Wm^{-2} from clouds and about 40 Wm^{-2} from the Earth's surface. Of the total long-wavelength radiation escaping from the Earth's surface (396 Wm^{-2}), about 333 Wm^{-2} radiation returns to the Earth surface after being absorbed and reemitted in all directions by GHG molecules and clouds, leading to warming of the Earth's surface and lower atmosphere (Kiehl et al., 1997; Houghton et al., 2001; IPCC, 2014).

While CO_2 is emitted through combustion of fossil fuel, deforestation, and desertification, CH_4 is discharged through rice cultivation, cattle rearing, biomass burning, coal mining, and emission of natural gases, landfill, and use of wood fuel. Nitrous oxide is normally released through agriculture, fossil fuel combustion, and use of catalytic converters in cars, while CFCs are produced by air-conditioners, freezers, solvents, and insulators (Hardy, 2003). In the year 1700, CO_2 and CH_4 concentrations rose respectively from 280 ppm (parts per million) and near zero to 403.3 ppm and 1,853 ppb (parts per billion) in 2016; the N_2O content increased from 280 ppb to 328 ppb during the same period (Scott et al., 2016). The sector-wise contribution of GHG to the atmosphere by various IOR countries is given in Table 5.2.

However, the effects of all of these GHG and aerosols on the Earth's climate depend in part on how long they remain in the atmosphere (Table 5.1). About one-third of the CO_2 emitted in any given year remains in the atmosphere for 100 years, but its real impact continues for tens of thousands of years. Methane lasts for approximately a decade before it is removed through chemical reactions. Particulate matters (PM), on the other hand, remain in the atmosphere for only a few days to several weeks. Therefore, some quick mitigating actions to reduce PM contribution can show results nearly immediately, while that for mainstream GHG may take decades. Observational evidence shows that heat-trapping gases (CO_2, CH_4, SO_2, N_2O, etc.), other than water vapor account for between 25% and 33% of the total greenhouse effect (Shindell et al. 2012; Lacis et al., 2013).

Among the several related activities making problems for the ocean (and posing real threats to the blue economy) are its acidification, unregulated emergence of

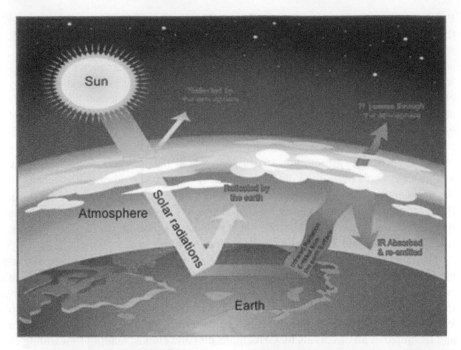

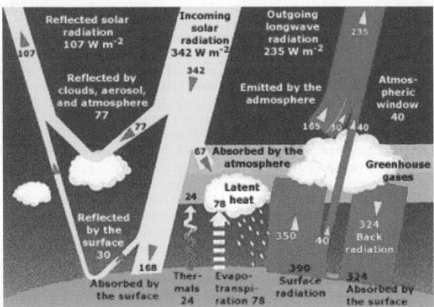

FIGURE 5.2   Earth's Annual Energy Balance and Warming of the Globe

TABLE 5.2
Global GHG Emissions

GHG Types		Sector-wise		Country-wise		Per-Capita	
1	2	3	4	5	6	7	8
CO_2 (1)	65	Power	25	China	30	Qatar	49
CO_2 (2)	11	Agriculture + Forestry	24	USA	15	Trinidad & Tobago	30
CH_4	16	Industry	21	EU	09	Kuwait	25
N_2O	06	Transportation	14	India	07	UAE	25
F-gas	02	Buildings	06	Russia	05	Saudi Arabia	19
		Other Energy	10	Japan	04	Australia	17
				Germany	2.2	USA	16.2
(1) = From Fossil fuel + Industry,				Iran	1.9	Canada	15.6
(2) = From Deforestation + Human				Saudi Arabia	1.8	UK	05.8
F-gas = Fluorinated gas like HFC, PFC, SF_6				South Korea	1.7	Africa + Asia	Least
				All Others	22.4	World Avg.	4.8

Source: After Boden et al. (2012); IPCC (2014).
Notes: Columns 2, 4, and 6 in percentage; Col. 8 in tons.

Sector-wise GHG Emissions in IOR Countries

	1	2	3	4	5	6	7
Australia	18	35	19	04	07	13	02
Bangladesh	–	33	–	17	02	38	10
India	10	–	24	28	33	–	05
Maldives	36	27	23	–	–	–	14
Pakistan	–	46	–	03	05	43	03
Sri Lanka	–	65	–	–	02	23	10
South Africa		84	–	–			
Kenya	–	31	–	–	4.6	62.8	1.4
Mozambique	–	09	–	59	1.5	27	4.0
Tanzania		08	–	73	0.5	17	1.6

Source: Compiled from Ranasinghe (2010); Sridhar (2010); Lohani and Baral (2012); IPCC (2014); Marcu et al. (2015); USAID (2015); WRI (2015); TNA (2016).
Notes: 1 = Tourism & Commercial activities, 2 = Energy & Electricity, 3 = Transportation, 4 = Residence (+ LULUCF), 5 = Industry, 6 = Agriculture, 7 = others (Waste + Fisheries), LULUFC = Land Use, Land Use Change & Forestry. All figures in percent.

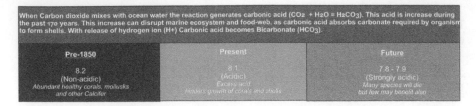

When Carbon dioxide mixes with ocean water the reaction generates carbonic acid (CO₂ + H₂O = H₂CO₃). This acid is increase during the past 170 years. This increase can disrupt marine ecosystem and food-web, as carbonic acid absorbs carbonate required by organism to form shells. With release of hydrogen ion (H+) Carbonic acid becomes Bicarbonate (HCO₃).

Pre-1850	Present	Future
8.2	8.1	7.8 - 7.9
(Non-acidic)	(Acidic)	(Strongly acidic)
Abundant healthy corals, mollusks and other Calcifer	Excess acid hinders growth of corals and shells	Many species will die but few may benefit also

FIGURE 5.3   Mechanism of Ocean Acidification (After www.biology.ucdavis.edu)

aquaculture/ mariculture industries, and ocean pollution. The ocean is known to hold more than 95% of the Earth's living resources and hosts more than 0.2 million known species of plants and animals. It generates half of the world's oxygen, and also serves as one of the world's most important natural "carbon sinks." In fact, it absorbs more than a quarter of the CO_2 released into the atmosphere by human activity. While this greatly counteracts atmospheric GHG, an increase in CO_2 absorption in ocean changes the chemical makeup of seawater through a process called "acidification" (Fig. 5.3).

When CO_2 is absorbed by seawater, chemical reactions occur that reduce seawater pH, carbonate ion concentration, and saturation states of biologically important calcium carbonate minerals (Hardy, 2003). In fact, the natural pH of ocean water (8.2, slightly alkaline) has been changing since the beginning of the industrial revolution in the mid-nineteenth century. It however declined rapidly by 0.1 units—equivalent to a 30% rise in acidity, which has serious biological consequences. As consumption of carbonate ions in seawater impedes calcification (carbon dioxide + water + carbonate ions = 2 bicarbonate ions), the organisms whose shells are composed of calcium carbonate ($CaCO_3$, basically chalk, such as oyster, shellfish, gastropods, pteropods, coral reefs, etc.) would erode and dissolve slowly as the pH of the ocean water becomes acidic. These changes would further hinder growth and reproduction for other organisms, and would impact the entire food chain. In turn, it would affect commercial fishing and tourism industries, causing a loss of more than US$1 trillion. What is worse is that acidification will make the sea surface warm, making it less effective in absorbing atmospheric CO_2. Ocean acidification prompts some plankton species to grow faster, while slowing growth for several others, thus disrupting competition and putting the entire ocean ecosystem at risk. It is estimated that ocean acidity will increase by 170% by the year 2100 compared with preindustrial levels, if present CO_2 emissions continue. Ocean acidification occurs more rapidly in the colder waters of the Arctic and Antarctic (IPCC, 2014).

Change in absorption level of CO_2 in seawater following acidification will impact fishery resources in a major way. Fish is a major source of protein and contains all essential amino acids for humans. Globally, aquaculture provides food, nutrition, and economic development, and produces an average of 10 kg of fish per person each year. Almost all fish produced through aquaculture is destined for human consumption. Aquaculture is also the fastest growing food sector economy and about

FIGURE 5.4 The Aquaculture Industry Is Becoming Hugely Popular and Economical (Courtesy Wikipedia).

350 million jobs worldwide are linked to fisheries (Fig. 5.4). Aquaculture experienced an average annual growth of 5.9% between 2001 and 2015. In 2015, aquaculture generated about 76.6 million tons of fish, worth roughly US$157.9 billion at first sale, along with 29.4 million tons of plants (US$4.8 billion), and 41,000 tons of non-food products such as pearls and shells, valued at US$208.2 million (FAO, 2018).

Several IOR countries are increasingly taking up aquaculture as a major activity. Besides accomplishing the UN's Sustainable Development Goal (SDG) #14 (conserving and sustainably using the oceans, seas, and marine resources), the aquaculture industry can also fulfill several other SDG, amongst others, SDG #1 (no poverty), #2 (zero hunger), #3 (good health and well-being), and #8 (decent work and economic growth). However, if the aquaculture industry is not managed carefully, the greed to make a quick buck could cause outbreaks of diseases, coastal degradation, and serious environmental risks.

Pollution could cause physicochemical imbalances in oceanic waters, impacting on carbon absorption and bringing about a change in climate. Of all the harmful polluting components, plastic contributes about 60–90% to marine pollution. More than 220 million tons of plastic are produced each year. In 2006, the UN Environment Programme (UNEP) estimated that every 2.6 square kilometer of ocean contained 46,000 pieces of floating plastic, and by the year 2050 the amount of discarded plastics may outweigh the amount of fish in our oceans. A plastic bottle can last up to 450 years in the marine environment. Plastic (including micro-plastics) chokes to death an estimated one million sea birds and an unknown number of sea turtles each year by clogging their digestive tracts (UNEP, 2016; Fig. 5.5).

A summary of possible climate change impact in the IOR on varied areas like economy, agriculture, food production, sea-level rise, rainfall, extreme events, etc.

FIGURE 5.5   Representative Photo of Plastic and Oil Spill Pollution in Sea (Courtesy Wikipedia).

is collated in Table 5.3. Although most of the Indian Ocean countries are alive with threats from climate change, and many of them have announced intended nationally determined contribution (INDC) as per UNFCCC (United Nations Framework Convention of Climate Change) through appropriate climate mitigation policies and actionable measures, there appears to be a lack of comprehension of the entire gamut of the problem. An analogy would be saving oneself from torrential rains with an umbrella that has several holes (Mukhopadhyay, 2019).

Another example ironically comes from a recent report on traveling of world leaders (www.statista.com). A list of top ten CO_2 emitters in 2018 (through their personal and official aerial tour) finds Shinzo Abe of Japan at the top emitting

TABLE 5.3
Summary of Climate Change Impact in the IOR

Sl	Impact Summary
1	**Enhanced Temperatures** More CO_2 in atmosphere means more growth of plants, but also more demand on water. Rain-fed wheat grown at CO_2 concentration of 450 ppm demonstrated yield increases with temperature increases of up to 0.8°C, but declines with temperature increases beyond 1.5°C (Xiao et al., 2005). Again, in India wheat yields are predicted to decline by 6–9% in sub-humid, semiarid, and arid areas with 1°C increase in temperature (Sultana and Ali, 2006), while even a 0.3°C decadal rise could have a severe impact on important cash crops like cotton, mango, and sugarcane. In Sri Lanka, half a degree temperature rise is predicted to reduce rice output by 6%, and increased dryness will adversely affect yields of key products like tea, rubber, and coconut (MENR, 2000). In the hot climate of Pakistan, an increase of 2.5°C in temperature would translate into much higher ambient temperatures in the wheat planting and growing stages.
2	**Rainfall & Water Management** Water resources are inextricably linked with climate. Tendencies of increase in intense rainfall events spread over few days are likely to impact water recharge rates and soil moisture conditions. Of 2,323 glacial lakes in Nepal, 20 have been found to be potentially dangerous with respect to GLOFs (Glacial Lake Outburst Floods). The most significant such event occurred in 1985, when a glacial lake outburst flood caused a 10–15 m-high surge of water and debris to flood down the Bhote Koshi and Dudh Koshi rivers for 90 km, destroying the Namche Small Hydro Project in Nepal (Raut, 2006). Rapid depletion of water resource is already a cause for concern in many countries in South Asia. About 2.5 billion people will be affected with water stress and scarcity by the year 2050 in South Asia alone (HDR, 2006). With the islands of the Maldives being low-lying, the rise in sea levels is likely to force saltwater into the freshwater lens. The groundwater is recharged through rainfall. Although the amount of rainfall is predicted to increase under an enhanced climatic regime, the spatial and temporal distribution in rainfall pattern is not clear (MEC, 2005).
3	**Extreme Events** South Asia suffers from an exceptionally high number of natural disasters. Between 1990 and 2008, more than 750 million people—50% of the region's population—were affected by a natural disaster, leaving almost 60,000 dead and resulting in about US$45 billion in damages. Several studies showed that generally, the frequency of occurrence of more intense rainfall events in many parts of South Asia has increased, causing severe floods, landslides, and debris and mud flows, while the number of rainy days and total annual amount of precipitation has decreased (Mirza, 2002; Lal, 2003). An increase in the frequency of droughts and extreme rainfall events could result in a decline in tea yield, which would be the greatest in regions below 600 m (Wijeratne, 1996). With the tea industry in Sri Lanka being a major source of foreign exchange and a significant source of income for laborers the impacts are likely to be grave. On an average during the period 1962–88, Bangladesh lost about 0.5 million tons of rice annually as a result of floods that accounts for nearly 30% of the country's average annual food grain imports (Paul and Rashid, 1993). Again, excessive water withdrawals can exacerbate the impact of drought. Changes in the frequencies of extreme events will have an impact on land degradation processes such as floods and mass movements, soil erosion by both water and wind, and on soil salinization.

(continued)

TABLE 5.3 (Continued)
Summary of Climate Change Impact in the IOR

Sl   Impact Summary

4   *Crop Yield, Agricultural Productivity*
     Agriculture is the mainstay of several economies in South Asia. It is also the largest
     source of employment. The sector continues to be the single largest contributor to
     the GDP in the region. As three-fifths of the cropped area is rainfed, the economy
     of South Asia hinges critically on the annual success of the monsoons (Kelkar and
     Bhadwal, 2007). In South Asia, there could be a significant decrease in non-irrigated
     wheat and rice yields for a temperature increase of greater than 2.5°C that could incur
     a loss in farm-level net revenue of between 9% and 25%. One study points out that in
     Bangladesh, production of rice and wheat might drop by 8% and 32%, respectively,
     by the year 2050. Studies show that a 0.5°C rise in winter temperature could reduce
     wheat yield by 0.45 tons per hectare in India. Other studies suggest that 2–5% decrease
     in Indian wheat and maize yield potentials for temperature increases of 0.5–1.5°C
     could occur.
     For countries in South Asia, the net cereal production is projected to decline at least
     between 4% and 10% by the end of this century under the most conservative climate
     change scenario. In Sri Lanka, Somaratne and Dhanapala (1996) estimate a decrease in
     tropical rain forest of 2–11% and an increase in tropical dry forest of 7–8%. This study
     also indicates that increased temperature and rainfall would result in a northward shift
     of tropical wet forest into areas currently occupied by tropical dry forest. Droughts
     combined with deforestation increase fire danger (Laurance and Williamson, 2001).
     GWCC will also impact crop pests and diseases.

5   *Fisheries*
     A strong El Nino Southern Oscillation will cause declines in fish larvae abundance in
     the coastal waters of South Asia. There is potential to substantially alter fish breeding
     habitats and fish food supply and therefore the abundance of fish populations in Asian
     waters due to the response to future climate change to the following factors: ocean
     currents; sea level; sea-water temperature; salinity; wind speed and direction; strength
     of upwelling; the mixing layer thickness; and predator response.

6   *Sea-Level Rise*
     Low-lying mega coastal cities will be at the forefront of impacts (Fremantle, Perth,
     Yangon, Singapore, Jakarta, Kuala Lumpur, Kuwait City, Sharm El Sheikh, Karachi,
     Mumbai, Chennai, Male, Colombo, Maputo, Dares-Salam, Aden, Port Said, and
     Dhaka-Khulna); vulnerable to the risks of sea-level rise and storms. A rise in sea
     level would raise the water table, further reducing drainage in coastal areas. All these
     effects could have possibly devastating socioeconomic implications, particularly for
     infrastructure in low-lying deltaic areas. The population of Maldives mainly depends
     on groundwater and rainwater as a source of freshwater. Both of these sources of
     water are vulnerable to changes in the climate and sea-level rise. With the islands of
     the Maldives being low-lying, the rise in sea levels is likely to force saltwater into the
     freshwater lens (MEC, 2005).

Source: Compiled from Sivakumar and Stefanski (2011), IPCC (2014), UNFCCC (2015), Mukhopadhyay
et al. (2018b), Mukhopadhyay (2019).

maximum CO_2 (14,442 tons), followed by Donald Trump (USA, 11,487), Moon Jae-In (South Korea, 11,461), Xi Jinping (China, 8,280), Emmanuel Macron (France, 7,645), Vladimir Putin (Russia, 7,616), Narendra Modi (India, 7,477), Angela Merkel (Germany, 7,325), Giuseppe Conte (Italy, 6,394), and Recep Tayyip Erdogan (Turkey, 5,088).

The IPCC report (IPCC, 2014) finds beyond reasonable doubt that since 1990, each of the last three decades has been successively warmer, as the average increase in the temperature of the Earth's surface has been 0.85°C. Of this, the warmest decade has been between 1996 and 2005. In fact, oceans absorb about 80% of increased heat in the atmosphere (Hardy, 2003). As a result, sea level has risen globally since 1870 by about 200 mm (= 20 cm; Fig. 5.6), and the effects will impact all aspects of economic activities in the ocean. Rise in sea level is normally caused by thermal expansion (57%), melting of glacier/ice caps (28%), and melting of polar ice (15%; Ramanathan and Carmichael, 2008). Shift in surface wind pattern, and gravitational field of the Earth could also cause regional fluctuations in sea level (Friedlingstein et al., 2001; Brohan et al, 2006).

Additionally, changing patterns in rainfall, and melting snow and ice are altering freshwater systems, which affect the quantity and quality of water available in many regions of South Asia for household use and agriculture. Cloud burst, extreme rainfall within a short time period, and extended drought are occurring in regions that have no record of such incidences in the past. Phenomena like ocean acidification, oxygen minimum zones, degradation of coral reefs and mangroves, increase in natural disasters, and pollution are the likely products of climate change, which in turn could immensely impact food security, tourism potential, and trade and commerce globally. Climate change could alone cause a loss of US$10 billion in annual revenue to global fisheries by 2050 (Ortuno and Dunn, 2016).

5.1.2 Geo-Hazards

The vagaries of nature are displayed by hazards in various intensities and types, some of those caused by water such as floods, storms, cyclones, typhoons, and tsunamis,

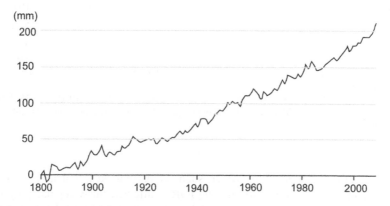

FIGURE 5.6 Global Average Variation in Sea Level

FIGURE 5.7   Geological Features of the Indian Ocean

Note: Marine gravity of the Indian Ocean from satellite altimetry. Inverted Y of red dots at the center are mid ocean ridges giving birth to new oceanic crust, while red curved line on the northeast corner is the Andaman-Java-Sumatra subduction zone where old crust gets consumed.

TABLE 5.4
Major Disasters in IOR—South Asia

Type of Disaster	Number of Disasters		Loss of Life	
	1985–97	1998–2009	1985–97	1998–2009
Droughts	6	9	300	200
Earthquakes	27	35	13,166	155,233
Extreme Temperature	28	41	4,011	10,021
Floods	141	235	32,610	26,111
Storms	97	99	165,360	23,056

Source: South Asian Disaster Resource Inventory (SAARC) and Asia Regional Integration Centre (SAARC, 2016).
Notes: Total number of disasters in South Asia during the years: 1985–9 = 96, 1990–4 = 135, 1995–9 = 134, 2000–4 = 166, 2005–9 = 199. Total cost of the property lost during natural calamities and disasters was US$44,787,984,000.

and a few others unrelated to water such as earthquakes, volcanic eruptions, landslides, and rock-falls and so on. The IOR is affected fairly regularly by many of these hazards. The activity at Barren Island (a part of Andaman Island group), a dormant volcano that first erupted in 1787, was followed by eruptions in 1789, 1795,

TABLE 5.5
Glaciers, Glacial Lakes, and Potentially Dangerous Lakes in South Asia

Country	Glacier Numbers	Glacier Area (sq. km)	Glacier Ice Reserves (cu. km)	Glacial Lakes Number	Glacial Lakes Area (sq. km)	Glacial Lakes Potentially Dangerous
Bhutan	677	1,317	127.31	2,674	106.77	24
India—Himachal Pradesh	2,554	4,161	387.35	156	385.22	16
India—Uttaranchal	1,439	4,060	475.43	127	2.49	0
India—Teesta River basin	285	577	64.78	266	20.20	14
Nepal	3,252	5,324	481.32	2,323	75.64	20
Pakistan—Indus River basin	5,218	15,041	2,738.50	2,420	126.32	52
TOTAL	15,003	33,344	NA	8,799	801.83	203

Sources: Ives et al (2010).

1803–4, and 1852 and after a long hiatus it erupted in 1991 that lasted for six months. The volcano once again erupted as recently as in 2017 (Global Volcanism Program, 2013). However, earthquakes are quite common in the Himalayan collisional belt and along the Andaman-Java-Sumatra subduction stretch, where the Indo-Australian tectonic plate is shoving below the Burma Plate (Fig. 5.7).

Considering the long coastlines of India, obviously the perils of hazards are far more from the seas. Year after year the east coast of India is more affected either by severe storms or cyclones, as compared to the west coast. These events cause large-scale flooding in the coastal states. An account of extreme events/disasters that visited the IOR (with South Asia in particular) is tabulated (Table 5.4). Glaciers, glacial lakes, and potentially dangerous lakes that might burst and cause devastating floods are also documented (Table 5.5).

Another water-related geo-hazard is tsunami. The December 26, 2004 tsunami, generated following a 9.2 magnitude earthquake at the Sumatra subduction zone, devastated several areas of Tamil Nadu, Andaman and Nicobar Islands, Indonesia, and Sri Lanka. As many as 200,000 people lost their lives and the repercussions of this tsunami are still felt by millions of people, both in India and in the neighboring countries. Following this mega event, a tsunami warning center was established by the Indian National Centre for Ocean Information Services (INCOIS, Hyderabad), which is a wing of the Indian ESSO (Earth Science System Organisation, Ministry of Earth Sciences, India).

On the west coast also, several earthquakes occurred following the subduction of the Arabian Plate beneath the continental Eurasian Plate. This compression is what formed the Zagros Mountains, which stretched for over 1,500 km from Turkey to Baluchistan through Iran. Located only 100 km off the south coast of Pakistan, the seismic activities near the Makran Trench led to the earliest known tsunami, which is dated to be during the fifteenth century (Regard et al., 2010) followed by two major earthquakes in the recent past. On November 27, 1945, a 8.0 strong earthquake coupled with tsunami occurred along the Makran Trench that killed 4,000 people in Pakistan, while a weaker earthquake of 6.3 intensity occurred on February 8, 2017 but there was no casualty (David Jacobson, February 8, 2017, United States Geological Survey, personal communication). An increase in awareness and better coordination between the littoral states are needed to enhance the capability to manage earthquake impact and also of the early tsunami warning system.

At national levels, many IOR nations have established ministries and institutional structures for environmental protection. Some of these are very focused (Sri Lanka, Bangladesh, Seychelles, Mauritius, Australia, Singapore) and some are not (India, Maldives, Pakistan, Kenya, Iran).

In summary, to mitigate the impact of climate change, the IOR countries may consider adopting the following: (a) ingrain the concept of the blue economy in their policy strategy; (b) bring all activities in the ocean (industrial, commercial, navigation, environmental affairs, etc.) under one umbrella authority to streamline licensing, governing, and coordination issues; (c) exchange documents to arrive at common consensus regulations in consonance with international agreements; (d) ratify such specific legislations; and (e) establish all-weather observatories in strategic areas (for example, in Java, Sundarbans, Male, Yemen, Cape Town) to acquire real-time dynamic data of atmosphere, biosphere, lithosphere, and hydrosphere (ABLH observatories).

5.2   GEOPOLITICAL INTIMIDATION

Besides the climate change impact, the other important aspect that can influence the blue economy concept is the geostrategic positioning of IOR nations. Geopolitics can have a telling effect on harmony, peace, and progress in the region. Most of the IOR nations are ill-equipped to respond to any sea-borne calamity, environmental disaster, or armed intrusion. The failure to challenge the terrorists on November 26, 2008 in time and to stop their entry into the city of Mumbai to create mayhem confirms the vulnerability of Indian shores and neglect of naval security forces. In both the short and long terms, India was found horrendously unprepared for any contingency that may occur in her waters. It is not that the IOR countries spend less on national security, but the culture of Command, Control, and Coordination (CCC) was missing.

Such a lackluster casual attitude is despite the fact that if any of the major ports in the IOR get closed even for a week, the economy of the entire region would be in turmoil. The authorities in these countries seem to believe more in adjustment and people's resilience, rather than developing standard operating procedure (SOP) in effectively countering such natural and man-made threats (Bharatan, 2011). Hence, besides the negative impact of climate change, the other major speed-breakers for a smooth blue economy paradigm have been the political manipulation, lack of effective security, and militarization of the IOR under one pretext or another.

About 100,000 ships transit annually through the Indian Ocean carrying about 66% crude oil, 55% bulk container, and 33% bulk cargo (Indian Trade Portal, 2019). However, shipping has been increasingly insecure in this ocean with an upsurge in piracy, hijacking of ships, and pilferage of materials. These incidents are reported more in and around choke points in the Indian Ocean. Among the 30 straits and channels present in the Indian Ocean, the important choke points are: Strait of Malacca between Malaysia-Indonesia and Singapore, Strait of Hormuz between Iran and Oman, Babel-Mandeb between Yemen and Djibouti, Mozambique Channel between Mozambique and Malagasy Republic, and the Suez Canal connecting the Mediterranean Sea and the Red Sea (Fig. 5.8, Table 5.6). A strategic control over these points can give a country the most valuable leverage to oversee navigations in the ocean, and in the process control the economy of the IOR nations.

Around 40% of the world's offshore oil production comes from the Indian Ocean. Accordingly, more than 80% of the world's seaborne trade in oil transits through three Indian Ocean choke points, of which about 40% passes through the Strait of Hormuz, 35% through the Strait of Malacca and 8% through the Bab el-Mandab Strait (Table 5.6). In addition, about 50% of the world's seaborne container traffic, and 33% of the world's seaborne bulk cargo involving some 100,000 ships pass through this ocean and its adjacent waterways.

The volume of transportation through the choke points in the IOR, the plethora of violent terrorist organizations, and the military bases of three major naval powers in this ocean region—the USA, China, and India—are shown in Table 5.6. The 2011

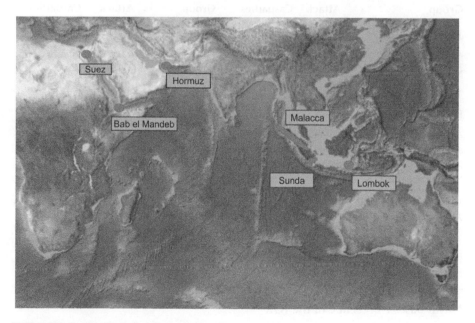

FIGURE 5.8   Choke Points in the Indian Ocean

Note: Presence of these choke points have made the Indian Ocean geopolitically and militarily controversial.

TABLE 5.6
Choke Points, Security Risk, Terrorist Organizations, and Military Bases of Major Powers in the IOR

Table 5.6 (A)
Choke Points and Security Risk in the IOR

Choke Points	Brief Description
Strait of Hormuz	About 17 million barrels of oil are transported per day through this choke point. The oil comes from Iran and Persian Gulf countries and is transported to Europe, the USA, Japan, India, Korea, and China; this is about 40% of oil transported through the Indian Ocean.
Strait of Malacca	About 15.2 million barrels of oil are transported per day through this point. The oil is sailed from Gulf countries to Pacific Rim countries, China, Japan, and Korea. This is roughly 35% of oil transported through the Indian Ocean.
Bab el-Mandab	About 3.40 million barrels of oil are transported per day through this choke point from Gulf countries to Europe and North America. This is approximately 8% of oil transported through the Indian Ocean.

Table 5.6 (B)
Terrorist Groups Operating in the IOR with Casualties in 2012

Group	Attack	Casualties	Group	Attack	Casualties
Taliban	525	1,842	Boko Haram	364	1,132
Al Qaida/ISIS[1]	249	892	Al Qaida/ISIS[2]	108	282
Naxalite/Maoist	204	131	Al Shabaab	121	278
Tehrik-e-Taliban (TTP)	103	510	PKK	80	83
FARC	71	122	FLNC	58	00

Additionally, Jaish-e-Mohammed, Lashkar-e-Toiba, Hizbul Mujahedeen, and Haqqani groups are dreaded terrorist organizations. These groups are actively supported morally, logistically, and financially by the Pakistan Army and its intelligence wing (ISI) to operate in the Indian State of Jammu & Kashmir and in Afghanistan against Indian, US, and NATO interests

Source: US Department of State, National Consortium for the Study of Terrorism and Response to Terrorism, May 30, 2012.
Notes: 1 = Operation in Iraq, 2 = in other Arab countries, Naxalite/Maoist in India, TTP in Pakistan and Afghanistan, PKK= Kurdistan Workers Party (Iraq), FARC = Revolutionary Armed Forces of Colombia, FLNC = Corsican National Liberation Front.

Table 5.6 (C)
Naval Bases / Presence in the IOR

USA	Australia, Bahrain, Diego Garcia, Egypt, Indonesia, Kenya, Kuwait, Oman, Qatar, Singapore, UAE
China	Bangladesh, Myanmar, Pakistan, Sri Lanka, Djibouti
India	Mauritius, Maldives, Iran

Failed States Index highlights that 11 out of the world's 20 most unstable states are located in the IOR. According to the World Bank in 2010, out of a global GDP of US$63 trillion the IOR accounts for US$6 trillion, compared to US$13.2 trillion for the Asia Pacific region, US$5.1 trillion for Latin America and the Caribbean, US$2.4 for the Middle East & North Africa, US$1.9 trillion in the Arab World, and US$1.1 trillion for Sub-Saharan Africa. A list of terrorist organizations having their presence in the IOR appears in Table 5.6.

The USA has been the sole strategic and military power in the Indian Ocean since the dawn of this century. Operating from a tiny (25 km × 11.5 km) V-shaped naval base Diego Garcia, a part of the British Indian territory leased to the USA and located almost at the center of Indian Ocean, the USA keeps an eye on the military and commercial activities in the region and secures its interest. The USA is also believed to have used this base for launching air attacks over Iraq during the first Gulf War in 1991, the war in Afghanistan in 2001 and in Iraq in 2003, and during the operation that liquidated the dreaded terrorist Osama Bin Laden in Abbottabad in Pakistan in 2011.

The USA wants to maintain its decades-long supremacy in the Indian Ocean by countering the emerging economic giants of Asia viz. India and China, and wants to ensure the freedom of navigation, conducting military operations, monitoring Iranian military deployments and deterring Iranian aggression, undertaking maritime security operations, such as counterterrorism, countertrafficking (humans, drugs), and counterpiracy. The Fifth Fleet of the US Navy (also known as the Central Command of the US navy and headquartered in Manama, Bahrain) is responsible for deployment of naval forces in the Persian Gulf, Red Sea, Arabian Sea, and parts of the Indian Ocean. Meanwhile another major power China is busy flexing its muscles in the region as a challenge to the USA.

Among the IOR countries, Iran has the highest number of naval assets, followed by India, Indonesia, and Pakistan (Table 5.7). India is the only country in the region to have an aircraft carrier. India also leads in most of the other naval facilities, such as in having 13 frigates, 11 destroyers, and 139 patrol vessels. Iran has the maximum number of submarines, while Indonesia owns maximum corvettes, and patrol vessels equal to India. However, the IOR countries have very little to compare in naval warfare with the major world powers who have their presence in this ocean. For example, while the USA and France have 24 and 4 aircraft carriers respectively, the UK, China, and Russia have one each. However, China has the maximum numbers of total naval assets, frigates, corvettes, submarines, and patrol vessels put together, indicating a determined effort by Beijing to rule over both the Indian Ocean and the South China Sea.

5.2.1 China's Presence in the IOR

Although not bordered by the Indian Ocean and seemingly unconnected, China of late has been pumping enormous foreign direct investment (FDI) into the IOR countries in a bid to control the economy of these nations. China's initiative is also to counter the strategic supremacy of the USA, and the moral authority of India in the Indian Ocean. China has expanded its presence in the IOR, both economically

TABLE 5.7
Major Geostrategic Military Powers in the IOR

Countries	1	2	3	4	5	6	7	8
Major IOR Players								
Australia	47	02	10	02	00	06	13	06
India	295	01	13	11	22	16	139	01
Indonesia	221	00	08	00	24	05	139	11
Iran	398	00	06	00	03	34	88	03
Pakistan	197	00	09	00	00	05	11	03
South Africa	30	00	04	00	00	03	31	02
Israel	65	00	00	00	04	06	37	00
Vietnam	65	00	09	00	14	06	26	08
Bangladesh	89	00	06	00	06	02	26	04
Saudi Arabia	55	00	07	00	04	00	09	03
Major World Powers								
USA	415	24	22	68	15	68	13	11
China	714	01	52	33	42	76	192	33
France	118	04	11	12	00	10	23	17
Russia	352	01	13	13	82	56	45	47
UK	76	01	13	06	00	10	22	13

Source: /www.globalfirepower.com/countries-comparison.asp; Mukhopadhyay et al. (2018b).
Notes: 1 = Total Naval Assets, 2 = Aircraft Carrier, 3 = Frigates, 4 = Destroyer, 5 = Corvettes, 6 = Submarine, 7 = Patrol Vessel, 8 = Mine Warfare.

and militarily. This expansionist attitude has been fueled by the enlargement of Chinese economy by 55 times between 1979 and 2016, i.e., from less than US$200 billion to approximately US$11,000 billion. In 2009, China became the world's largest exporting nation, and by 2013 it became the world's largest trading nation. China is well-set now to overtake the USA as the world's largest importer of fossil fuels.

Moreover, the thirst for energy and resources, and to ensure safe navigation for their merchant ships (as more than 70% of China's imports are through this ocean), China is modernizing its navy. And probably to avoid raised eyebrows from any quarter China is offering excuses to redevelop its ancient trading routes to central-west Asia and Africa to enter into the Indian Ocean. China has variously named this initiative as One Belt, One Road (OBOR), the Belt and Road Initiative (BRI), the Silk Road Economic Belt (SREB), or the 21st Maritime Silk Road (MSR; Fig. 5.9). This expansionist approach by China, as alleged by many countries, is unsettling the peace and tranquility in the region, which works against the basic tenet of the blue economy.

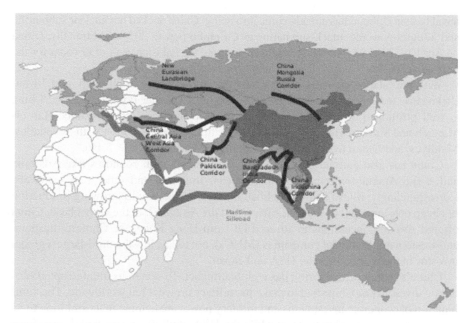

FIGURE 5.9   The Proposed OBOR/BRI/MSR Framework of China

Note: Black lines = Belt = Road and rail corridors; blue lines = Road = sea route. BRI connects China to central and west Asia, and Africa.

With saturation in industrial growth looming large and with its growth rates starting to moderate from double digits to around 6% currently, China has started exporting steel, cement, and engineering skills to other countries under the provisions of OBOR, and reestablishing in the process a silk route for trade and commerce. But the world remains suspicious of China because its market is opened selectively in return for technology. For example, Indian pharmaceutical companies complain about the lack of access to the Chinese market, when approximately 80% of the raw materials (formulations, pills) for such medicines are imported from China.

The OBOR project includes Belt (land corridor of road and rail) and Road (shipping lanes; Fig. 5.9). The Belt will connect (a) Beijing to Irkutsk via Ulan-Bator, (b) Xi'an to Moscow via Astana and Kazan, (c) Xi'an to Anaklia via Khorgos and Tashkent, (d) Xi'an to Duisburg via Tehran, Ankara, and Istanbul, (e) Xi'an to Gwadar, and (f) Kunming to Dhaka and Kuala Lumpur. The Road (shipping lanes), on the other hand, would connect Quazhou to Venice through Haikou, Singapore, Kolkata, Colombo (also probably Male), Gwadar, Mombasa, Djibouti, Suez, and Piraeus.

The Chinese navy has 255,000 servicemen and women, including 10,000 marines and 26,000 naval air force personnel. After the USA, China has the second-largest navy in the world in terms of tonnage. China's naval presence in the Indian Ocean includes building 18 naval bases in the IOR countries to surround India (Fig. 5.10). These bases will be, amongst others, at Djibouti, Yemen, Oman, Kenya, Tanzania, Mozambique, Seychelles, Madagascar, Pakistan, Sri Lanka (yet to be finalized), and Myanmar. Its naval base in Djibouti has already come into being. China docked her nuclear submarine and a diesel-powered attack-submarine at Colombo port in 2014. Additionally, China on April 14, 2015 won the leased rights to operate the Pakistani port of Gwadar for the next 40 years.

With the establishment of its first overseas military base in Djibouti, and Jiwani (Pakistan), and having obtained the leased right of Gwadar Port in Pakistan, China is well positioned to militarily influence the two important choke points: Bab el-Mandeb and Strait of Hormuz. Additionally China is in negotiation with Seychelles, Maldives, and Oman to expand its economic presence (Figs. 5.9, 5.10).

China has also entered into long-term agreements with countries like Australia and Saudi Arabia for import of fossil fuels, iron ore, and other raw materials. Correspondingly, it also had to establish markets for its rapidly growing production of cheap consumer goods in Asia, Africa, and the Americas. In the early phase, China offered several concessions to attract FDI from Hong Kong, the Chinese diaspora, and western multinational companies (MNCs), but today, it is the third-largest global investor in the world, after the USA and Japan.

China's economic expansion has a global impact. To secure its uninterrupted economic investments, China had to make its military prowess felt worldwide. The combination of economic investments and military power are helping in its ambition to be recognized as a superpower. Meanwhile the USA, with its unwise withdrawal from climate negotiations (COP-21, Committees of Parties) gave China an easy option to take the leadership role for a clean environment worldwide. China is further taking

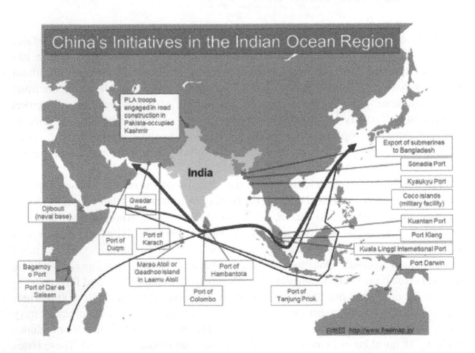

FIGURE 5.10 Chinese Initiatives in the IOR (After Ahmad (2018); Tasir (2018); Nagao (2018)).

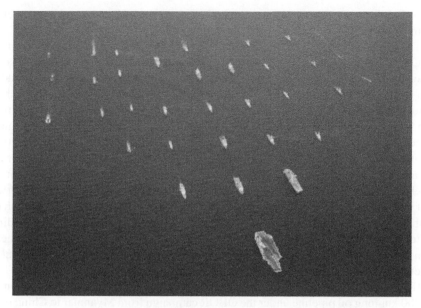

FIGURE 5.11 Advancing Chinese Naval Fleet in the IOR with Aircraft Carrier (PLAN Portal).

the lead in anti-piracy operations in the Gulf of Aden to ensure free flow of trade and secure access to energy supplies.

China's growing naval presence in the Indian Ocean is of major global concern (Fig 5.11). In fact the Indian Navy has been prodded into action to counter this presence, which is a part of what is known as Beijing's "string of pearls" strategy, launched early this century to expand its geopolitical influence by building ports and other civilian maritime infrastructure along the Indian Ocean periphery. When joined together, the ports being developed by China in Pakistan (Gwadar), Bangladesh (Chittagong), Myanmar (Kyaukpyu), and Sri Lanka (Hambantota) would form a "string of pearls" around the neck of the Indian subcontinent, and are feared by New Delhi to be part of a Chinese encirclement strategy. These ports are all currently commercial facilities and used for civilian purposes only. In the near future, though, they may serve as resupply and recreation stations for Chinese naval vessels and their crew, and would certainly enable the Chinese Navy to extend its reach into the Indian Ocean. China has also established a military surveillance facility on Myanmar's Coco Island near the Malacca Strait, helping to facilitate the entry of Chinese naval ships into the IOR.

China's presence grew steadily following its participation in anti-piracy operations off the coast of Somalia in 2008. It seems eight ships of the Chinese Navy were regularly deployed in the Indian Ocean, including nuclear-powered submarines (NDTV, 2017). China also has sold ten submarines to India's closest neighbors: eight to Pakistan and two to Bangladesh. As mentioned earlier, China established a naval base in Djibouti, which also includes a free-trade zone. This naval base provides logistical support for China's naval fleet, protects its commercial fleet from piracy and other threats in the region, and can harbor as many as 10,000 naval personnel at a time.

The China-Pakistan Economic Corridor (CPEC), a part of the bigger canvas of China's Silk Route, is considered by many as a threat to the Indian Ocean security. The CPEC connects the rail and road corridor from Xinjiang to Gwadar at a cost of US$60 billion, mostly as a loan to Pakistan. The port of Gwadar in Baluchistan province of Pakistan could be used by China in a demanding situation as a large-scale naval mobilization to threaten India's security, risk global sea lanes of commerce (as it is close to the point of origin of the world's oil trade exiting from the Arabian peninsula), and challenge the USA for regional naval superiority. Gwadar is connected in the north to Pakistan-occupied Kashmir and China through newly constructed rail and road, thousands of kilometers long. It is also reported that China's growing investment in Pakistan, predominantly to fund a new batch of four new nuclear reactors by 2020, with four more reactors in the pipeline (adding up to a total power capacity of 7,930 MW by 2030) can be viewed with apprehension as a disguise for a possible uranium enrichment facility to make N-bombs, when required.

If CPEC runs through the northwestern border of India, the recent construction of an 800 km-long oil pipeline under the OBOR campaign from Myanmar to China runs along the northeastern boundary of India, and directly connects China to the Indian Ocean. Also China has built a port in Kyaukpyu (in Myanmar) to transport gas from the Gulf, and laid oil and gas pipelines to move Myanmar's own gas, and imported oil and gas through Myanmar to its Yunnan province, circumventing the Malacca Straits. Myanmar has proved to be a stable strategic partner of China, as also about 70% of Myanmar's arms is supplied by China.

Similarly, in Sri Lanka, the state-run Chinese investment companies have subsidized billions of dollars in the development of the port city in Colombo. Additionally, China has signed a US$1.1 billion 99-year lease with the Island government to manage the Hambantota port. The work for the first phase in 2008 cost more than US$300 million of which 85% came as a loan from Exim Bank of China. China is further constructing a new city near Colombo port, which also handles approximately 75% of the transshipment of India's trade. The problem is about 82% of GDP of Sri Lanka goes in loan repayments. China is ever ready to offer money on loan, but the Island government has however no clue how to get out of this debt-trap laid by China. China is further building roads, rail lines, and power projects in Bangladesh, and is seeking association with port-building activities and on their upgrading.

The Maldives remained a traditional Indian security partner until 2017, when the Chinese reached an agreement with the Maldivian government to establish a joint ocean observation station on the Mukunudhoo Atoll. Although set up for R&D purposes, the station could serve as a vantage point on a crucial Indian shipping route. The Chinese also brokered a Free Trade Agreement (FTA) with the Maldivian government around the same time. The establishment of the R&D station, together with China's previous acquisition of 17 other Maldivian islands in the same area, has raised concerns regarding the lack of transparency about the purpose of these acquisitions. Chinese cooperation with the Maldives expanded in early March 2018 when China sent a combat naval force in support of the pro-Chinese Maldivian president during a declared state of emergency. However, the follow-up election in the island brought back an India-supported leader as the president of the Maldives.

Like many Asian countries, China has been successful in drawing many poor African countries into its fold for economic engagement, in terms of infrastructure and commercial building projects. Only recently Tanzania has signed a deal with China to help improve and expand its main port at Dar-es-Salaam. In addition, a variety of infrastructure projects received technical and financial help from Chinese firms. In a similar fashion, Kenya has also been deeply involved in China's OBOR. The most significant project related to this involvement is the building of the Standard Gauge Railway (SGR) connecting capital Nairobi to its most viable port Mombasa, costing several billions of dollars.

The OBOR also facilitates Chinese firms to engage in construction work across the globe on an unparalleled scale. However, the concern that OBOR loans may put several countries into a debt-trap to China who might use it as a weapon in polit-ical disputes to extract strategic concessions is coming true. For example, in 2011, China wrote off an undisclosed debt owed by Tajikistan in exchange for 1,158 km^2 of disputed territory. China also stopped salmon imports from Norway after it awarded the Nobel Prize to Chinese dissident Liu Xiaobo in 2010. In 2012 China discontinued the supply of rare earth elements to Japan, and ceased tourism to and banana imports from the Philippines. The most vulnerable countries in terms of falling into a debt-trap to China are Djibouti, Kyrgyzstan, Laos, the Maldives, Mongolia, Montenegro, Pakistan, and Tajikistan.

Beijing's US$995.6 billion (another estimate put the figure at US$1,200–1,300 billion) OBOR initiative is a state-backed campaign for global dominance. The OBOR touches 71 countries that account for half the world's population and a quarter of global GDP, with China planning huge construction work in more than 60 of these countries. As many as 15 IOR countries are among the designated partners of OBOR. The rest, except India and Oman, however participated at the high-level inaugural meeting of the Belt and Road Forum for International Cooperation (BRFIC) held on May 14–15, 2017 in Beijing. The meeting was attended by 29 foreign heads of state and government, and representatives from more than 130 countries and 70 inter-national organizations. In summary, China is showing almost equal attachment and interest with the Indian Ocean as that in the South China Sea, in terms of commerce, trade, and military activities.

5.2.2   INDIAN PRESENCE IN THE IOR

China's push into the Indian Ocean raises the specter of full-fledged military skirmishes at short notice between India and China. Even with a trivial misunderstanding or when the Indian Ocean sea lines of communication are being blocked by Chinese warships during a contingency such armed conflicts are a distinct possibility. But there are ramifications even during peacetime, as Beijing's amplified presence may prompt some countries that had been friendly toward New Delhi to shift allegiances.

In a sense, China may be practicing what ancient military strategist Sun Tzu held up as being the supreme art of war: "to subdue the enemy without fighting."

In contrast to the belligerent initiatives of the USA and China, India on the other hand has so far taken up the cudgel in winning the hearts of the IOR nations by getting involved in building their capacities in health, education and skill development, imparting training, cultural exchanges, and strengthening its military installations (a sort of soft diplomacy). Agreeing that India needs to further raise its profile in a changed circumstance, it is responding to China's advances largely in two ways through: (a) military presence and (b) training and capacity building.

5.2.2.1 Military Presence

India has had non-military supremacy in the Indian Ocean over several decades in the past. As for the maritime security India sees itself as the natural preeminent regional power (Sakhuja, 2011). However, presently the key reason for India's militarization in this ocean is to counter the growing Chinese presence and influence. In late 2017, the Indian Navy adopted a new plan for deployment of warships that was aimed at countering the increasing presence of China in the Indian Ocean. Under the new plan, a dozen of mission-ready warships are deployed across four strategic regions, such as the Malacca Strait and Andaman Sea; North Andaman Sea and Bay of Bengal (Bangladesh and Myanmar); Lakshadweep Islands and Maldives; Madagascar; the Persian Gulf and the Arabian Sea (TOI, 2019b).

This facility overseeing the crucial Malacca Strait through which about 60,000 ships pass each year is led by the Andaman and Nicobar Command (ANC). This, India's easternmost naval base, has not seen much military upgrading since its establishment in 2001. However, the Indian Navy is giving the ANC now top priority and probably will have a division-level force, with an S-400 surface to air missile system, fighter squadron, more airstrips (headway has been made in that regard), and major warships. In addition, the Anti-Submarine Warfare (ASW) has been upgraded keeping in view the increase in Chinese submarine excursions. India has deployed two ASW aircrafts. In addition, India has deployed the Arihant-class nuclear-powered ballistic missile submarine in the Andaman waters for anti-piracy activity, safe navigation of merchant navy, and to strengthen its global image and status in the Indian Ocean.

To display its strength, India has taken several steps since 2001. It has modernized its bases on the Andaman and Nicobar Islands (near the Malacca Strait), established naval surveillance facilities in Madagascar and Mauritius, and launched a military satellite to enable communication even while far offshore. India's warships are visiting IOR countries frequently on goodwill tours (50 such tours were made in 2014 alone), including friendly port calls by an aircraft carrier to the Maldives and Sri Lanka. In 2016, the Indian Navy hosted an international fleet review, in which over 100 warships from 50 countries participated (Fig. 5.12).

In October 2017, India launched seven "missions of engagements" to display its presence, strength, and goodwill. The missions are: (a) Malacca Strait, (b) Andaman and Nicobar Islands, (c) northern Bay of Bengal and northern Andaman Sea, and, to the west, to the (d) Strait of Hormuz, (e) coast of Somalia, (f) Sri Lanka and Maldives, and (g) Mauritius, Seychelles, and Madagascar. To facilitate travel of

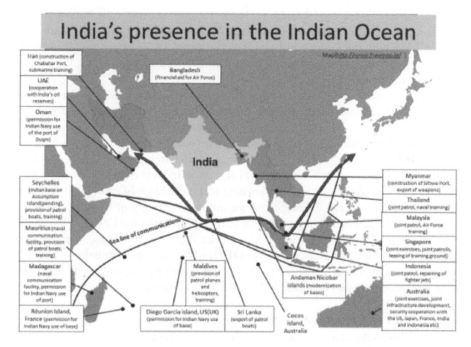

FIGURE 5.12   India's Initiatives in the IOR (After Ahmad (2018); Nagao (2018)).

warships on goodwill visits, India secured access rights to Diego Garcia from the USA in 2016, to the port of Duqm from Oman, and to Réunion Island from France. And in 2018, it reportedly reached agreement with the government of Seychelles to build a base on Assumption Island. Since May 2018, India's fighter jets are stationed in the Andaman and Nicobar Islands to counter any security threat to the Indian Ocean (Nagao, 2018).

The production and acquisition of warships (destroyers, frigates, and submarines) by China since early this century clearly outnumbers that of India. To counter this, and as a retaliatory measure, in 2017 India upgraded her military partnership with Singapore to allow the Indian Navy to use Changi Naval Base for logistical and operational support, and for rearmament to the Indian fleet to operate both in the South China Sea (SCS) and IOR. In addition, Sri Lanka recently reached a 40-year agreement to lease her Hambantota airport to India (remember the Hambantota seaport is leased to China). Again, India helped in the development and modernization of Iran's Chabahar seaport, and this would be a countering tool to China's Gwadar seaport in Pakistan. In 2016, India also inked a trade corridor deal with Afghanistan and Iran to facilitate transport of men and materials and carry out trade and commerce through Chabahar port to Central Asia. More importantly for India, proximity of Chabahar to the regular oil trade coming out of the Persian Gulf provides geostrategic value for India in the way Gwadar port would do for China. In February 2018, India signed an agreement with Oman that includes an understanding of military cooperation and, more significantly, exclusive military access to the Port of Duqm. This is highly significant given the fact that this port is geostrategically proximal to Persian Gulf shipping lanes, and India would receive all logistical and operational support for any sort of military operation in the northwest IOR, including the Gulf of Aden off the coast of Djibouti.

Perhaps the most noteworthy strategic move that India has made, however, comes with regards to agreement to invest half a billion dollars and build a naval installation on Assumption Island in Seychelles. Assumption Island, which is among the southern-most points of Seychelles just north of the Comoros and southeast of Dar-es-Salaam, provides India's navy with the ability to strategically monitor important shipping routes in the Mozambique Channel. This base will also allow India to engage in its anti-piracy campaign as well as combating illegal fishing and trafficking. In addition, it will also provide India with a critical launch pad for military oversight of Chinese economic and strategic developments in the western IOR. Additionally, the Logistics Exchange Memorandum of Agreement (LEMOA) between the USA and India, coupled with Defense Technology and Trade Initiative (DTTI) which permits both India and the US military forces to use each other's bases and other infrastructure, is going to help both countries immensely and maintain the balance of power in the region.

5.2.2.2 Training and Capacity Building

Stressing the fact that the entire IOR region is the product of same ancient heritage and culture, India's second approach has been to provide capacity-building assistance for countries in the region, in the form of training of personnel, financial support for new infrastructure (Built-operate-transfer/Turnkey/Grant), and trade-tourism-culture.

The first type of assistance includes accepting foreign students to undergo educa-tion and training in India, repairing equipment, or loaning tracts of land to conduct military drills. Malaysian fighter jets, for instance, were brought to India for ser-vicing, and Singapore has a long-term arrangement for the use of India's training grounds. Assistance provided in the recipient country usually consists of dispatching instructors, sometimes together with the provision or export of equipment. For example, Indian military personnel have been stationed in the Maldives, Seychelles, and Mauritius to train local troops in the use of patrol planes, helicopters, and boats that were supplied by India to those countries.

Financial assistance has increased conspicuously in recent years. In April 2017, India agreed to furnish US$500 million in aid for the Bangladesh Air Force, and this amount is expected to be used to purchase spares for fighter jets and to conduct studies in the selection of next-generation fighters. Gaining ground in Dhaka would not only keep check on the Chinese invasion, but would also prevent Bangladesh from purchasing jets from China, denying the Asian giant from gaining a permanent foothold at an air base in Bangladesh.

5.2.3 RELATIONSHIPS

Beyond the military and strategic manipulations, the trust between countries matters a lot in maintaining a warm, peaceful, and mutually beneficial relationship in the IOR. It is intriguing to note that IOR experiences two extreme types of understanding among nations, where one relationship based on mistrust and vendetta can drain both the countries of finance and goodwill, while other relationship developed over trust and cooperation could bring about mutual progress and prosperity, in terms of economic and social development. While the Indo-China and Indo-Pakistan relationships fall under the first category, the second category is exemplified by the India-Bangladesh, ASEAN, and IORA relationships. We now discuss these issues briefly, and aim to show how trust and cooperation among countries could achieve an enduring ecosystem for growth of the blue economy.

5.2.3.1 China–India Face-Off

A race between the two largest Asian military powers, India and China, to gain control over important shipping lanes, proximal to Middle East energy supplies, and access to the African continent (the last global economic frontier) has thus both economic as well as militarily strategic components. While commenting on the growing militarization of the Indian Ocean power game, Nagao (2018) felt that the Maldives unrest (declaration of a state of emergency in February 2018) triggered a near-showdown of military might in the Indian Ocean. India was reportedly preparing to send troops to restore political stability to the island nation, and at around the same time 11 Chinese naval vessels entered the Indian Ocean. In addition, China already had 3 ships carrying out anti-piracy operations off the coast of Somalia, meaning that 14 PLA (People's Liberation Army of China) Navy vessels were deployed in the Indian Ocean. The sizeable Chinese presence prompted India to send a fleet to head off the Chinese incursion, after which the Chinese warships retreated (Reuters, 2018).

Japan, like the USA, is also worried over the safety and security of the Indian Ocean, particularly with the deployment of a large number of warships by China following its OBOR campaign, which is to materialize by the year 2049. In this scenario, deployment of Japanese ships to defend the Indian Ocean may be necessitated, which is a vital sea lane for Japan too. In fact, although the Japan-USA security alliance is focused on the East and South China Seas, Japan has been dispatching naval vessels to the Indian Ocean since 2001 to defend shipments of her oil from the Middle East. Insiders feel that the time has come for both Washington and Tokyo to help New Delhi undertake the bulk of the security task in the Indian Ocean. Some security experts have speculated that India is building an "undersea wall" of sensors along the coastline of the Bay of Bengal, stretching as far as Hainan Island, to detect Chinese submarine activity (Singh, 2016). Such an initiative would bolster Japan's national security, and the Japanese government would do well to lend its full and active support to New Delhi. Japan could share shipbuilding expertise and provide ships and aircrafts not only to India but also to other friendly countries in the region. In fact, Japan has already given two US-2 air-sea rescue amphibious aircrafts to India, and is considering to give two patrol boats and a used P-3C maritime surveillance aircraft to Sri Lanka (Nagao, 2018).

China's string of pearls strategy (Fig. 5.10, Fig. 5.13) and its effort to generously finance infrastructure or other facilities in many IOR countries may make Japan and other major countries irrelevant in this region in the near future. Japan's official development assistance and investments by Japanese companies has played a major role so far in the region's economic development (Watanabe, 2019). That must continue.

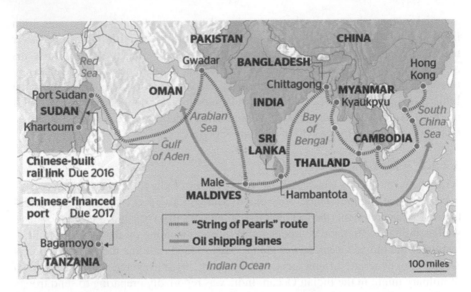

FIGURE 5.13   China's Specific Expansion Plan for the IOR (After People's Republic of China Government Portal).

However, it is argued that Japan could extend a helping hand by offering solutions to the problems these countries are facing. This additional help may not necessarily be in the form of military, but a potentially highly effective approach would be the closer civil-sector cooperation and scientific collaboration to elevate the quotient of goodwill. Japan with its long time-tested experience in disaster management could extend assistance to disaster response centers of many IOR countries offering relief and rehabilitation in the event of natural calamities. For example, the Asia Pacific Alliance for Disaster Management (A-PAD), an internationally active and acclaimed Japanese NGO, could tie up with local NGOs of each country to carry out the relief and rehabilitation (Watanabe, 2018).

Surprisingly, the countries that require the greatest assistance in the area of disaster management already have close ties with China, namely Sri Lanka, Bangladesh, Myanmar, Indonesia, Vietnam, and the Philippines. The first three countries are counted as being part of China's string of pearls strategy, while the latter three ironically have territorial disputes with Beijing in the South China Sea.

The race for strategic control over the Indian Ocean between China and India is only speeding up and the construction of a naval base by China in Djibouti appears to be extremely strategic. Besides attending to piracy threats, Chinese can launch from here direct military operations in the western IOR, if the need arises (Brewster, 2014). Additionally, the element of secrecy that shrouds almost all the activities of China (be it acquisition of 17 islands in Maldives, or the nature of Gwadar port facility, or similarity between R&D centers in Maldives with that in South China Sea, which can also be used for military application) is creating enormous apprehension in the minds of other major players in the world viz. the USA, Japan, Australia, South Africa, France, Britain, and India. Another concern that may spell doom in the continuation of China-funded economic activities in poor IOR/OBOR countries, has been the building up of a potential debt-trap. This would make loaned countries vulnerable to Chinese interference in their internal politics and economic decisions. In fact, what India started softly several decades ago to help build capacity in IOR countries through training, education, skill development, and cultural exchanges, is now suddenly threatened to be made redundant out by mega-scale grants from China.

Again there are significant differences in the military and commercial capabilities of these two countries (Table 5.7). For example, India has 28 premodern-era shipyards, which have failed the aspirations of a Navy in need on several occasions. In contrast, an aggressive China has over 800 shipyards and even countries like Vietnam and the Philippines produce quality ships at competitive prices. India could make this industry open to investment by private sectors and modernize her shipbuilding and ship-repairing facilities in good numbers. The Maritime Agenda 2010–20 pegs an investment of US$2.85 billion by 2020 for dredging to make ports usable for large naval and commercial vessels. The dredging industry may need quick private investment. In contrast China has created two islands—Spartly Island Chain and Fiery Cross Reef—through the dredged material, even violating the provisions of UNCLOS, the UN regulatory body (Kopela, 2017).

Again, while China is fast moving toward its goal of 5–6 aircraft carriers, India's goal of 3 is constantly missing the deadlines. For India, *INS Vikramaditya* is presently serving, while *INS Vikrant* will be made available only in 2023, and *INS Vishal* is still at the design phase. In contrast, aircraft carrier Liaoning (commissioned in 2012) and the second Shandong are serving the PLA (Chinese) Navy. China has 68 submarines and India has only 14 (and is adding another 15 quickly). However, the gap speaks volumes, if not the whole story, about China's undersea capabilities (Table 5.7). Nevertheless, India's long-standing naval partner the USA could help it develop capacity in this regard. In fact, India's naval policy has been inconsistent, slow, plagued with lack of coordination, and does not honestly espouse great faith. India must move quickly with dexterity and in an integrated manner in achieving a sea-based deterrent system.

India and China are no aliens to one another when it comes to conflicts. In 1962 and in 1967, they engaged in a short but violent war in the Himalayan terrain. A feud over various territorial claims between these two nuclear nations has been going on since 1980. And as recently as in 2017 they had the Doklam border dispute in Bhutan and skirmishes in eastern Ladakh in June 2020 in which tens of soldiers of either side lost their lives. In addition, the long pending border problem in Arunachal Pradesh and Jammu & Kashmir, the OBOR that encircles India with Chinese presence, the in-equal trade licensing, and flooding of the Indian market with cheap electronic goods are detrimental to Indian industries. All these activities have soured the relationship between these two countries. In future, both countries are expected to go all out for military, commercial, and economic initiatives in the Indian Ocean. And if this security dilemma persists, which is in fact likely only to escalate, confrontation among the navies of these two countries cannot be ruled out. Such hostility may also travel on land along the vast territorial border that these two countries share. Hence, the next few years will be critical in determining whether the world would see a peaceful competition or disastrous collision for regional supremacy in the Indian Ocean (Brewster, 2014; Mukhopadhyay et al., 2018b; Watanabe, 2019). However, because of such nefarious actions the worst victim will be the implementation of the blue economy in the Indian Ocean.

Despite several stark security anomalies like the above, astonishingly China and India have collaborated on the global platform in the areas of trade, climate change, in the formation of Asian Infrastructure Investment Bank (AIIB), and the BRICS Bank (the Brazil-Russia-India-China-South Africa bloc). The mutual visits of top leaderships to each other's countries also helped reduce distrust. But there have been frequent instances where the relation appears to remain inconsistent and incompatible. Although the Indian economy is only 20% the size of China's, the elevated moral, democratic human right, and high spiritual position of India are making China look small ethically.

Insisting on application of multilateralism and rules-based politico-military order, Mohan (2018) saw a vision for a new foreign policy approach in the Indian Ocean. Going beyond the USA-China-India scenario, Mohan found immense economic and strategic importance for Europe, as the Indian Ocean is the gateway to enormous Asia-Pacific markets. It seems about 35% of Europe's exports go to Asia as this continent is highly dependent on unhindered maritime highways or sea lines of communication. And yet, European society and media are more focused on the South China Sea rather than on the fast-changing military and strategic landscapes in the Indian Ocean.

China's entry in the Indian Ocean region, through economic investments, political influence, and military presence under the OBOR campaign is significantly altering regional dynamics. The Chinese Defense White Paper (2015) spoke about China's ambition to become a major maritime power, indicating its interest beyond the South China Sea, and particularly securing its trade routes and access to natural resources in the Middle East and Africa. Over the years China has also assured to its hinterland a continuous supply of natural gas from Myanmar, and oil from the Gulf. Most of the IOR countries import a range of finished products from China, which is also emerging as a major investor in manufacturing and infrastructure, as well as an arms supplier to ASEAN nations. Yet, many of these countries remain critical to its South China Sea policy for conflicting territorial claims (Mohan, 2018).

Incidentally, India's reach and influence in the IOR have received a beating despite being ideally located to represent a common culture, history, and civilization with many African, South Asian, Gulf, and Southeast Asian countries. For example, India is the largest exporter to only one country in the entire IOR region (Sri Lanka), while China has been the largest exporter to as many as 22 countries. Similarly, India exports goods worth US$5,500 million and imports US$639 million from Bangladesh, whereas China-Bangladesh bilateral exports have recently reached US$15 billion. While exports to Sri Lanka from both China and India remain the same, China imports goods worth only US$259 million from Sri Lanka compared to Indian imports of US$848 million. The scale of investment also differs drastically. Compared to Indian investment in Australia (US$2.4 billion), and in Malaysia (US$0.4 billion), China has invested US$10 billion and US$3.1 billion, respectively in these countries. India, however, invested more in Singapore (US$5.9 billion) compared to China (US$1.1 billion).

Vietnam and the Philippines have however been very vocal in decrying China's expansionist strategy, particularly in territorial claims in the South China Sea. The latter even won a favorable maritime boundary judgment from the UNCLOS tribunal against China, but has lately caved in to accept US$24 billion in Chinese investment in infrastructure. Even the former (Vietnam) permitted US$4.15 billion Chinese investment in 2016, imported goods worth US$66 billion from and exported only US$25 billion worth of goods to China. In contrast, India has invested only US$1 billion in Vietnam in 2016 and the bilateral trade remains at US$8 billion. Chinese constitute 1.2 million population in Australia (5.6% of the total Aussie population), while India has only 0.46 million by ancestry (~2% of the country's population). Australia however is slowly getting wary and vocal of China as the latter has started taking over Australian hi-tech companies.

In South Asia also, China is consolidating its position. As mentioned earlier all South Asian countries except India and Bhutan participated in the 2017 OBOR conference. The CPEC could be a threat to India, as it passes through Pakistan Occupied Kashmir (POK). This would ensure the presence of Chinese troops close to the Indian western border and may embolden Pakistan to continue with terrorist activities against India. The deal signed between China and Sri Lanka with regards to Chinese investment for the development of Hambantota port might face some difficulty due to recent changes in the government in Colombo. This agreement might also alert the island nation to a possible debt-trap. China is also aiming to invest in building a port in Bangladesh, after it consolidated its engagement with the supply of a couple of submarines. A Chinese company replaced an Indian company in modernizing an airport in Maldives, and is developing a tourist resort on an island over a 50-year lease. Meanwhile, China is all set to pursue deep-sea mining of minerals in the Southwest Indian Ridge and probably hydrocarbon in the Indian Ocean off Madagascar.

Threatened by these developments, India responded by focusing on the Security and Growth for All in the Region (SAGAR) strategy in the IOR region, and stressed on strengthening the Bay of Bengal Initiative for Multi-Sectoral Technical and Economic Cooperation (BIMSTEC). India is also in league with Japan, the USA, Australia, and South Africa to play a balanced commercial and military role to counter the Chinese asymmetric policy of expansionism and crony-capitalism-based commercial activities in the region. Understandably, India and Japan are increasing their collaboration in the areas of maritime security, and infrastructure development of the Andaman archipelago. The civil nuclear cooperation between India and Japan, and involving Japan in Indo-US Malabar military exercises are some of the other areas of closeness (Ghosh, 2017). Moreover, India signed with the USA the Logistics Exchange Memorandum of Agreement (LEMOA), which will pave the way for a balanced growth. India's commercial shipbuilding and ship repair sectors, complementing the Sagarmala project (Fig. 2.5) of port development have the potential to take economic transformation forward. Unfortunately, when China is establishing communication with other IOR countries, India is busy developing her own ports and harbors.

India meanwhile is trying to expand its commercial base through the Port of Chabahar (in Iran) to directly access the huge market of Central Asia by road through Afghanistan. India is also competing with China for access to Seychelles, Maldives, and Oman. India, in fact, is spending about US$46 million in foreign aid in Seychelles to improve its coastal defense and airstrips. However, the project has run into rough weather as some locals at the alleged instigation of China and probably abetted by the change of government in the Island country has raised the issue of environmental degradation.

As China's economic interests begin to span the entire globe, it was inevitable that Beijing would try and secure them eventually through its own military means. That is what all great powers do. As the world's second-largest economy (aggregate GDP of US$13 trillion) and the second-largest annual defense budget (US$250 billion), China has both the impetus and the resources to acquire foreign military bases (Mohan, 2018). The political impact of OBOR can be assessed from the fact that an influx of workers, materials, and money on loan from China to several IOR countries has placed many of these nations in Chinese debt traps. Beside large-scale land acquisition and related environmental degradation, such loans will have clear security and political ramifications. In fact, the national politics and governance of Sri Lanka, Pakistan, Myanmar, and Maldives have already experienced such interventions. Such intrusion will only increase and spread in future, as China's economic investments elsewhere might also give undemocratic, weak, and corrupt governments a new lease of life.

China, it seems, wants to settle the border dispute with India quickly. Leaders of both the nations have publicly advocated that these two time-honored civilizations have sufficient wisdom to resolve all the pending disputes. Meanwhile, India and China are resuming their annual two-week-long hand-in-hand military combat joint exercise in December 2019 (frozen after the Doklam border scuffles in 2017–18) to focus on transnational counterterror operations, establishment of a joint command post, humanitarian assistance, and disaster relief. However, the unprovoked intrusion of the Chinese Army in Indian Territory in Ladakh in May–June 2020 and limited armed skirmishes with loss of life on both sides could threaten the chance of these two countries reaching any quick understanding.

Yet, to counter the Chinese aggressive stance in the Indian Ocean, India should have a five-pronged response: (a) quickly enhance its naval and other military power in the region; (b) strengthen military partnership with the USA, Japan, Australia, and France; (c) win more friends and partners in the IOR to establish naval bases; (d) set up trading, economic, and tourism exchange facilities in various IOR nations to ensure balance growth; and (e) increase offering of soft help to IOR nations in terms of training, capacity building, education, R&D, and heritage and cultural exchanges to bring people of these countries close to India. India's US$3 trillion economy is growing, and getting globalized, with a desire to reach US$5 trillion dollar by 2024. In this regard, the White Paper, titled "China's national defence in the new era" issued by the Chinese Navy in July 2019 has tried to justify building military facilities in foreign locations for promoting international security and protecting China's overseas interest. Hence, India needs a well-oiled efficient integrated marine/naval organizational structure to ensure its security imperatives beyond its borders to the IOR and probably to the Indo-Pacific region for an effective progression of the blue economy.

5.2.3.2   Pakistan–India Face-Off

As China desires to establish military bases in friendly IOR nations, Pakistan was only more than ready to accommodate the request. Being the closest ally, Pakistan quickly integrated to China's military strategy. Given this background and armed with nuclear weapons, Pakistan has not been maintaining the best of relationships with her eastern neighbor, India (Mohan, 2018). In fact, they have fought so far four full-fledged wars: in 1947, 1965, 1971, and 1998, but with India winning all. However, the acrimony remains, particularly along the Line of Control (LOC) in Jammu & Kashmir. It is hard to believe that Pakistan would allow any opportunity— be it diplomatic or military—to needle India go to waste.

In reply to this, in November 2017, India deployed its second Arihant-class nuclear-powered SSBN (ship submersible ballistic nuclear submarine). Currently, India is also constructing two more Arihant-class submarines. Meanwhile, the Indian Space Research Organization (ISRO) has a dedicated GSAT-7 satellite, which is used by the Indian Navy as a multi-band military communications satellite. Apart from India's second operational nuclear-powered submarine, it has 13 diesel-electric ones, among which about half are in service in the Indian Ocean. Pakistan believes that Indian ambitions, along with growing economic, industrial, and naval capabilities including a range of missiles are posing serious challenges not only for Pakistan's maritime, energy, and economic security but also for its conventional and strategic capabilities (Khan, 2018). However, India's nuclear doctrine (2003) has agreed for three principles: No First Use (NFU), Inflict Massive Retaliation (IFU), and have Credible Minimum Deterrence (CMD).

In contrast, Pakistan's navy at present operates five French diesel-electric submarines—three purchased in the 1990s and two dating from the late 1970s. In May 2012, Pakistan established its Naval Strategic Force Command (NSFC) that is the custodian of Pakistan's sea-based developing capability to strengthen its CMD and maintain strategic stability in the region. In November 2016, Pakistan established a Very Low Frequency (VLF) communication facility that provides a secure military communication link, hence enhancing the flexibility and reach of operations including the use of submarines. Pakistan has also developed Babur III SLCM (450 km strike range; Khan, 2018).

The problem of Pakistan essentially comes from their decades-long support to Islamic militancy and terrorism, and exporting these to neighboring countries to run a proxy war. This terror proxy has since been ingrained in their foreign ministry strategy. Almost all global security and financial institutions have reprimanded Pakistan for this nefarious activity against humanity and have taken punitive actions, yet Pakistan choose to deny them. It was however quite embarrassing for Pakistan's prime minister to admit recently (July 2019) the presence of 30,000 to 40,000 terrorists on Pakistani soil who have fought in Afghanistan and in the Indian part of Jammu & Kashmir.

Interestingly, much of the militarization effort by India in recent years in the Indian Ocean is seen by Pakistan as against them, although these are globally believed as India's stand to counter the advance of China in this ocean. However, as indicated earlier Pakistan could act as a dark horse in the battle for supremacy between India and China in the IOR, and could alter the balance of power in favor of China. Hence, India, the USA, and Japan must take this aspect seriously.

5.2.3.3 Bangladesh–India Friendship

In contrast to the estranged relationships of India with China and Pakistan, as an example of what a superlative cooperation and sustained collaboration among the countries of IOR could achieve, we figure out the Bangladesh-India relationship. These are the two closest neighbors, for about 4,000 km of boundary of Bangladesh is surrounded by five Indian States (West Bengal, Assam, Meghalaya, Mizoram, and Tripura). These two countries have long-standing problems, such as settling the sovereignty of hundreds of *chhit-mahals* (enclaves), redrawing the international and maritime boundaries, and Teesta River water sharing (Fig. 5.14).

Bangladesh took the first step to improve relations with India by closing down the training camps of terrorists on her soil and insurgents' active in India's northeast, and even handing over some of its leaders who had taken refuge in Bangladesh. Later, Bangladesh

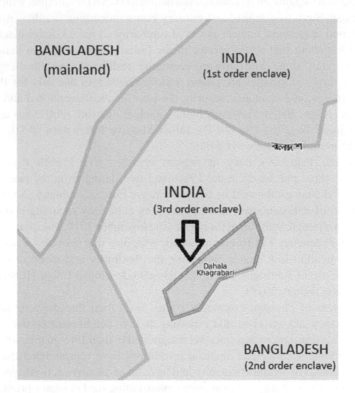

FIGURE 5.14 Redrawing the Border between Bangladesh and India

Notes: Dahala–Khagrabari (Enclave #51) was the world's only third-order enclave before India ceded it to Bangladesh in 2015. It was a piece of India within Bangladesh, within India, within Bangladesh. Fewer than 7,000 square meters (0.70 ha; 1.7 acres) in area and was the site of a jute field. 28 smaller enclaves existed within the overall complex.

conferred the country's highest civilian award to the late Indira Gandhi (former Indian prime minister) for her role in the liberation of Bangladesh. Bangladesh had also moved purposefully to recover her secular democracy. Such a status that defines its fundamental difference with Pakistan was lost soon after the assassination of Sheikh Mujibur Rahman (the father of Bangladesh) in 1975. This deviation was further consolidated by successive military coups and the anti-Indian policies pursued by the present main opposition party, who were in power most of the time after the country gained independence in 1971. The opposition party is infested with Razakar and those who ironically supported Pakistan during the independence struggle in 1971. Razakar was an anti-Bangladesh Bengali paramilitary force formed, financed, and abetted by the Pakistan Army in the then East Pakistan, now called Bangladesh, during the Bangladesh Liberation War in 1971. The term Razakar is now synonyms with "traitors" as they committed numerous atrocities (genocide) on their own Bengali brethren.

Rational political will solved the seven-decade-old vexed enclaves (*chhit-mahals*) by redrawing the international boundary amicably and without using military might. At the time of partition of India in 1947, East Pakistan (now Bangladesh) had 102 enclaves of Indian Territory, while India had 71 Bangladeshi enclaves. A joint census in 2010 found 51,549 people residing in these enclaves: 37,334 in Indian enclaves within Bangladesh and 14,215 in Bangladeshi enclaves within India. Following the Land Boundary Agreement in 1974 that got revised in 2015, India received 51 Bangladeshi enclaves (covering 28.77 km^2) in the Indian mainland, while Bangladesh received 111 Indian enclaves (covering 69.44 km^2) in the Bangladeshi mainland. Although India, in fact, lost around 40 km^2 to Bangladesh, it was conceived as a goodwill gesture to strengthen neighborly relations.

Again, Bangladesh won the case against India of maritime boundary in the international court and India did not prolong the problem as it did not ask for a review. Both the countries were again very close to inking the Teesta River sharing in the early part of this decade, but inadequate coordination between the federal and state governments in India hampered the two sides from reaching a much-awaited agreement. Yet, India and Bangladesh have agreed to consolidate their friendship further by exploring new prospects of oil and gas, and environmental concerns in the Sundarbans (Hossain, 2012), and furthering S&T collaboration in the field of oceanography.

In fact, saving the environments of the Sundarbans and the home of the Royal Bengal tiger could be the focus of Bangladesh-India cooperation. Mangroves in the Sundarbans are a dominant carbon sink, storehouse of unique flora and fauna, and also the source of livelihood for millions on both sides of the border. Some of the major rivers flow through both the nations, so also the countries share one language, culture, literature, and thought process. The warmness between India and Bangladesh is now back after a low from 1975 to 1996, and then again from 2001 to 2006, with top leaders of both the countries having found values in cooperation. This is a positive take away for the stability, peace, and harmony for blue economy in the IOR.

5.2.3.4   ASEAN Cooperation

The Association of South East Asian Nations (ASEAN) is another example from the IOR that displays commendable mutual respect, trust, and confidence among members for the progress of the people of their respective countries and for humanity in totality. Formed on August 8, 1967, the ten-member ASEAN comprises Brunei, Cambodia, Indonesia, Laos, Malaysia, Myanmar, the Philippines, Singapore, Thailand, and Vietnam. The ASEAN countries agreed to enhance peace, stability, and regional resilience by promoting political, security, economic, and sociocultural cooperation among its member countries.

ASEAN promised to preserve Southeast Asia nuclear free, and to facilitate free flow of goods, services, investment, capital, professionals, and to create a stable economically integrated and highly competitive common market. It further aims to strengthen democracy, human rights, good governance, and alleviate poverty. An example from East Asia Sea countries is given in Section 1.2. This association has been extremely successful and has set an example for other nations and cooperative organizations to follow.

The resource economy estimates of the countries of the East Asia Sea (EAS) region is another example of successful blue economy paradigm. EAS comprises Cambodia, China, Indonesia, Malaysia, Korea, Philippines, Thailand, Timor Leste, and Vietnam. Both ASEAN and EAS give formidable importance to the ocean as it regulates climate and weather by constantly transporting heat from the equator to the poles. Oceans produce about half of the oxygen in the world and also store 50% more carbon dioxide than the atmosphere. The blue carbon component (mangroves, salt marshes, and sea-grass) in the EAS region removes carbon dioxide ten times more than a tropical rain-forest and store three to five times more carbon and in the process make our life bearable on this Earth. In terms of money, the estimated blue carbon value in the EAS has been US$111 billion from mangroves and another US$77–95 billion from sea-grass. Moreover, coral reefs and mangroves reduce impacts of waves by dissipating respectively 97% and 66% of the wave energy. The 1,400 oil and gas offshore fields in the EAS region produce 2 million barrels of oil per day (equivalent to US$34 billion), catch 40% of the world's fisheries (= US$35 billion), and 80% of the world's fishes from aquaculture (contributing US$100 billion). EAS witnesses about 90% of world naval trade passing through its waters involving considerable income besides providing jobs to thousands. Accounting for about 26% of the world tourist footprint, EAS collects a revenue of about US$200 billion (PEMSEA, 2018).

5.2.3.5   Regional Collaboration

IOR countries have formed several multilateral regional economic groups among themselves and on a global level. For instance, 24 IOR countries are a part of the 57-member Organization of Islamic Conference (OIC). The Arab League has 14 IOR states out of a total membership of 22 states. The G-20 includes 5 IOR countries. The Next-11 is made up of 5 IOR states. The Group of 77 comprises 48 IOR states out of a total membership of 132 states. Out of the 13 states that are members of the G8+5, 2 are IOR countries. Out of the 48 states that are categorized by the UN as the world's Least Developed Countries, 21 are located in the IOR. Out of 119 states, the Non-Aligned Movement is composed of 47 IOR nations (Cordesman and Toukan, 2014).

A somewhat new regional trade partnership agreement in the Indian Ocean has emerged recently. Although in negotiation since 2012, the Regional Comprehensive Economic Partnership (RCEP), created to make things easy for increasing inter-IOR trade, witnessed some rough weather, as India refused to sign the agreement due to its heavy leaning/tilting toward China. Planned to have all ten ASEAN countries (Brunei, Cambodia, Indonesia, Laos, Malaysia, Myanmar, the Philippines, Singapore, Thailand, and Vietnam) and six FTA countries (India, China, Japan, Australia, New Zealand, and South Korea), the RCEP accounts for a population of 3.4 billion people with a total GDP of US$49.5 trillion, approximately 39% of the world's GDP.

Along these lines, 22 IOR nations have formed the Mauritius-headquartered Indian Ocean Rim Association (IORA). This association is an example of warm, trustworthy relationships among nations of the IOR. Established on March 7, 1997, this 22-member body has been a unique coordinating organization. This has been one of the first organizations to work out a detailed plan for the blue economy (Table 5.8).

5.3   HUMAN MINDSET

This chapter has so far discussed the enormous intervention of climate change (a natural affair) and geopolitics (a man-made affair) on the existing economic models, and on the blue economy activities. The third significant threat to the blue economy however comes from "within." It is the mindset, belief, and the perception of human beings on how success and progress are defined. We would show here that socio-economic developments for a long time have been ill-defined and the concept of genuine progress has been deliberately ignored for commercial benefit. Erosion of values and ethics is a real challenge

5.3.1   UNBOUND ASPIRATIONS

The human mind is guided largely by aspiration and the follow-up strategy or mechanism to accomplish such goals. Aspirations involve a strong desire for personal advancement and to achieve something higher and greater. Aspirations (also often referred to as desire/hope/ambition) motivate people to dream, live, and to keep trying. These are vital to the growth of a human being and also for society. They keep the clock of development moving.

Aspirations are not constant, and change with time. Most people in general aspire to earn money, settle quickly, adopt for a better lifestyle that may include affordable and reliable healthcare, education, smooth communication facilities (rail, road, smartphone, internet, etc.), and have a comfortable, peaceful life. Some others aspire to do as much work as possible for society. A few others would aspire to become powerful—financially, politically, academically, or socially—irrespective of their strength. All have different aspirations in their life.

The major drawback in this regard however has been that people very often remain unable to figure out what they want exactly in life, or what they deserve, or how skillful are they in achieving such desires. Hence, in many cases, the aspiration would be drawn disproportionately to the skill and education of the person concerned. As a result, the difference between aspirations and accomplishment remains wide. And when people start believing in their undeserved aspirations and try to achieve such unfounded wishes standing on the right or wrong side of the law the world becomes a dangerously different place to live in.

TABLE 5.8
Blue Economy Declaration of the IOR Countries

Declaration of the Indian Ocean Rim Association (IORA) on the Blue Economy in the Indian Ocean Region
Jakarta, Indonesia, 8–10 May 2017

WE, the Ministers and representatives of the Member States of the Indian Ocean Rim Association (hereinafter referred to as "IORA"), the Commonwealth of Australia, the People's Republic of Bangladesh, the Union of Comoros, the Republic of India, the Republic of Indonesia, the Islamic Republic of Iran, the Republic of Kenya, the Republic of Madagascar, Malaysia, the Republic of Mauritius, the Republic of Mozambique, the Sultanate of Oman, the Republic of Seychelles, the Republic of Singapore, the Federal Republic of Somalia, the Republic of South Africa, the Democratic Socialist Republic of Sri Lanka, the United Republic of Tanzania, the Kingdom of Thailand, the United Arab Emirates and the Republic of Yemen attended the Second IORA Ministerial Blue Economy Conference (BEC-II) in Jakarta, Indonesia, on 8–10 May 2017;

RECALLING
• the 1982 United Nations Convention on the Law of the Sea (UNCLOS) and other international conventions and instruments related to the activities in the oceans and seas;
• Goal 14 of the Sustainable Development Goals (SDGs), to conserve and sustainably use the oceans, seas and marine resources;
• the Recommendations and the Declaration of the First Ministerial Blue Economy Conference in Mauritius on 2–3 September 2015;
• the Jakarta Concord on Promoting Regional Cooperation for A Peaceful, Stable and Prosperous Indian Ocean, signed in Jakarta, Indonesia, on 7 March 2017;
• Relevant UNGA Resolutions, including 61/105, 64/72, 66/68, 69/292;

RECALLING ALSO the intention to implement the IORA Action Plan of 2017–2021 as adopted by the Council of Ministers' (COM) Meeting in Jakarta, Indonesia, on 6 March 2017;

RECOGNISING that oceans, along with coastal and marine resources, play an essential role in human well-being and social and economic development;

STRESSING the need for the IORA Member States to harness the potential of the Blue Economy to promote economic growth, job creation, trade and investment, and contribute to food security and poverty alleviation, whilst safeguarding the ocean's health through the sustainable development of its resources;

CONCERNED about the disparities in economic development of the IORA Member States, including in skills and human resource development, research and development, business opportunities, resource allocation; technology and innovation and its impact on the public and private sector, including the Small and Medium Enterprises (SMEs);

AWARE OF the need to promote communication and maritime connectivity in the Indian Ocean region;

STRESSING the need to promote observation, protection, conservation and sustainable use of ocean resources so as to continue to meet the needs of the present without comprising the opportunities of future generations;

TABLE 5.8 (Continued)
Blue Economy Declaration of the IOR Countries

REAFFIRMING that research and investment are required to address key challenges of the IORA and to provide solutions and create a friendly business environment to attract investors in the Blue Economy in the Indian Ocean region;

RECOGNISING the importance of promoting entrepreneurship, innovation and SMEs, with a special focus on promoting youth and women's engagement in the sustainable development of the Blue Economy;

MINDFUL OF the increasing challenges, both natural and human factors, such as overexploitation of resources, increasing marine plastics debris and nutrient pollution, illegal, unreported and unregulated (IUU) fishing, over fishing, destructive fishing, crimes in the fisheries sector, biodiversity loss and its impacts on blue carbon stocks, illegal mining and the impacts of global climate change and natural disasters;

ENCOURAGING the IORA Member States to move towards integrated and ecosystem-based approaches in the management of marine resources to maximise sustainable economic yield from the ocean, including through utilising the appropriate management tools such as marine spatial planning, marine protected areas, etc;

RECOGNISING the importance of public-private partnerships in the development of and cooperation in the Blue Economy;

ENCOURAGING sharing of information, experiences, expertise, best practices and technology in Blue Economy related cooperation among IORA Member States and Dialogue Partners;

ACKNOWLEDGING the outcomes of IORA Blue Economy events on various related topics, including marine aquaculture, marine tourism, postharvest processing, seafood safety and quality, maritime connectivity, port management and operation, ocean observation monitoring, forecasting and seabed minerals and hydrocarbons;

HIGHLIGHTING the importance of collaborating and cooperating with relevant stakeholders, including regional and international organisations for the advancement of the Blue Economy in the Indian Ocean region;

ENCOURAGING the IORA Member States to mainstream ocean-related issues in their national planning and policy-making process based on their priorities;

EMPHASISING the need to foster support and financing opportunities, as well as promote transfer of technology, capacity building and skills development, for local fishery entrepreneurs and coastal communities that are directly dependent on the sea, including through triangular cooperation;

REAFFIRMING IORA's role and commitment in the development of the Blue Economy, through the sustainable use, management, observation, protection and conservation of marine resources in the Indian Ocean region;

REITERATING the commitment to establish the IORA Working Group on the Blue Economy which would enhance cooperation to promote the Blue Economy.

We, the Blue Economy Ministers/Head of Delegations of the Member States of the Indian Ocean Rim Association;

HEREBY DECLARE AS FOLLOWS:

That the Member States of IORA will be guided by the following principles when developing and applying blue economy approaches to sustainable development and the enhancement of socio-economic benefits, particularly of the coastal communities, in the Indian Ocean Region:

(continued)

TABLE 5.8 (Continued)
Blue Economy Declaration of the IOR Countries

1. The Blue Economy should ensure the sustainable management and protection of marine and coastal ecosystems to avoid significant adverse impacts, including by strengthening their resilience, and taking action for their restoration in order to maintain healthy and productive oceans, and achieve inclusive economic growth in the Indian Ocean region;
2. The development of IORA's Blue Economy priority sectors namely: Fisheries and Aquaculture; Renewable Ocean Energy; Seaports and Shipping; Offshore Hydrocarbons and Seabed Minerals; Deep Sea Mining, Marine Tourism; and Marine Biotechnology, Ocean Observation, Research and Development, should be carried out in an environmentally sustainable manner;
3. IORA Member States are encouraged to pledge their voluntary commitments, including in implementing capacity building programs, in the concerted effort to strengthen cooperation in the blue economy;
4. IORA Member States are encouraged to develop their Blue Economy sectors, based on their priorities, that could contribute to boosting their economic growth and contribute to job creation and poverty alleviation;
5. IORA Member States, in collaboration with Dialogue Partners, should encourage the financing of ocean economy infrastructure and development projects, including development and investment in Economic Development Zones, as well as investment and exploration of new technologies for Blue Economy Development;
6. IORA Member States and Dialogue Partners should enhance cooperation and collaboration to promote: research and development; networking; technology transfer; sharing of information, data and best practices; exchange programmes and expertise; and networking across the Indian Ocean region for the sustainable development of the Blue Economy;
7. IORA Member States should adopt ecosystem-based approaches to sustainably manage and use their marine resources, while protecting and conserving the marine environment;
8. IORA Member States are encouraged to consider the full range of technologically advanced solutions as well as local wisdom and traditional knowledge, as appropriate, in the context of adaptation and mitigation strategy to confront climate change effects on societies;
9. IORA Member States, in collaboration with Dialogue Partners, should promote capacity building, including collaboration of ocean observation training and scientific capacities, and skills development in the Blue Economy sector, through reinforcing collaboration and networking with relevant regional/international organisations and institutions in the Indian Ocean region;
10. IORA Member States, in collaboration with Dialogue Partners, need to address challenges and key issues related to the Blue Economy, including overexploitation of resources, marine plastics debris pollution and nutrient pollution, biodiversity loss, IUU fishing, illegal mining, climate change, and its impact on marine resources and ecosystems;
11. Collaboration between IORA Member States and Dialogue Partners in various aspects, including financing and development of Blue Economy activities and projects, and technology transfer should be strengthened so as to ensure balanced economic development in the Indian Ocean region;
12. Cooperation among IORA Member States, Dialogue Partners and relevant stakeholders in: carrying out marine scientific research; sharing, collecting, and managing data and information; and the implementation of concrete projects on emerging ocean science and blue economy issues;

TABLE 5.8 (Continued)
Blue Economy Declaration of the IOR Countries

13. The development of effective legal, regulatory and institutional frameworks and ocean management policies should be enhanced as appropriate, for informed decision and policy-making, which are crucial steps toward structuring and guiding its growth;
14. Sustainable development of the Blue Economy should be in accordance with the 1982 United Nations Convention on the Law of the Sea (UNCLOS);
15. IORA Member States are encouraged to promote public-private partnerships and the involvement of business communities in developing the Blue Economy, including infrastructure development and transfer of technology in various blue economy sectors such as: fisheries and aquaculture; ocean observation; renewable ocean energy; seaport and shipping; deep-sea mining and marine tourism, including cruise tourism;
16. IORA Member States should cooperate to promote efficient monitoring and inspection programme to prevent maritime trade of uncertified/unauthorized chemicals and pesticides;
17. IORA Member States consider, if deemed necessary, supporting the establishment of an IORA business travel card to ease business travel on blue economy businesses and collaborate with member countries who are ready to do so;
18. IORA Member States, in accordance with international laws and consistent with existing obligations, should perform environmental impact assessments before engaging in relevant deep-sea mining activities and fulfil relevant obligations to ensure effective protection of the marine environment from any harmful effects of deep-sea mining;
19. IORA Member States are encouraged to adopt and implement transparency and traceability measures to strengthen the application of sustainable fishing practices by regulating harvesting & ending poverty, the fight against IUU Fishing destructive fishing and crimes in the fisheries sector, provide access to small scale artisanal fisheries to marine resources & markets and protect food security;
20. The empowerment of women and youth to participate in the development of the blue economy is essential through better access to education, training, technology and finance. Women and youth should be encouraged especially by supporting MSMEs and small scale fisheries, to be equitably included in sustainable economic growth;
21. The proposed Working Group on the Blue Economy would consider programs, activities, pilot projects and studies for regional cooperation in the Blue Economy;
22. IORA Member States to consider developing a Master Plan on the Blue Economy to identify and prioritize concrete projects and tangible areas of cooperation, to promote the blue economy as a driver for socio-economic development;
23. The sustainable development of the IORA priority sectors of the Blue Economy in the Indian Ocean Region would contribute to: food security; poverty alleviation; the mitigation of and resilience to the impacts of climate change; enhanced trade and investment; enhanced maritime connectivity; economic diversification; job creation and socio-economic growth;
24. IORA Member States and Dialogue Partners should increase the economic benefits derived from the Blue Economy to Small Island developing States (SIDs) and least developed countries (LDCs) from the sustainable use of marine resources, including through sustainable management of fisheries, aquaculture and tourism;

(continued)

TABLE 5.8 (Continued)
Blue Economy Declaration of the IOR Countries

25. Collaboration among IORA Member States, Dialogue Partners, research institutions, industries and public-private partnerships should be enhanced so to create an environmentally sound business environment and attract foreign investment that would accelerate the commercialisation of ongoing research in exploring data and in creating new products derived from marine and maritime data resources;
26. IORA Member States, in collaboration with Dialogue Partners, are encouraged to carry out pilot projects and set-up modern and accessible technologies to effectively develop the Blue Economy in a sustainable manner.

ADOPTED by the Blue Economy Ministers / Head of Delegations of the Member States of the Indian Ocean Rim Association on 10 May 2017 in Jakarta, Indonesia.

In this regard the experiment made by Dr. Dylan Seltarman, a professor of Psychology at the University of Maryland (USA), appears apt. Dylan created a moral dilemma to his class students with a cruel ethical choice. Since 2008, he has been offering his students to opt for extra-gratis marks between 2 and 6 in their exam paper over and above their actual marks, but with a rider. If more than 10% of the students in the class opt for 6 grace marks then none of the students would get these additional marks. And it so happened that only once in the last ten years have students got grace marks. It means that more than 10% of students desired to have the full 6 grace marks. The students were very clearly greedy to obtain more undeserving additional marks. If they would have been satisfied with small grace marks below 6 (i.e., less greedy), the entire class would have benefited throughout the years. But the greed of 10% of students denied benefit to all.

A similar thing is happening in society these days (Fig. 5.15). A few persons' greed and a few countries' ego are preventing the world from turning into a harmonious platform. Success is often considered commensurate to earnings of a person/company/country, in terms of money, social status, and property. Similarly, progress of a society is measured in terms of development of roads, buildings, communication facilities, waste disposal amenities, etc.

A study has recently been made worldwide by some US and Swiss scientists to see how many people would return a wallet full of cash to the owner without tampering. The most honest were Switzerland and Norway (70% would return the wallet full), and one of the least honest countries is China (only 20% would return the wallet with cash). Complete dishonesty was shown by Mexico and Peru. India is little better, positioning ten slots above China. Overall, 51% people the world over would return the wallet full, while another 40% would return it empty. The classical economic theory hypothesizes that the greater the temptation, the less likely people are to be honest. But this study found exactly the opposite. Even with a higher amount in the wallet people did not necessarily take out the money. The study concludes that more people chose to be honest even when dishonesty had the greater reward (TOI, 2019c).

FIGURE 5.15 Skewness of Human Mind

Notes: The human mind will tend to do anything that comes for free. On the left panel, people continue to park their vehicles at a place not marked for parking but against a penalty that appears not-forceful enough. But when such a facility is made officially available against a genuine payment (right panel) nobody turned up.

The social equations change with time, and many of the beliefs and theories of the present day are expected to be redundant in the near future, particularly in a blue economy environment. Research and technological innovations will be the key tools to cope with such an existential changeover and to meet the ever-increasing human aspiration. Time has come to broaden the skill, education, and human capacity to make research and teaching meaningful to society. In this regard, it is essential to enliven the progress made in the fields of science and technology with equal intensity of arts, ethics, values, and egalitarianism to enable sustainable, balanced, and purposeful humane growth in society. Science and Arts are complementary to each other. The more these two harmonize the better for a blue economy society.

If undeserving aspirations can be legitimately termed as greed, people have a tendency to fulfil such greed through a method that might not be an acceptable legal procedure. Such improper acts additionally steal the legitimate aspirations of other deserving individuals, insult rules and regulations, create bad blood in society, and disturb the sustainable blue economy mandate.

Noted philosophers like Lord Buddha, Rabindranath Tagore, Swami Vivekananda, Mahatma Gandhi, and others knew that there would be clashes between evil and ethics, and erosion of the latter with time. They were profound philosophers of humanism, universalism, religious liberalism, self-less service, practical spirituality, and essential unity in diversity. The solution they proposed to arrest the fall of values and lack of ethics is to fall back on nature through a psychophysical process. They suggested that the growth must come naturally from within (the way a human body develops from childhood to maturity), from the dictates of higher consciousness through yoga, the art of cosmic union (Kundu, 2019). When the world is moving in the wilderness with unsustainable egoism, religious factionalism, and myopic nationalism, the clarion call of these philosophers to assimilate the pluralism and walk toward inclusiveness can give humanity hope and the blue economy a chance for success.

Similarly, the harmony that exists between the individual and Nature represented by the Earth, water, light, air, fruits, flowers, etc. is a note to complete the symphony of joy, peace, and fulfillment. Clearly, the activities of any organization must make sense to the staff that superiority of human beings does not rest in power of possession but in power of unison and empathy. The activities of all governments in the blue economy regime must continue to be in consonance with this line of philosophy. One must aim to create a fearless nature-centric enabling environment, which is essential for the growth of the blue economy.

5.3.2   GDP vs. Genuine Progress

Reaching a trillion-dollar economy has been the aspiration of all IOR countries. While the USA has been a trillion-dollar economy (TDE) since 1969, China could get there only in 1998, and India is hoping to reach the mark 5 TDE in 2024. The USA is now a 21.34 TDE, China 14.22 TDE, India 2.97 TDE, and Brazil 1.96 TDE (TOI, 2019c). Connected to the economy, another conflict of interest exists between GHG emission and growth in GDP, as the factors that makes GDP climb up have been amidst much criticism. The GDP is considered as a figure of economic progress without realizing the cost for such a bloated claim. In this regard, the Genuine Progress Indicator (GPI) is proposed by several socioeconomists to depict the real picture of progress (Fig. 5.16; Stiffler, 2014).

GPI, for example, differentiates between desirable and undesirable economic activities. It suggests that the cost of damage reached to the environment (land, river, coast, forest) and society (culture, demography, crime) owing to industrial and infrastructure growth must be subtracted from GDP data. Once such social, environmental, and economic costs (totaling about 70%) incurred by a nation to achieve high GDP are subtracted the *genuine progress* emerges, which remains less than only 30% (Fig. 5.16). For example, the per-capita GDP of the USA increased from US$13,500 in 1950 to US$35,000 in 2005, while the per-capita GPI during the same period has increased only from US$6,500 to US$10,000.

Compared to just three billion a half-century ago on this planet, the footfall has now reached seven billion people. Today, average per capita income is US$10,000, with the rich countries averaging around US$40,000 and the developing world around US$4,000. That means that the world economy is now producing around US$70 trillion in total annual output, compared to around US$10 trillion in 1960. The economy of China and India is growing at around 6–8% annually, while Africa is averaging roughly 5% (World Bank, 2014a).

Rapid economic growth also helps in alleviating poverty. As the world economy grows at 4–5% annually; it will be on a path to double its size in fewer than 20 years. Today's US$70 trillion world economy will be at US$140 trillion before 2030, and US$280 trillion before 2050, if we extrapolate considering today's growth rate.

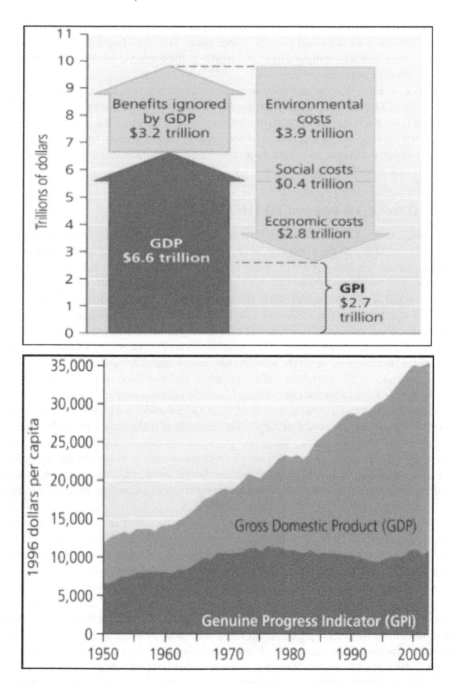

FIGURE 5.16   Gross Domestic Product versus Genuine Progress Indicator

Unfortunately, our planet will be unable to physically support this exponential economic growth if we let greed take the upper hand. With the disastrous effect of climate change impact looming large, the world will be unable to feed 9–10 billion people in the near future.

Hence, as change is the only consistent phenomenon in this Universe, the age-old definition and beliefs with regard to success and development deserve a modification. It is not how much one has earned (property, money, fame, etc.), but what percentage of his/her earning s/he is ready to contribute/give back to society should be the criterion to evaluate the success. This measure could be established as a marker to a new set of development paradigm.

5.4   TOWARD A BALANCED THEATER

It seems George Orwell (a twentieth-century English novelist, essayist, journalist, and critic) visualized the commencement of a cold war at the end of World War II. While nuclear weapons would deter any direct confrontation, he was however apprehensive that continuing efforts by the superpowers to score points in international affairs could make the world thirst for peace. In contrast to what could be seen as cold war between the USA and USSR (Soviet bloc, now presented largely by Russia), Orwell projected involvement of a third power, China, which looked very much unlikely then (Wertheim, 2019). But by the end of the tenure of the President Obama administration in 2016, Washington feared that China was deviating from the agreed path of liberalization and cooperation, and moving toward dominating the world. China's claim for the entire South China Sea (demeaning claims of East Asian countries) and its rapid expansion in the Indian Ocean under the pretext of OBOR are seen as threats by the world at large. No ideological difference can be the cause, because the USA is no more interested in exporting democracy, while China also does not dream for the universal triumph of communism. It is simply the icy battle for trade, commerce, and military hegemony. Some say that US sanctions on Russia and its increasing tariff on Chinese goods could bring Moscow and Beijing onto the same page against the USA (Wertheim, 2019).

While securing peace in the Indian Ocean, military power of a country may not be all, as armed strength needs to be ably backed by technological development vis-à-vis economic growth. China, for example, has invested several billions of dollars in recent years to develop the military and civilian applications of emerging technologies such as 5G, semiconductors, microchips, artificial intelligence, and quantum computing. Such investments transformed the country from an assembly line of low-tech manufactured goods into a preeminent economic and technological power manufacturing high-tech goods. India must take note of such an aggressive China whose economy is five times that of India, and also its military armed with the new high-tech weapon systems, like cyber warfare and robot technology.

India is often perceived as a regional power, but a closer look reveals that it is in a disadvantageous position vis-à-vis China in South Asia. The first reason is that Indian governments never had the political will, economic strength, or military desire to pursue their regional power ambitions. South Asian countries could always play the China card in order to evade India's influence. Second, although

India's new South Asia policy with focus on trade and connectivity has improved regional cooperation since 1991, China remains an economically more attractive and politically more reliable partner for India's neighbors (Wagner, 2016).

In such a demanding situation India can hardly afford to fall back further, lest she gets marginalized in the world and in the IOR power matrix (Gupta, 2010). Hence, India's political leadership, defense establishment, scientific and academic institutions, and industry must work together to accelerate the pace of her development quickly and squarely in these directions.

India is proud of her liberal ideology, democratic governance, and pluralistic civil society. This idea needs to be spread out to the IOR nations. Such a scheme of things that includes promoting liberal democracy in the neighborhood and taking pragmatic steps for larger stability in this region could as well be ideal for the blue economy to take a strong root in the IOR. In fact, if all these are followed honestly by the IOR nations, the region will be much more stable, transparent, peaceful, and prosperous. The push for a no-nonsense democratic set-up in Nepal by India's civil society, as against the subtle support of monarchy by the Government of India until a decade ago shows how civil society could influence the foreign policy of India. India's philosophy of unconstrained ideology, egalitarian governance, and diverse civil society, however, came under critical examination in recent days through two major IOR cum global initiatives. This occurred first in Afghanistan, and then in Iran (Gupta, 2010).

Political consensus reached recently by four countries—the USA, China, Russia, and Pakistan—on bringing peace, prosperity, and political stability to Afghanistan is a major initiative that will impact the IOR in terms of harmony and consistency. The consensus has agreed on four points—withdraw foreign forces from Afghan soil, prevent Afghanistan from becoming a haven for terrorism, end the current violence, and launch an intra-Afghan dialogue including Taliban that will define a new political arrangement in Kabul (Mohan, 2019).

Surprisingly India was not invited to the meeting despite being a major player engaged for the last few decades to help build Afghanistan. India wanted the peace negotiators to focus on three issues—the fair condition of the ceasefire, destroy Taliban's sanctuaries in Pakistan, and stop belittling or delegitimizing the existing Afghan government to make room for untrusted Taliban. None of these three issues raised by India were answered by the negotiators. Although the exclusion of India might not be a welcome step, for the larger interest of peace and stability in the region India has even agreed for such a dispensation. While India would pray for peace in Afghanistan, one should be aware of the fragility of this peace consensus.

The second challenge to Indian philosophy came from Iran, as the USA has put sanctions on India importing oil from Teheran for the latter's alleged desire to develop nuclear capability against the wishes of major powers in the world. In July 2019, Britain impounded an Iranian oil tanker in Gibraltar, claiming that the vessel was carrying oil to Syria in violation of the European Union's sanctions. Iran is working on retaliation. The Persian Gulf is slowly attaining pointless and avoidable military build-up. Although India and Iran are great partners and allies, Delhi is trying to cope with such sanctions on oil import from Teheran, sacrificing both friendship with Iran and economic advantage. India must look ahead at potential challenges, especially if the stand-off escalates in the Persian Gulf and threatens larger flows of oil.

At a time when the Indian economy is slowing down, this could have significant consequences (Raghavan, 2019). Shedding off its decade-long inhibition, India must now rise to the occasion to enthusiastically mediate among its closest allies—the USA (+ EU + Israel) and Iran, and help them reach a negotiated settlement. It needs to be remembered that the Shia-dominated Iran is one of those few countries who had the courage and guts to fight against the majority Sunni terrorists (ISIS, Al Qaeda, Taliban, JeM, LeT, etc.) backed by Saudi Arabia and Pakistan (and indirectly by China). Any harm to Shia-Iran will embolden these dangerous global Sunni militancy. An understanding with Iran, therefore, can counter global terrorism, and such understanding will have a positive impact in solving the Afghanistan problem and to a great extent the Middle East quagmire, and in the process would further secure the IOR and the world.

It is believed that the world's geopolitical CG (center of gravity) may shift in future from the Pacific Ocean to the Indian Ocean. Besides being the global center for trade, commerce and energy flows, the IOR arc from Indonesia in the east to the lower tip of Africa to the west via the Middle East (except India and Sri Lanka) represents largely a violent, archaic, male-dominated Islamic way of life, with patriarchal belief. At the same time, the region also has the world's largest concentration of fragile or failing states—from Yemen and Somalia to Pakistan and the Maldives. Hence, the challenges in IOR get multiplied every day. Failed states produce more transnational terrorists, pirates, lawbreakers, unemployment, and human rights violators. No wonder threats to navigation and maritime freedoms are increasing (Chellaney, 2015). Security in the Indian Ocean, therefore, is always a pressing concern given the increasing importance of its maritime resources and sea lanes.

Although the IOR represents one of the poorest parts of the world, it houses at the same time some of the established economies (most of the Gulf countries) and some emerging economies like Indonesia, India, Iran, and East Africa. Yet the natural disasters occurring in the IOR present a high humanitarian risk. The IOR is again a front-line area to receive the brunt of change in the climate. Maldives, Mauritius, and Bangladesh are quite vulnerable to sea-level rise. Consequently, the sustainable development of oceanic resources, which form the heart of the blue economy, is extremely important for the Indian Ocean. The seas can probably meet the burgeoning humankind's needs, and shrinking resource reserves (demand-supply deficit gap) in terms of food, medicines, energy, employment, and socioeconomic development. Hence, it is essential to first formulate an approach that is acceptable to both environmentalists and business.

For example, IOR may first deploy the latest technology to generate adequate reliable data on resources and to understand forcing parameters in terms of atmosphere-land-ocean interaction. Second, IOR nations must get into in-depth and sustained dialogue among themselves for a multisectoral understanding to prepare a practical and implementable blueprint for blue economy. Bhatia (2017) insisted for track-2 (or 1.5?) dialogue, involving officials and independent experts from select countries to carefully recalibrate geopolitics and security-related issues. Third, realizing that the ocean does not recognize geopolitical boundaries, IOR nations must make best use of transboundary resources and new opportunities. Hence, a cooperative approach among the IOR is extremely essential.

In fact, the IOR may achieve sustained growth and balanced development by increasing trade and investment, promoting small and medium enterprises, liberalizing trade rules, rationalizing barriers, improving connectivity, and opening supply chains to hinterlands and land-locked countries. Deepening cultural linkages with the people in the region, backed by diplomacy, could be one such tool to build the Indian Ocean as a frontier of human understanding, peace, and harmony.

In this direction, the 22-member countries in the Indian Ocean Rim Association (IORA) exhibited significant dynamism in the past few years as the trade in the region increased by over four times from US\$302 billion in 2003 to US\$1.2 trillion in 2012. A broad range of activities to enhance the regional and international cooperation in security and governance has been suggested by the working group of IORA (IORA, 2017; Table 5.8).

Indeed, to promote sustained growth and balanced development in the IOR, IORA has focused on: (a) Maritime Safety & Security, (b) Trade & Investment facilitation, (c) Fisheries management, (d) Disaster Risk Assessment, (e) Tourism & Cultural exchange, (f) Academic, Science & Technology, (g) Blue Economy, and (h) Women's Economic empowerment. The IORA calls for sharing of skills, data, and knowledge among IOR nations; integrated regional-scale planning; adoption of marine-specific policies; robust governance; and collaboration between governments and industry (Jakarta Declaration).

Another country perturbed greatly with the aggressive inroads of China in the Indian Ocean is Indonesia, the de facto leader of ASEAN. To ensure a purposeful Indo-Pacific cooperation and to keep up the balance of power in Asia, Indonesia and India have agreed to safeguard the free movement of people, goods, and services through the Strait of Malacca, one of the busiest shipping routes between the Indian Ocean and the Pacific. In fact, freedom for navigation, availability of port infrastructure, and unhindered access to markets are mandatory for this purpose. India's interest in joining the Malacca Straits Patrol (MSP)—a four-nation arrangement between Indonesia, Malaysia, Singapore, and Thailand—is being considered actively pending operational and technical approvals (Kaura, 2018).

And despite direct warning of China, Indonesia has recently given permission for Indian access to northern Sumatra's Sabang port, enhancing the Indian Navy's ability to maintain a forward presence in the Straits of Malacca. Delhi and Jakarta have agreed to elevate their relationship to the level of a comprehensive strategic partnership by accelerating economic and security cooperation in the maritime domain, and in the process furthering the development of the Indo-Pacific concept (Kaura, 2018). Similarly, the Quadrilateral (QUAD) grouping of India, the United States, Japan, and Australia is widely perceived as a counterbalance to increasing Chinese geo-economic and geopolitical assertiveness (Fig. 5.17).

However, Panda and Parameswaran (2018) find that India's notion of regional order in the Indo-Pacific should maintain a common distance with Beijing and Washington. This is despite the fact that four major democracies in the world— the USA, Japan, Australia, and India formed the Quadrilateral Security Dialogue (QSD or Quad) in 2007 as an informal strategic dialogue forum. Considered by many

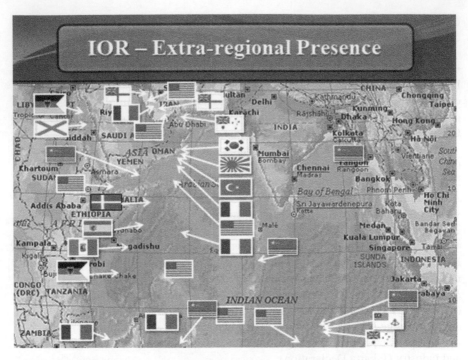

FIGURE 5.17   The International Presence in the Indian Ocean (After MEA Portal).

as the "Asian NATO" in making, these like-minded democracies have the potential to dramatically change the region's security landscape. The Quad has provided India a platform to address Asia's power asymmetry with China while positioning Indian interests more prominently between US-led and China-led schemes of politics in the Indo-Pacific. Yet, India maintained Quad as only a consulting forum (not strategic) on Indo-Pacific region and appears not to offer this grouping a deliberate strategic intent as that might be construed as anti-China. The three other members of the Quad are somewhat wary of India's such ambiguous diplomacy.

In fact, India wants to see Quad working for the overall economic development of the Africa-Asia-Pacific region, leaving a room to accommodate China and Africa in the scheme of things. The USA, on the other hand, considers Quad more maritime-centric and has committed US$113 million investment funds for infrastructure, energy, and technology. The renaming of the Pacific Command as the Indo-Pacific Command also indicates the importance that the USA is attaching for Indo-Pacific strategic domain. The USA is particularly encouraging both India and Japan to play major roles in the implementation of the Quad concept. Although formed to keep the status quo in the post-Cold War era, the Quad is now seen as a tool to balance the power equation in Asia (Matsuda, 2020). In the near future, all IOR governments must quickly undertake economic reforms encouraging more private investment in a mixed economy environment. The other important aspect should be a careful handling of geopolitics. In this regard, getting the Quad working in the Indian Ocean must be a priority.

India, however, appears to be little hesitant to go full-steam with Quad as she would prefer to view China as a maritime dialogue partner to avoid any direct confrontation. This hesitancy was reflected with India differing from other three members of Quad on drafting an official press release of the Quad meet. This is despite the fact that among Quad countries, India will be directly threatened by the OBOR schemes of China. New Delhi possibly wants to remain strategically autonomous and keep up an exclusive relationship with China and Quad. India probably finds an alliance with China more comprehensive as it includes economic and infrastructure development in the OBOR countries (Panda and Parameswaran, 2018). But India must not be oblivion to the fact that a majority of contracts for such development are going to Chinese infrastructure companies only, and China is encircling India rapidly and comprehensively.

The entire spectrum is generating bad blood between Beijing and the receiver countries. This asymmetric behavior of China may call for enhanced counter-asymmetric conduct by other players in these regions. China must see reason and must understand that her ambition to become a global superpower "quickly" banking solely on military and economic power might fail and would not be sustainable. The elements of ethics and values appear to be missing in the entire context of the Chinese pattern. China must take what it deserves and must avoid creating any unnecessary challenges, and refrain from generating tension in the region.

Another aspect that might have contributed to the phenomenal growth of China is the lack of economic and social reforms in most of the Afro-Asian countries. Absence of any tangible economic reform (other than ASEAN) made these countries continue in fiscal depravity and poverty. In such a depressing situation, fresh offers by China showcasing its rapid growth, however lethal that may be, unsettled the Afro-Asian countries and literally lured them to toe the Beijing line.

In this regard, India is seen as falling short of playing its role in the IOR region effectively. Banking on a rich history of an ancient civilization, India should have propagated an alternate ethical approach for balanced growth. India may propose a doctrine based on the wisdom of four great philosophers (Buddha-Tagore-Vivekananda-Gandhi, or BTVG; see Chapter 8 for details) to lead the effort to bring in rational sensibilities and balance growth through ethical governance. The near absence of India in the theater is proving costly not only for the region, but also for the world's peace and prosperity.

The growing competition between China and India presents a strategic opportunity for the USA. It must expand its relationship with India. The USA has generally viewed India as a strategic partner to offset China and should be interested in counter-balancing China by enhancing the capabilities of the Indian Navy. Also, the USA and India could make efforts to rope in Australia, France, South Africa, Japan, Iran, UK, and Israel in their bid to make the Indian Ocean a more balanced theater of cooperation, growth, and peace. Moreover, as indicated earlier, India could play an important role to help normalize the relation between the USA and Iran.

In summary, the time has come for all the IOR countries to get down to the business of contributing to maintaining peace and sustainable activities in the Indian

Ocean. To start with, the IOR nations are to develop their own blue economy policy that includes traditional sectors, such as, fisheries and aquaculture, port development, marine tourism, and shipping. Later, an SOP commensurate to ground reality may be formulated, empowering in the process the local marine authorities to deal with any emergency. Later these policies must be integrated into a tangible framework of the India Ocean action policy to achieve the objectives of the blue economy.

6 Opportunities Ahead

Science, technology, and innovation should play major roles in a successful blue economy architecture. Additionally, as providing employment to millions of youths is extremely essential for harmonious socioeconomic growth in an area like the IOR, labor-intensive ventures will be more appropriate than capital intensive ones. Hence, we discuss in this chapter a few possible new ventures that could shape up a blue economy paradigm keeping innovation and employment at the core.

The IOR countries must especially vie for "blue growth"—a term normally used for an array of responsible economic and industrial growth through innovation. A wide variety of new ventures could be planned under the blue (growth) economy to substantially maximize the economic support from the ocean in a sustainable manner. The IOR must get encouraged with the progress made in blue growth areas by a few countries in the world in the recent past. For example, China's blue economy amounted to US$963 billion in 2014 (i.e., 10% of GDP), while that of the USA accounted for US$258 billion in 2010 (i.e., 1.8% of GDP) and of Indonesia for 20% of GDP. The blue economy prospect suggests that in 2030, two out of three fish catch will come from "fish-farms" (World Bank, 2013), the offshore wind facility would have expanded by nearly ten times, while the marine trade would enlarge four times by 2050 (International Transport Forum, 2015).

6.1 EMERGING NEW VENTURES

Besides the large traditional sectors such as tourism, fishery, trade, shipping, and services that continue to expand swiftly, the sustainable exploitation of ocean resources like mineral, oil & gas, and offshore renewable energy (such as wind, tides, waves, biomass, ocean thermal energy), marine engineering, deep ocean water applications, naval architecture, ship repairs, coastal engineering, marine pharmaceuticals, marine surveillances and security, marine spatial planning, conservation, pollution, and climate change studies are rapidly emerging with new approaches and innovations

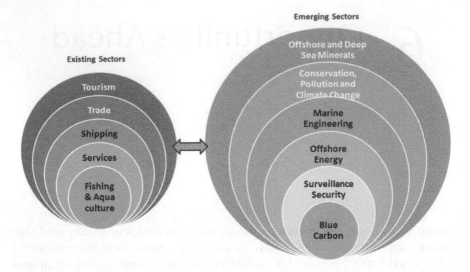

FIGURE 6.1   Existing and Emerging Sectors in Blue Economy Paradigm

(Fig. 6.1). From 37% in 2012, aquaculture will contribute nearly two-thirds of total global fish consumption by 2030 (World Bank, 2013). The offshore energy production is predicted to rise from 8 GW in 2014 to nearly 29 GW in 2020 (International Energy Agency [IEA], 2014).

The ecosystem resources and services significantly vary from place to place and season to season. After mapping the ocean resources in detail, and by striking a balance with the precise exploitation methods, the following conceptual inspirations can be mulled over for betterment of resource enhancement and subsequent harvesting in a healthier way (Spalding et al., 2016).

- As mangrove is proved to be a good habitat to nurture a substantial fish population, countries like Madagascar, Mauritius, Seychelles, India, Bangladesh, Singapore, Thailand, Sri Lanka, and Australia may follow the Indonesian example. Innovative measures in developing and conserving mangrove areas could go a long way to promote fish catch.
- One single hectare of sea grass and salt marsh area supports the generation of about 30,000 fish per year and 235 kg of shrimp, as well as 170 kg of blue crab. Based on the given geographical and ecosystem condition, the figure may go up. In view of this, there is an urgency for the IOR countries to have a roadmap and work out an action plan in this direction. Moreover mangroves, seagrass, and wetlands sequester large amount of atmospheric CO_2.
- It is a fact that coral reefs are potential areas for fish catch. When such areas are treated for ideal conservation measures, various fish species habitats will be encouraged, in turn, promoting a high potential zone for fishing. Healthy coral reefs can generate 5 to 10 tons of fish/km^2/year.

6.2 FOOD, MINERALS, ENERGY, INFRASTRUCTURE

The new innovative ventures to develop the four sectors—food, minerals, energy, and infrastructure, under a blue economy regime are briefly presented here.

6.2.1 MARINE FOOD

6.2.1.1 Pelagic Fishing

With the advent of new ideas and scientific innovations, marine living resources can be effectively expanded to new ventures to better address both the food security and the economic returns (Table 6.1). The new ventures may include hatchery-based seed production to plan for large pelagic fishing, as well as mariculture of value-based species such as sea bass, grouper, marine eel, pompfret, mullet, etc.

Another emerging venture is the deep-sea fishing. Countries like the USA, China, Australia, and Japan are successfully operating this activity. The IOR countries' fish catch from the deep sea is limited. However, as fish catches can only be expected to increase in the future, more fish processing centers are to be planned accordingly, along with augmentation of cold storage facility and dedicated terminals in shipyards to handle fish catch. There is also a promising potential for fish oil products. These downstream industries increase job potentials as well as contribute to the economy.

TABLE 6.1
Established and Emerging Ventures in Marine Fisheries

Activity	Subcategories	Established Industries	Possible New Ventures
Food - Marine Living Resources	Sea food harvesting	Primary Fisheries, Secondary Fisheries, Trade Seafood products, Trade non-edible sea food products	Investigate large pelagic fish harvesting, long line fishing of tuna and allied fishes, Hatchery based seed production, Mariculture of sea bass, grouper, marine eel, pompfret, and mullets, Increase soft shell crab farming from hatchery produced seeds, Deep-sea fishing, More seafood processing plants, cold storages and terminals in shipyard, fish product industries like fish oil.
	Aquaculture, Mariculture	Marine Bio-technology	Multispecies aquaculture of seaweed, marine algae, sea cucumber, octopus culture, cage culture, shell fish breeding
	Marine Bio-technology	Bio-prospecting	Pharmaceuticals, chemicals, and food

With ever-increasing new biotechnological advancements, aquaculture has gained with newer innovations in culturing marine algae, seaweed, octopus, and shellfish breeding. The cage and line culture has been widely demonstrated successfully. Of late, the application of bio-prospecting on pharmaceutical and cosmetic industries from sea products has a flourishing potential. In addition, a few more emerging approaches are narrated below that could be applied to most of the IOR countries.

6.2.1.2 Mariculture

Offshore aquaculture (mariculture) is forecasted to grow exponentially to meet the ever-growing need for seafood. Countries like India, Mauritius, Sri Lanka, and other IOR countries, especially those that have a larger maritime regime, put forth their action plan to expand the offshore aquaculture (LiVecchi et al., 2019).

The IOR countries need to have innovative initiations to further improve brackish and marine aquaculture practices. Breeding and farming of high-value aquaculture species such as Asian sea bass or *Lates calcarifer*, mullets, pomfrets, and shell-fish (mud crab) could be hopeful prospective resources. For example, soft shell crab farming is being practiced with new technology in a few places in Bangladesh (Hussain et al., 2017).

6.2.1.3 Community-Based Aquaculture and Seafood Management

Community-based aquaculture is a promising substitute to reduce pressure on captured fisheries and marine biodiversity. Madagascar, due to its geographic isolation and arid climate, has found aquaculture the best alternative for a sustained local economy. The isolated coastal communities could be brought together to enhance their economic status in the main island through the sea farming of red seaweed (*Kappaphycus alvarezii*: widely used in cosmetics as a texturing agent), sea cucumbers (*Holothuria scabra:* as a health food and aphrodisiac), etc., which have high international trade potentials.

Another example of a community-led seafood management strategy was shown by the coastal communities in Popisi village, Banggai Island, Central Sulawesi, Indonesia where the highly productive fisheries sites were closed for three months (October 2018 to January 2019) and the entire facility ventured into octopus fisheries. The Bajo, a nomadic seafaring people, still living in stilt houses around the coastal waters of Indonesia, were trained for octopus fisheries. Banggai Island's marine territory covers about 13,000 km^2. In this protected coral reef habitat, the development of octopus fisheries is an activity with high potential, growing rapidly and producing quick catches as well as high income. This scalable innovative project highly influences the conservation of endangered the Banggai Cardinal Fish in coral areas by the Indonesian Nature Foundation (LINI) along with the Blue Ventures (blog.blueventures.org).

6.2.1.4 Commercial Marine Algae Farming

Commercial-scale farming of marine algae such as seaweed, blue-green algae and micro-algae can be a source for biofuels as they have appreciable contents of more polysaccharides and less lignin. Moreover, these marine algae are considered widely in the pharmaceutical, cosmetics, and food processing manufacturing sectors. Being less carbon intensive and an efficient way to sequester carbon, the biofuels will attract more significance in the next decades. More research on marine algae is discussed under the section Science and Technology.

6.2.1.5 Aquacelerator

The aquacelerator is an endeavor being initiated by the Australian Department of Foreign Affairs and Trade (DFAT) in association with SecondMuse, to support and scale aquaculture in the Indian Ocean by inspiring new ideas and approaches (www. huffpost.com). For example, the temporary closure of fisheries and subsequent shifting to octopus fishing in Indonesia has made an immense impact on ecological conservation and rejuvenation of cardinal fishes. This new approach helped the local community to find an alternative income through octopus culturing and fishing.

The vision of the SecondMuse is to fabricate twenty-first-century economies through innovations and new approaches (www.secondmuse.com). In Australia, collaboration between the Great Barrier Reef Foundation and Tiffany & Co, has helped rehabilitate the ecosystem of the Great Barrier reefs (www.outofthebluebox. org). In 2018 during the sixth annual conference of Australian and Indonesian Youth (CAUSINDY), *SecondMuse* proposed two projects, one in Bali on seafood innovation and another in Surabaya on Ocean Plastic Prevention Accelerator with the funding from DFAT (www.secondmuse.com).

6.2.1.6 Aquaponics

Aquaponics is the combination of aquaculture (raising fish) and hydroponics (the soil-less growing of plants) that grows fish and plants together in one integrated system and symbiotic environment. In Kenya, with an investment of about US$15.37 million, the aquaculture and aquaponics technology development and their innovation transfer program through training are implemented (Blue Economy bankable projects, 2018). In addition, with US$2 million, another program to rehabilitate fish habitats, farming, breeding and harvesting leading to fish processing and value addition has been recently executed to link with the market. With additional investment of US$380 million, a new container terminal to handle 1.5 million TEUs per annum has been completed at Mombasa. In addition, some of the berths have been converted into terminals. Further to the above-mentioned developments, other new ventures that have dawned in Kenya include establishment of desalination plants, development of mariculture farms for crab, prawns, and seaweed, rehabilitation of the mangrove ecosystem, construction of a fish port with landing berths, cold storage facility, development of local fishing fleets, establishment of cruise ship terminals, and construction of fish auction markets and necessary infrastructure.

6.2.2 MARINE MINERALS

Though the non-living sector is one of the widely recognized activities, it is not yet fully developed in the IOR countries. A few countries, like India, Malaysia, Australia, and South Africa, are involved in mineral exploitation in the coastal areas. In most of the IOR countries, the offshore and seabed mapping for marine mineral resources

TABLE 6.2
Established and Emerging Ventures in Marine Mineral Development

Activities	Subcategories	Established Industries	Possible New Ventures
Minerals—Non-Living Resources	Mining for minerals	Sea salt and its components	Research on salt composition, salt refining, and marketing using innovative techniques.
		Offshore mining for mineral sands	Intensify exploration and mapping, innovative offshore mining, mineral processing, and value addition
		Seabed mining for phosphorites, sulfides, tin, etc.	Exploration and estimation and mining technology development
		Gas hydrates	Technology development
		Hydrothermal exploration	Exploration, research on quantity, quality and valve, mining and marketing
		Manganese nodules	Exploration, mining and marketing, economic feasibility
	Oil and gas	Energy resources	Intensify exploration and exploration with innovative approaches
	Desalination	Freshwater extraction	Intensify, research on innovative approaches

is yet to be carried out in detail. The emerging ventures in this sector are outlined in Table 6.2.

Occurrence of minerals is specific to water depth and ocean physiography. For example, placer deposits occur in near shore and offshore regions (up to 40 m depth average). Sands for construction purpose and ceramic graded clays are richly housed in offshore region. Resources in continental margin includes phosphorite deposits at around 1,000 m depth; gas hydrate at around the same depth; and oil & natural gas at the shelf and slope regions. In the deep-sea region, cobalt-rich crusts occur at the summit regions of seamounts (~2,500m depth); hydrothermal deposits at 3,500m depth, and manganese nodules at depth of about >5,000m.

6.2.2.1 Salt Deposit

Salt and its products form an essential part of flavoring, preserving, and packaging food and fish commodities. Other than common salts, epsom, gypsum, and supplementary varieties of salts are also in demand in chemical and cosmetic industries.

Innovations in salt research will further elevate the essentiality of various products. Since most of the IOR countries are provided with wider salt and mud plain areas, there could be greater scope of increased salt production for downstream industry needs.

In most of the Island Countries and major islands, fresh water is a targeted and high-rise commodity for which desalination processes are under an immense demand. Nearly 150 counties are adopting to convert saline water (sea water, brackish water, and less saline water) to fresh water. From 25% as of today, Singapore trusts to provide fresh water from desalination to about 30% of her citizens by 2060. It uses the ceramic membrane technology and constructed largest plant (Choa Chu Kang Waterworks) in the world with a capacity of 181,800 m^3/day (www.globalwaterintel. com).

6.2.2.2 Heavy Mineral Sand Deposits

The placer minerals in offshore regions are significant due to their high economic value, usages in high-tech industries, large occurrence, and available technological capability. Global distribution of placer deposits is remarkable, and more so in IOR countries' coastal regions (Fig. 6.2).

Beach sands rich in heavy minerals and offshore placer deposits are actively exploited by IOR countries, particularly India, South Africa, Indonesia, Sri Lanka, and Thailand. Placer minerals are used in high-tech industries such as electronic, aerospace, petroleum, paint-making, biomedical, and refractory. In this digital era, each and every high-tech industry needs a titanium metal which is at least two times stronger and half as light than steel. Also, titanium metal is non-corrosive and chemically non-active in normal condition. These physicochemical characteristics qualify

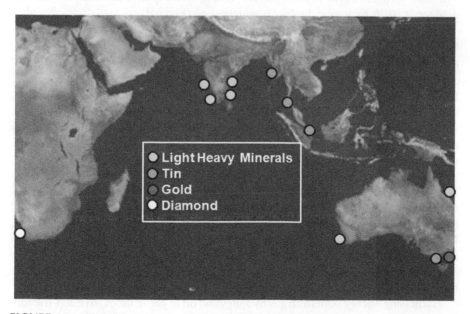

FIGURE 6.2 Global Occurrences of Major Offshore Placer Deposits

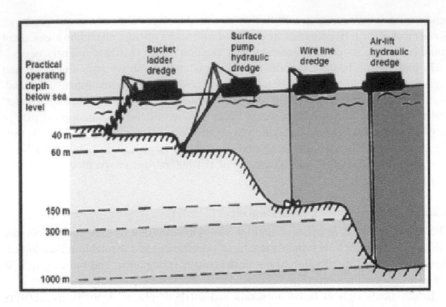

FIGURE 6.3   Offshore Mining Technologies at Various Depths in Ocean

the titanium metal to be sought by the high-tech industries. The raw minerals for this titanium metal are ilmenite, rutile, and leucoxene that are widely found in coastal sands, especially on the beaches and offshore areas.

In shallower ocean up to 60 m depth, usually, the bucket ladder dredger or hydraulic dredger is used to mine placers. With increasing depths, either the wireline dredger or airlift hydraulic dredger is widely used (Fig. 6.3).

6.2.2.3   Continental Margin Resources

Gas hydrates in offshore areas encompass a good deal of prospects in future green energy. A huge amount of reserves have been discovered in a few African countries, India, and Australia. Since this has a towering potential of economic importance, other Indian Ocean countries can invest in exploration of this resource. Additionally, the technology for harvesting and storing of gas hydrates along with the techniques to generate power from hydrates in sustainable means are not available yet in the market and may need intense research in the future.

The oil and gas reserves in Indian Ocean countries are yet to be fully assessed and realized. Exploration and production of oil and gas in offshore areas continues to grow rapidly as a high-value commercial interest to nations. Huge reserves of hydrocarbons are explored and being tapped in the offshore areas of Saudi Arabia, Iran, India, and Australia. About 40 percent of the world's offshore oil production comes from the IOR. Besides some African countries, India, and Australia, the other countries such as Seychelles, Bangladesh, Somalia, Pakistan, Sri Lanka, etc., are investing a high budget for offshore exploration activities.

Harnessing energy both from gas hydrates and oil & gas in innovative eco-friendly ways would attract a high degree of research and innovation in the future.

6.2.2.4 Deep-Sea Resources

Because of ever-growing demand for strategic metals, the focus of the international community has fallen on deep-sea manganese nodules occurring at a water depth of more than 4,500 m. The mining of polymetallic nodules from the seabed remains a tempting, but technologically challenging option. The new ventures involving exploration and mining have greater scope for development and economic contribution to every country. A recent economic appraisal and strategy for mining of nodules from the Indian Ocean Nodule Field—one of the four economically potential areas in the world oceans—suggests that in contrast to an overall perception of non-viability of nodule mining, a fair degree of economic feasibility and commercial sustainability to mine the deep-sea manganese nodules exists in the IOR (Mukhopadhyay et al., 2019). The other deep-sea resources like cobalt crust atop major seamounts, and hydrothermal sulfides from the mid-oceanic ridge system need more research and constrained exploration to identify and assess the extent and richness of mineralized zones.

6.2.3 RENEWABLE ENERGY SECTOR

The IOR nations could produce more than the required energy from sunlight of which the entire region is especially blessed. The solar radiation reaching the earth's surface in one year provides more than 10,000 times the world's yearly energy needs. Hence, the IOR and the world should have started obtaining at least 30% of their energy from renewable sources by 2020.

Further, biomass can be used as energy source. Fast-growing trees can be grown for fuel, so also rotten plants and manure can be used to produce methane gas. There is no strong argument against biomass energy including the alleged unavailability of land. Hydrogen, the most basic and ubiquitous element in the universe, could also play an important role in solving energy scarcity. Instead of any harmful CO_2 emissions, when burned hydrogen gives out only heat and pure water. When millions of end-users connect their fuel cells to publicly owned hydrogen energy webs (HEWs), they can begin to share a new decentralized form of energy generation. Wind power too is an important part of the strategy to generate renewable no-pollution energy. Every megawatt-hour produced by wind energy avoids the production of an average of 1,220 pounds of carbon dioxide (www.audubon.org). Many of the IOR countries can harness power effortlessly and the generated power could be distributed through IOR power grids or regional power grids.

Generation of electricity from ocean energy is a reality, and with appropriate technology being added regularly there is not very far to go to transform this energy generating process into being economically viable. This future emerging technology could be a laudable step and a game-changer toward the low-carbon green energy commencement. Since 1974, about 26 countries have been focusing on and investing in ocean energy (IEA, 2011). In Europe, a variety of concepts for ocean energy conversion with nearly 200 devices have been developed with a target to generate 100 GW of electricity by 2050 (Science Daily, 2018). In this context, new ventures in tide and Ocean Thermal Energy Conversion (OTEC) reserve appreciable promises, so also the International Solar Alliance, formed by countries located along the Equator.

TABLE 6.3
Established and Emerging Ventures in Blue Energy Sector

Blue Economy Activity	Subcategories	Established Industries	Possible New Ventures
Renewable Energy	Offshore energy: Wind, wave, tide stream, water current, OTEC	Offshore wind as an alternative green energy	Innovation and development of new technologies in tide, OTEC etc.

TABLE 6.4
Ocean Energy Patents Filed by the Various Countries

Country	Wave	Tide	OTEC	Total Ocean
Global	717 (52%)	592 (43%)	53 (4%)	1,368
United States	118 (17%)	109 (18%)	32 (60%)	242 (18%)
United Kingdom	118 (17%)	87 (15%)	06 (11%)	228 (17%)
Germany	71 (10%)	82 (14%)	03 (6%)	154 (11%)
Rest of Countries	410 (56%)	314 (53%)	12 (22%)	744 (54%)

Source: European Patent Office (EPO) (2016).

Likewise, the IOR needs a comprehensive Ocean Energy Roadmap and implementation strategies for the next ten years (Table 6.3).

The public funding for ocean energy R&D has increased over the decades, as has the number of filing of patents (Table 6.4). As many as 717 patents were filed globally in ocean energy research among which wave and tidal stream research is on an increasing trend.

Unfortunately, until recently, the IOR countries have yet to fully realize the potentiality of harnessing energy from the ocean. The prospect of generating energy in the IOR has been estimated to be 40,000 MW from wave energy, 9,000 MW from tidal energy, and 180,000 MW from OTEC. Along the Indian coast, the average wave potential is calculated as 5–10 kW/hour (Sannasiraj, 2019).

Wave Energy: After nearly four decades of global research with the advancement of artificial intelligence and learning algorithms, various new innovative designs have been invented based on the use of orbital velocities of water particles to convert wave energy to electricity. Continuing research in innovative mechanical systems with dielectric elastomers could be the future emerging technology (Science Daily, 2018).

Tidal Energy: The development of tidal energy converters especially floating tidal devices is demonstrated at a semi-commercial stage in Europe. With an advantage of not requiring a heavy foundation system, nowadays, the third generation of tidal energy converters produce energy from a tidal flow using sails, kites, or simulating fish swimming motion (Science Daily, 2018). The tidal energy has been attempted in high tide range coastal areas in the Gulf of Kutch and Gulf of Khambhat, Gujarat, and in Sundarbans, West Bengal. In spite of the location advantages, somehow, these two projects have failed to bring the expected results. This area of power generation, hence, needs more research and development.

OTEC: The OTEC technology is based on the water temperature differences between 29°C at the surface and 7°C at 1,100 m depth of water. With a huge potential of generating 180,000 MW, the IOR needs to lap up the opportunity. Though the capital investment of OTEC is a little bit more expensive, the future improved technologies may bring down the cost to affordable and economic limits.

6.2.4 MARINE INFRASTRUCTURE DEVELOPMENT

In view of rapid expansion of the blue economy sectors, the IOR countries have greater prospects to expand their shipping and cargo handling infrastructure. The modernization as well as building of new harbor facilities in suitable locations would enhance the capacity handling of cargo consignments in a professional way. Industrial zones around ports and coastal special economic zones normalize the cargo handling issue and timely delivery to the destinations. Since the logistics are the lifeline of the economic activity, roadways and railways play an essential role in easing the movement of goods in a competent way. In addition, mini air strips and mini harbors further augment the economic activities. The following new ventures are possible and need potential investments to intensify economic activities (Table 6.5).

6.2.4.1 Mechanized Floating Ports/Cities

The world's first offshore blue economy platform has been established by Australia. The platform has facilities for power generation renewable energy, aquaculture, and marine engineering activities. Further, in the northwest shelf of Australia, about 488 m long and 88 m wide an innovative offshore liquefied natural gas (comprising predominantly of methane, plus some ethane) terminal has been constructed that serves

---

TABLE 6.5
Established and Emerging Ventures in Infrastructure Development

Blue Economy Activity	Subcategories	Established Industries	Possible New Ventures
Marine Infrastructure Development	Ports and Harbors	Ports and Harbors	Modernization and building of new ports, Port-based industrial clusters and coastal economic zones, Construction dedicated and improved new terminals
	Hinterland connectivity and multimodal logistics	Logistics and connectivity	Link logistics to transport goods in good roadways and railways, Construction of new mini cargo airports/air strips

as a production platform as well as an offshore port, and also as operating plants. This innovative structure subsequently minimizes the laying of underwater pipelines that greatly reduces not only the cost of transportation but also the pollution thereto. Such innovations also enhance the economic returns of the port-based activities. Such innovative models that are beneficial to boost the blue economy can be appropriately implemented by other IOR nations.

Singapore is implementing a concept of Very Large Floating Structure (VLFS) technology which acts as an elaborate maritime space hub to support container port infrastructure with a 65 million TEU (TEU = Twenty-foot equivalent unit is an inexact unit of cargo capacity often used to describe the capacity of container ships) capacity. Also the floating charter cities are under planning, which represents the convergence of the blue economy with traditional infrastructure focusing greater economic potential (www.sgsme.sg).

6.2.4.2 Trade and Commerce

The maritime trade by the IOR countries has recorded an increasing level of business from US$302 billion in 2003 to US$1.4 trillion in 2012. The cargo transport is an essential activity in the blue economy that needs a well-planned infrastructure network (Table 6.6). Well-furnished new terminals with cargo storage, fast clearance and movements, planned services, logistics, increasing numbers of fleets in various capacities, skilled manpower, more berths, improved amenities, etc., are to be incorporated for better trade capabilities. Singapore has demonstrated this activity very well and has excelled in service-oriented activities that earned a huge economic return. As a part of new ventures, many countries have started passenger cruise liners and tourist cruise liners. Potentially both would earn capital and would simultaneously create considerable employment. While the passenger cruise activities reduce the pressure on land-based transports, the tourist cruises attract more tourists and foreign exchanges.

The Indian Ocean accounts for nearly 20% of global water coverage and has five major choke points along commercial sea lanes (see Chapter 5 for details).

TABLE 6.6
Established and Emerging Ventures in Marine Trade and Commerce

Blue Economy Activity	Subcategories	Established Industries	Possible New Ventures
Commerce and Trade	Coastal Transport	Shipping	Develop and intensify cruise liners for passenger transport, Fine tune waterways infrastructure, Increase the fleet and its facilities, Port infrastructure and services
	Trade and Cargo	Cargo Transport	New terminals with amenities for cold storages and up-to-date infra facilities, Increase the numbers of fleet

These choke points are the gateways for movements of cargo ships. With enormous opportunists in the Indian Ocean for trade and commerce through ship movements, safe and secured transport of goods through these choke points are most desirable.

The World Bank prepared a Logistics Performance Index (LPI) with 155 countries in 2012 to connect the trade logistics with the global economy (World Bank, 2012). The LPI has been calculated based on the following six parameters: (a) customs (efficiency of the clearance process, speed, simplicity, and predictability of formalities by border control agencies); (b) infrastructure (quality of trade- and transport-related infrastructure—ports, railroads, roads, information technology); (c) shipments (ease of arranging competitively priced shipments); (d) logistics (competence and quality of logistics services—transport operators and customs brokers); (e) tracking and tracing (ability to track and trace consignments); and (f) timeliness (frequency with which shipments reach the consignee within the scheduled or expected delivery time). The LPIs of some of the IOR countries are listed in Table 6.7.

The table shows that only a few IOR counties could make it to the first 50 in global ranking. Singapore tops the list followed by UAE (17), Australia (18), South Africa (23), Malaysia (29), and India (46). It clearly shows that other IOR countries have greater scope to introduce new ventures in the trade section by just concentrating on the six parameters of the LPI. Some of the countries are yet to join the club although they are located in the high potential trade zone in the Indian Ocean.

6.2.4.3 Smart Ports

The construction of smaller ports substantially reduces the trade pressure and congestion on the larger ports. This potential sub-activity encourages new ventures in terms of infrastructure developments, creation of sophisticated goods-handling facilities and other logistics and also one way it creates a considerable employment opportunity to the local people.

The faster clearance of cargo by the concerned authorities does ensure the fast delivery of the goods to the customer. Countries like Singapore and Malaysia have excelled in this act and accordingly they have policies to promote faster clearances by customs, without compromising on safety, security, or inspection using new digital methods.

Further, coastal waterways are good means of connecting neighboring coastal cities and certainly facilitate not only the cargo movement but also domestic transport between coastal urbans. New ventures in this category could include dedicated new passenger terminals with basic amenities and increasing the number of passenger liners. The passenger cruises are continuous money earners.

6.3 TOURISM AND RECREATION

Both maritime tourism and coastal tourism are professionally considered under the service sector and enjoy the benefits of the government care, support, and subsidy. While the maritime tourism is essentially based on only water-based activities (boating, yachting, cruising, and other water-based sports), coastal tourism covers beach base activities (surfing, swimming) and near-beach activities in the

TABLE 6.7
Logistics Performance Index with Scores and Ranks of Indian Ocean Countries

Country	Rank	Score	LPI %	Country	Rank	Score	LPI %
Australia	18	3.73	87.2	Pakistan	71	2.83	58.4
Comoros	146	2.14	36.5	Sri Lanka	81	2.75	56.0
India	46	3.08	66.4	Tanzania	88	2.65	52.9
Indonesia	59	2.94	62.2	Thailand	38	3.18	69.9
Iran	112	2.49	47.6	UAE	17	3.78	88.9
Kenya	122	2.43	45.9	Maldives	104	2.55	49.4
Malaysia	29	3.49	79.8	Yemen	63	2.89	60.3
Mauritius	72	2.82	58.2	Cambodia	101	2.56	50.0
Singapore	1	4.13	100	Vietnam	53	3.00	64.1
South Africa	23	3.67	85.5	Madagascar	84	2.72	55.1
Maldives	104	2.55	49.4	Oman	62	2.89	60.4

Source: World Bank (2012).
Note: LPI = Logistics Performance Index.

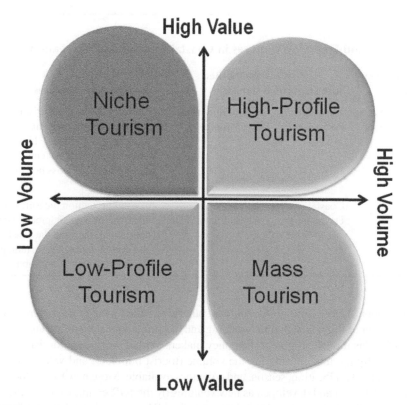

High Value

Niche
Tourism

High-Profile
Tourism

Low Volume

High Volume

Low-Profile
Tourism

Mass
Tourism

Low Value

FIGURE 6.4   Tourism Development Model(Ecorys (2013)).

coastal areas. The business model of Sun-Sand-Sea-Leisure Tourism is marketed as low-cost tourism as it is entirely season based. In order to encourage round-the-year coastal tourism, many strategies including winter tourism and rain tourism could be thought of with appropriate leisure and recreational products (Fig. 6.4, Table 6.8).

Based on the differentiation between a range of maritime and coastal tourism activities, the European Union (EU) has developed a relationship model to define the best equation of high-end eco-friendly tourism. The model has four components—niche tourism, high-profile tourism, low-profile tourism, and mass tourism (Fig. 6.4). The volume of tourists and amount of value have been considered in the light of location importance and amenities available. Through this model, it would be easy to assess the problems, challenges, characteristics, and opportunities for maritime and coastal tourism. Any IOR country would welcome high-profile mass tourism for a vibrant tourist economy. While high-profile tourism would ensure development of superb facility (lodging, boarding, travel, transfer), mass tourism would guarantee more footfalls (Boissevain, 1996; Bramwell, 2004).

Niche tourism is provided with specific value-added services. It draws lesser but specific groups of people with high spending capacity. It is also known as luxury

TABLE 6.8
Established and Emerging Ventures in Coastal Tourism and Recreation

Blue Economy Activity	Subcategories	Established Industries	Possible New Ventures
Recreation and Tourism Trade	Tourism and Recreation	Cruise lines and water sports	Intensify the beach tourism, construct coastal museums, water sports facilities, state-of-art aquariums, whale watching, coral diving, increase glass bed tourist vessels
	Coastal Development	Urbanization, Coastal Smart Cities	Coastal protection measures and infrastructure development, beach nourishing, increase the beach and coastal amenities including seaside resorts and restaurants

tourism (Poon, 1989). Low-profile tourism, otherwise called local tourism, represents a small number of tourists with little spending and has limited local visibility and demand (Fig. 6.4). This model enlightens the policy makers to plan the coastal tourism activity systematically by duly considering the volume (tourist numbers) and value (spending capacity) based on location, season, and amenities available. More new tourist locations could be identified and developed as new ventures by the IOR countries and some may be tagged as the niche and high tourism spots for high economic returns, while some others may be designed also for a combination of mass and high-profile tourism.

6.3.1 COASTAL TOURISM

The coastal tourism sector is an important GDP contributor and employment creator for the IOR counties with a range of interests from wildlife parks to beaches to coastal sports to cultural tourism. Especially, the island countries attract more tourists for their vacations and fun holidays. The potential of coastal tourism is ever increasing as people prefer to spend holidays for leisure, fun, and to enjoy the cultural essence of various sites. The island countries such as Seychelles (total contribution 60.8% to GDP; US$0.66 billion in 2011, and total employment 25,000, i.e., nearly one-quarter of the country's population: WTTC, 2012), Madagascar (total contribution 14.9% to GDP; US$1.37 billion in 2011, and total employment 577,000, i.e., 12.5% of the country: Ministry of Tourism, 2012), Mauritius, and Sri Lanka have excelled in the tourism sector.

The tourism departments of the IOR countries require to plan and design new tourism sites. After identifying the sites, other infrastructure facilities such as approach roads, transport, accommodation, restaurants, and hotels are to be developed. Value addition measures like museums, cultural centers, animal and botanical parks are established and integrated.

6.3.2 Discovery and Wildlife Tourism

Australia has promoted ocean-based wildlife tourism by providing whale and shark watching, swimming with dolphins, coral diving, etc. The other IOR countries as well have high potential of ocean discovery tourism. In many places along the coastline of India (particularly in Gujarat, Maharashtra, and Goa) dolphin watching and coral diving are being practiced. Some other areas do have similar prospects. Due care must be taken to make those areas tourist friendly by developing necessary logistics. In order to develop discovery tourism, countries such as Singapore and Mauritius have developed state-of-the art marine aquariums. Some of the small islands are being developed into tourist spots with educational and knowledge-based ocean life displays. Underwater aquatic museums, glass bottom cruise boats, etc. are contributing considerably to coastal tourism.

6.3.3 Cruise Tourism

Globally, cruise tourism is a US$40 billion industry. The number of people cruising across the globe has profoundly increased from 13 million in 2004 to 24.7 million in 2016. There were 448 cruise ships in operation in 2016 and there were an additional 65 new ones by 2019. Most of the cruisers (passengers) are from the USA (50%) and Europe (30%). This industry could grow in the IOR countries in a big way under the blue economy. For example, in Sri Lanka, cruise passengers have increased from 19,000 in 2013 to 60,000 in 2017. Inter-country cruises, inland passenger cruises, and other coast-based passenger cruises besides sightseeing tourist cruises are in high demand and craving for developments with a well-defined action plan.

6.3.4 Cultural and Archaeological Tourism

There are many coastal archaeological and cultural sites in the IOR countries that are least developed. For a set of educated high-profile and niche tourists the cultural attributes of an area attract the most. Especially, some of the coastal cities are dotted with temples and other cultural signatory structures. The new ventures in this space may include the development of cultural museums, libraries, cultural parks, and visits to excavation sites and artefacts' museums. By developing facilities to connect cultural assets and ancient archaeological structures, the potential of a tourist spot increases manifold. Such initiatives also help educate people to get an insight into the ancient cultural richness of the areas.

6.3.5 Hospitality and Amenities

Opportunities in new ventures in tourism development include services and hospitality. The accommodation and facilities in rooms need more investments for growth as the visitors are ever increasing. For example, in Madagascar, the 768 hotels and 9,325 rooms increased by 162% and 120% respectively between 2003 and 2012 (Ministry of Tourism, 2012). In Seychelles, in addition to 4,066 rooms of accommodation, live-aboard yachts had 1,097 beds in 2010 (STB, 2011). For cruise tourism

TABLE 6.9
Established and Emerging Ventures in Marine Ecosystem Services

Blue Economy Activity	Subcategories	Established Industries	Possible New Ventures
Ecosystem Services	Carbon sequestration	Blue Carbon: Climate mitigation	Conservation measures and innovations
	Coastal production	Habitat protection and restoration: Resilient Growth	Research and engineering innovative design and structures
	Waste disposal	Assimilation of solid waste: Wastewater management	R&D in innovative measures
	Biodiversity conservation	Protection of habitats: Conservation	Best practise action plans

development, the following facilities are justified: (a) dedicated terminal with infrastructural facilities; (b) closer access to city areas; and (c) special immigration facilities.

6.4   MARINE ECOSYSTEM SERVICES

Sustaining a marine ecosystem is both a challenge and an opportunity. The air-land-sea interaction must maintain a balance. Table 6.9 provides a glimpse of the existing and emerging areas in nourishing an ecosystem. The threats are manifold and could come from change in climate, from military and strategic activities, and pollution.

Ethical blue economy technology (EBET) could play a major role in de-polluting the ocean and atmosphere. Major pollution in the IOR is generated by coal and diesel-based power facilities. In contrast, hydro-power and nuclear power generate only a little amount of pollution. Nuclear energy, although considered as the most environmentally friendly, has the problem with its high cost of set-up, radiation, and n-waste. Coal, oil, and nuclear power appear to be no solution to the future economy and must be phased out as quickly as possible.

6.5   CAPACITY BUILDING

In order to propel toward sustainable blue economy growth in the IOR, capacity building would act as a fundamental pillar. In order to construct a vibrant blue economy of the Indian Ocean countries, the following five initiatives on capacity building could facilitate the ongoing as well as future activities into a structured and well-planned approach (Fig. 6.5).

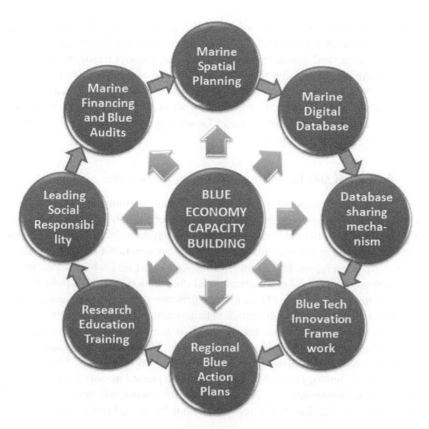

FIGURE 6.5   Flow Sheet on Capacity Building for Blue Economy

6.5.1   MARINE SPATIAL PLANNING (MSP)

MSP encompasses an integrated and comprehensive marine geographic and related business information. It includes administrative and ecological boundaries, spatial extension of marine conservation areas (marine parks, marine habitats, coral banks, etc.), seabed bathymetry, geology, geomorphology, and physical oceanographic details. MSP further includes locations of ports, harbors and other onshore details and their buffer zones, details of logistics in the coastal areas, and marine infrastructure details under the sea (pipelines, cables, shipwrecks, etc.). In fact, MSP classifies the maritime regions for its fullest potential for best suited sustainable activities. MSP addresses the marine policy and governance with the enclosures of both technical standards and information systems. In many Indian Ocean countries, MSP has not been taken up seriously. Some island countries like Seychelles and Mauritius are however the forerunners in developing MSP strategies.

6.5.2   MARINE DIGITAL DATABASE (MDD)

Inadequate and unsubstantiated databases in estimating the potential of oceanic resources plague the IOR countries considerably. Even for countries like Australia and India, the issue of known resources of the ocean is debatable. However, the resources in terms of total availability, generation cycle for living fauna, conservation, exploration, and alternative technologies need correct estimation. Such an integrated database essentially comprises a time-series assortment of biological, chemical, geological, biotechnological and physical parameters (data) collected from various oceanic environments (beach, near shore, offshore, terrestrial, and EEZ waters). The Indian Ocean Global Ocean Observing System (IOGOOS) could be strengthened to serve this purpose.

6.5.3 Blue-Tech and Innovation Strategies Framework

In fact, innovations are required in every sphere of blue economy activities, such as in the generation of non-polluting cheap energy from all conceivable natural resources, its transmission, and also in communication and transportation; in agriculture; in industries; in tourism; and in resource management. The prescription for a good intelligent society involves developing EBET. Such technology must be cost-intensive, intellectually cutting edge, and may need international cooperation. The IOR must focus on intense innovation and research of new ideas to generate power using natural resources. For example, EBET can technically enrich both agricultural reforms (collective farming and use of GM crops) and water management (linking of rivers, and recharging of aquifers), and could also create thousands of new types of jobs. After recognizing appropriate potential sectors for future developments through MSP findings, EBET (Ethical Blue Economy Technology) may be implemented in a phased manner through financial support from sponsoring agencies. A well-defined framework will help implement the innovation technologies that are the need of the hour for the Indian Ocean countries.

6.5.4 Education and Training on Skill Development

It is nice to see many IOR countries have established dedicated marine research institutes to carry out innovative investigation to promote sustainable use of oceanic resources. The NGOs and these academic/research institutions must play a significant role in imparting training to the local people and entrepreneurs with appropriate guidelines, skills development training in an understandable format, with practical demonstrations. The community-based activities especially in fisheries and aquaculture brought stupendous results in Madagascar and several other island nations in the Indian Ocean. The World Bank and other global institutions have been financing such community-based learning in well-structured layouts. Enhancing blue carbon activities are highly potential skill development actions being recommended and widely implemented in many Indian Ocean countries.

Based on type of marine ecosystem and nature of resource distribution (resulting from a detailed MSP), a region-wise best practice action plans could be worked out to achieve optimum level of returns. The region-specific action plans carefully designed and developed based on science input and technological back-ups will help the concerned countries to devise their own capacity-building plan, given their financial ability and State policies.

6.5.5 Marine Financing and Blue Audits

Financing the marine research projects and blue economic activities, both traditional and upcoming, are the critical issue as far as the policy of any government is concerned. Based on the potentials of blue resources and marine services, the financial outlay is designed keeping the economic and societal development in due concern. While considering blue financing, the rudiments of capacity building of the blue economy, i.e., MSP, the time series projections, innovations and new technologies, the carrying capacity of marine resources and societal acceptances are widely evaluated for economic returns, societal benefits and ecosystem conservation.

The World Bank-initiated Blue Bond initiative in Seychelles has been very successful and is fetching the anticipated and projected results in the development of economy of the nation. In a similar way, countries such as South Africa, Madagascar, Mauritius, Bangladesh, etc., are implementing the financial system for boosting their blue economy activities. Blue Auditing is another essential concept to be carried out at every stage of implementation of action plan. The interim assessments from the blue audit keep the various operational activities under an effective monitoring mode so as to fine tune the entire blue financing system in a competent and professional manner.

6.6 SCIENCE AND TECHNOLOGY

Rapid technological developments in terms of digital application have been swiftly growing in the blue economy paradigm. Robotics, digital video-surveillance, submersibles, drones, LIDAR, radars, etc. are deployed in marine research that enhance and enrich the economic activities in the sea. Marine environmental research, and new technologies in ocean energy could add a considerable degree of sustainability to the blue economy architecture (European Commission, 2012).

6.6.1 Solar Energy

Generating electricity from solar radiation is fast catching up and many nations in the world are cooperating on bilateral or multilateral levels to produce maximum energy out of solar insolation reaching the Earth. For example, the USA and Europe have entered into cooperation, followed by Belarus and Russia (in producing solar photo-voltaic cells [PV] and thermal), India and the USA (for solar PV, wind), South Africa and Europe (for biofuels and wind), and the USA and China (for solar PV). In this regard, greater economic integration and cooperation could be another solution to IOR acute resource shortages. The cost of manufacturing both solar cells and modules and other components has been falling steadily. As a result, the price of PV has fallen by an average of 5% per annum over the last 20 years (Greenpeace International, 2007).

Such advancement is however also going on in the IOR region. For example, China is presently producing 131,000 megawatts of solar power (1.8% of national need), the USA 51,000 megawatt (5.9%), and India 18,000 megawatts (2%). Volume-wise China produces 53% of global solar power, the USA 11%, India 10%, Japan 7%, and the rest of the world 13%. Hence, the economically well-off countries (including

China and India) pollute the environment by generating a large amount of electricity and emitting GHGs, but also keep up the balance by increasing solar power and Leaf Area Index (vegetation) coverage.

6.6.2 Aquatic Biotechnologies

Research and development on marine biotechnology and bio-prospecting with innovative ideas could invent new pharmaceutical drugs, chemical and associated products, industrial products, biomaterials, enzymes, and cosmetic materials from sea resources. Many requirements of present-day needs are being addressed in terms of the advancement of biomaterials, health care diagnostics, etc. Marine algae have much potential as far as their utilization in pharmaceutical application is concerned. They are the fastest-growing plants and can double their biomass daily. Out of 40,000 species, only 10% has been cultured. Moreover, the marine microorganisms constitute more than 10% of the total living biomass carbon of the biosphere. These bio-products are useful in the manufacture of drugs against cancer and dementia, and in various other products such as food additives, fertilizers, feed, cosmetics, detergents, etc.

6.6.3 Hybrid Technology

Advances in renewable energy, electric vehicles, and hybrid technology have led to significant reductions in emissions and waste disposal. Further improvements have been made in biofuels, organic photovoltaics, and hydrogen cars. The recent Decarbonathon competition conducted by the World Economic Forum Young Global Leaders initiative, alongside ENGIE and the National Physical Laboratory (USA), set out to find the most promising new ideas that could support the blue economy by reducing CO_2 emissions in cities. The competition selected four technologies that it thinks hold the most promise for the future.

Mobiliteam has been one such innovator. It has developed an air booster that reduces the energy consumption of electric vehicles by improving the efficiency of air conditioning systems, whilst having no effect on the passenger's comfort. Even in cool climates, air conditioning accounts for 5–10% of fuel consumption, meaning that there are economic as well as environmental incentives for manufacturers to fit the technology.

Bynd has been working to develop a car-pooling app that, unlike existing car-pooling services, is aimed at the regular commuter. According to the Campaign for Better Transport, 91% of car commutes are single-passenger journeys. Bynd aims to work with companies to develop an app that allows staff within the same business (or another nearby) to combine journeys and reduce the number of car journeys taken in cities.

The Traffic Energy Bar System (TEBS) takes a different approach. Instead of attempting to make cars more efficient, or reduce road traffic, it makes use of busy roads to generate energy for use elsewhere. TEBS is a system installed across areas where a high volume of traffic is slowing down, in which bars are pressed down by the wheels of each car as it moves over them, creating an up and down motion that generates electricity. It uses the waste energy from the cars slowing down, and harnesses it to power other systems in the city that require electricity.

The last innovation recognized as having big potential was Mutum that aims to reduce industrial and residential emissions. An idea borne out of the sharing economy, it aims to reduce overconsumption by making it easier to share things with others. For example, a typical electrical drill is only used for 12 minutes during its life-time. Mutum aims to show how such objects can be borrowed rather than bought. Overconsumption creates wasteful industrial processes through overmanufacturing, so reducing these emissions will help lower urban energy demand and subsequent GHG emissions.

These are just a few examples, but if the world desires to reach the ambitious pledges set through the Paris Climate Conclave (COP-21), more must be done, and new green technologies and continued innovation needs to be encouraged.

6.7 EMPLOYMENT GENERATION

Employment generation is a key component of human resource management to make the best situation for a blue economy to flourish. The blue economy, when first thought about, was supposed to make 100 innovations and creating in the process 100 million jobs within ten years (Pauli, 2010).

To establish a robust blue economy architecture, the IOR would need thousands of skilled professional staff along with innovative technology, to sustain the clean energy movement, an important component of ethical blue economy (EBE) aspir-ation. China has moved full-speed toward creating more clean energy professionals from PhD-level engineers to well-trained technical operators (Liu, 2010). The Chinese government and Chinese firms are using a number of strategies to attract and develop talent in clean energy. In less than a decade, dozens of universities set up clean energy faculties and scheduled to enroll undergraduates starting from 2011. Beijing is ready to seek help from outside. The China-EU Institute for Clean and Renewable Energy (ICARE) is a case in point. In fact, between 2011 and 2016, China generated about 600 postgraduates and retrained nearly 1,000 energy professionals for the local industry.

To groom more solar energy professionals, China's Zhang College provides a three-year vocational training program free of charge and offers scholarships to those from poor families. In 2009 alone, Hemin Solar poured more than US$2.5 million into education incentives. With such support, the number of students there climbed to several thousand in 2015, up from fewer than 100 in 2007, when the college had just started. The Belt and Road initiative of China also aims to usher in collaborative research among participating nations, mostly funded by China. The IOR may take a cue from Chinese experience and could form several dedicated institutes under the IOR to produce clean energy professionals and engineers and build its own pool of engineers to realize the solar power potential of the IOR.

Employment generation is mainly supported directly by fisheries, marine trans-portation, tourism, offshore oil and gas, defense and security, etc. In a global per-spective, nearly 820 million people are directly or indirectly associated with fisheries and related supply chain industries, as per the estimates of FAO. Among which about

15% of women are directly involved in fisheries activities (World Bank, 2016). In addition, an equal amount of job creations is supported by down-stream industries such as aquaculture, mariculture, fish processing, port activities, minerals, ship-building, etc. Besides, there is a huge opportunity of jobs from emerging sectors related to biotechnology application, blue carbon conservation, ocean energy production, coastal urban infra development, inland waterways, cruise liner passenger transportation, etc. Both in the skilled and unskilled categories, job generation could be doubled with proper planning and implementation.

6.8   SYNTHESIS

It is estimated that the majority of people around the world now live in urban areas—and the global urban population is expected to grow approximately by 1.84% every year. Such growth is a key driver behind the move to "smart cities" that aim to improve quality of life and efficiency of transport, energy provision and health care through technology. But as urban areas grow, GHG emissions are likely to grow. Hence, there is a great need to ensure that, as our cities become smarter, they also become greener.

Hence, innovations are required in every sphere of blue economy activities. The problem is that emerging green technologies can often struggle to secure investment, severely hampering their development and market uptake. On top of that, subsidies in the energy sector often create unfair market conditions by favoring established technologies, many of which are contributing to climate change rather than helping to address it. Ironically, government subsidies to the fossil fuel industry are four times more than to clean energy industries. Building confidence in new technologies is also very crucial to securing investment and market uptake.

Meanwhile the World Bank, UNFCCC, Asian Development Bank (ADB), and FAO are jointly helping coastal countries with finance and technology to promote effective governance, improve sustainable fisheries and aquaculture, make coastlines more resilient, establish coastal and marine protected areas, enhance coastal and ocean habitats, and develop knowledge, skill, and capacity building. The ADB and World Bank are also providing financial support for coastal infrastructure such as waste treatment, watershed management, and other activities that help reduce coastal and marine pollution.

Again, innovative employment and alleviation of poverty are closely linked, and are extremely important tools to appreciate the blue economy (ADB, 2014; World Bank, 2014b). In this regard, EBET must be seen as a huge opportunity to create new markets, jobs of completely different types, and growth potential of various ancillary industries and services. The companies and investors that seize these opportunities would be best positioned to thrive on a resource-constrained economy. In addition, such an initiative would simultaneously keep the environment healthy and individuals wealthy. Under the UNFCCC's Paris 2015 agreement the developed countries are to help the countries of the developing nations (including most of the IOR countries) with funds and transfer of advanced technology to mitigate the climate change impact.

There are a few best minor ways for every citizen to further the blue economy perspectives. The indicative (not exhaustive) ways are to be politically aware to make climate-conscious decisions during election voting, purchase electricity from renew-able green sources, make households energy efficient, buy energy- and water-efficient appliances, walk mostly or use bicycles/public transport, recycle/reuse/avoid useless purchases, do telecommuting and teleconferencing to avoid avoidable travel, and eat less red meat.

Therefore, it can be said that an emerging sector like the blue economy will require technological excellence to allow many scientific and commercial activities to reach their potential, particularly in generating clean/green energy, conserving ecosystems, and harnessing living and mineral resources. It seems there are more markets for new products with potential than anticipated.

7 Economic Projections

Understanding and measuring the contribution of the ocean economy to the national economy is difficult and still in formative stages, although many countries have attempted to quantify such an input. The main constraint to this exercise has been the data pertaining to ocean sectors, which are inadequate, inconsistent, and not comparable internationally. Unless detailed and centralized databases are made available, any attempt to quantity and model the ocean economy will not yield the expected results. Quantifying the contribution of the ocean sector becomes more complicated when we perceive the blue economy rather than the ocean economy. The blue economy encompasses many aspects of human, social, and ecological factors, which are not easily quantifiable, and not included in the estimation of the ocean economy. Hence, establishing a standard statistical framework to evaluate the ocean economy of different countries is a necessity (Wang, 2016).

Appropriation of ecosystem services to trace biophysical interdependence on the ocean economy was made a decade back in Auckland, New Zealand (Patterson et al., 2009). The study found limited ability of economists to capture ecological interdependence, although it is a fact that the economic activities depend largely on ecosystems. The first reason for such inability is that many of the benefits provided by the natural resources are not marketed appropriately and therefore do not fetch a market price. The second reason very often is that the economic activity and its ecological implications do not happen at the same time and at the same place. This makes assessment of ecological impact of economic activity difficult in order to study the sustainability issues.

Wai Ming To and Peter Lee (2018) applied four-parameter logistic models to assess the trends and to predict the future values of China's maritime economy. The study found that China's primary and secondary sectors of maritime economy (fishery and cargo shipping) have already passed their maximum growth rate. However, the tourism and transport sectors have potential to grow further. The study discussed qualitatively various implications of growth in the maritime sector to the marine ecology and the sustainability in China.

A bibliometric method to review the existing literature on marine econometrics suggests that quantitative analysis of marine economy is still very weak (Kedong Yi and Xuemei Li, 2018). In the late 1990s and particularly since the 2000s, research achievements in the field of traditional marine economics have been increasing gradually. This study also highlighted the deficiency of relevant data to quantitatively assess the contribution of ocean to economic growth. We here aim to provide the future prospects of the blue economy of 11 of the Indian Ocean Region (IOR) countries. These countries are Australia, Bangladesh, India, Indonesia, Kenya, Mauritius, Oman, Seychelles, South Africa, Sri Lanka, and Thailand. While doing so we consider three major ocean-related economic activities of these countries, namely, fisheries, shipping, and tourism. The basic data for all tables and figures in this chapter were sourced from FAO Fishery Statistics (FAO, 2018) and World Bank (2017), which were treated later as per the requirement by the authors.

7.1 BLUE ECONOMY DETERMINANTS

At the macro level, it is believed that the length of the coastline, and the extension of the exclusive economic zone (EEZ), mangroves, coral reef, and marine protected areas (MPA) determine the size of the blue economy. In addition, the technological advancement, ecological, and ethical considerations also contribute in determining the size of the blue economy.

The IOR is endowed with a long coastline, EEZs, coral reefs, and mangroves. How these resources determine the economic contribution of the ocean to the overall economic growth assumes a lot of significance and has far-reaching implications in planning for ocean development in this region. Hence, in this section an attempt is made to work on the following two hypotheses:

i Do length of the coastline and size of EEZ determine the scope of GDP in IOR countries?
ii Do MPA and mangroves influence the fish catch in the IOR countries?

Do Length of the Coastline and Size of EEZ Determine the Scope of GDP in IOR Countries?
In order to test the hypothesis of whether the area along the coastline and under the EEZ determines a country's economic growth, the following log-log econometric model is developed and tested.

$\ln Y = \beta_1 + \beta_2 \ln X_1 + \beta_2 \ln X2 + u,$

where, Y = the GDP of IOR
X1 = length of coastline in km
X2 = area of EEZ in km^2
ln = natural log
u = error term

The data to test the above hypothesis were obtained from as many as 30 IOR countries (World Bank, 2017; FAO 2018). The variables used in the model are GDP in million US$, coastline in km, and EEZ in km^2 (Table 7.1).

TABLE 7.1
Summary Statistics of the Variables Used

Variable	Mean	Median	SD	C.V.	Skewness	Ex. kurtosis
Coastline	4,548.7	1,329	10,621	2.3349	3.9469	15.338
EEZ	1,200,300	233,480	2,308,800	1.9236	2.5067	4.9423
GDP	336,770	115,290	561,030	1.6659	2.9710	9.4116

Significance at 1% level

Notes: Coastline in km, EEZ in km^2, GDP in million US$, SD = Standard Deviation, CV = Coefficient of Variation, Ex. kurtosis = Excess kurtosis. The mean and median are the two commonly used measures of central tendencies. The mean indicates the average value of a distribution and median, the middle value of a distribution. The standard deviation reveals to what extent the other values are dispersed from the mean value. The higher the standard deviation, the higher will be the dispersion. While the standard deviation is expressed in terms of the units in which the variable is measured, the coefficient of variation (mean divided by SD) is independent of units and is generally expressed in percentages. The skewness is a measure of symmetry or absence of symmetry. When the distribution is not symmetrical, it is called skewed. The skewness shows where the tail of the distribution is located. If coefficient of skewness is negative, the tail of the distribution lies on the left side and if skewness is positive the tail of the distribution lies on the right side. Kurtosis reveals whether data are heavily or light tailed as compared to the normal distribution. A high value of kurtosis shows heavily tailed distribution or distribution is with outliers. In excess kurtosis the coefficient of kurtosis is 0 if it is a normal distribution, positive value indicates higher kurtosis and negative value lower kurtosis.

The average coastline area is 4,549 km. The coefficient of variation of 2.33 reveals high inequality in the coastline area of the countries in the IOR. The distribution of coastline also shows a high level of skewness. The average EEZ area of the IOR countries is 1,200,300 km^2. The coefficient of variation of EEZ is high but less compared to the coastline. The mean GDP of IOR countries is US$336,770 million. Table 7.2 shows the results of the model estimated.

The estimated model is satisfactory, in terms of F value and adjusted R^2. While F value gives overall validity of the model in question, R^2 gives information on how good the fitted regression line is. The results indicate that the In-EEZ (In-EEZ is EEZ in log form) is not an important factor in determining the GDP as the coefficient has been found statistically insignificant (p value > 0.10). However, the coefficient of In-coastline (log of In-coastline) is statistically significant implying that longer the coastline, the higher could be the GDP. The results of the model are convincing as larger EEZ area need not lead to higher GDP unless huge investment is made to tap the resources available. On the other hand, longer coastline can enhance tourism, fishing, etc., thereby contributing to the GDP. The result reveals that if the coastline of a country is 1% more, the GDP of that country would be higher by 1.09%. The results also imply that ocean plays an important role in the economic growth of the IOR countries.

Do MPA and Mangroves Influence the Fish Catch in the IOR Countries?

An econometric model is estimated and tested here with fish production as the dependent variable, and the areas under MPA, coral reef, and mangrove coverage

TABLE 7.2
Impact of Coastline and EEZ on GDP of the IOR Countries

Dependent Variable: ln-GDP

Variables	Coefficient	Standard Error	t Ratio	P Value
Constant	7.25021	2.22921	3.252	0.0031 ***
ln-coastline	1.09403	0.314926	3.474	0.0017 ***
ln-EEZ	−0.317243	0.209240	−1.516	0.1411

Adjusted R^2 0.267553, F (2, 27) 6.296642, P-value (F) 0.005695

Notes: ln= natural log, Col. 1 = independent variables, Col. 2 = ordinary least square (OLS) estimates of β coefficients of the model, Col. 3 = standard error of each coefficient, Col. 4 = "t" ratio (value of coefficient divided by the standard error) is the test statistic used in the OLS-based regression models. The significance of the estimated β coefficients depends on the value of t ratio. Col. 5 = Probability values, $P < 0.05$ implies that the β coefficients are significant at 5% level and $P < 0.01$ implies that the β coefficients are significant at 1% level. Any P value > 0.10 implies that the coefficient is not significant. The higher the t ratios, the lower will be the p values and more likely that the coefficients would become significant. In OLS regression models, R^2 is the goodness of fit and indicates the explanatory power of the model and it varies between 0 and 1. The higher the R^2, the better the explanatory power of the model. The adjusted R^2 is the R^2 adjusted for degrees of freedom. In multiple regression models, it is more appropriate to consider the adjusted R^2 rather than R^2. F statistic is used to test the joint significance several regression coefficients. The F statistic is based on degrees of freedom of the numerator (k-1, the number of coefficients estimated minus one) and degrees of freedom of the denominator (n-k, the number of observations in the model minus the number of coefficients estimated). F (2, 7), in Table 7.2 means that the degrees of freedom for the numerator is 2 and the degrees of freedom for the denominator is 27. The critical value of F and the p values change as and when the degrees of freedom of numerator and denominator changes. The p value of F shows the significance of F value at 1%, 5%, 10% levels or not significant at as in the case of p values for t ratio.

as independent variables. The objective of this model is to understand the impact of these independent variables on the fish production. Data on coral reefs is obtained from Area Statistics, World Atlas of Coral Reefs, UNEP (WCMC, coral.unep.ch/atlaspr.htm) and data on MPAs (as percentage of territorial waters) is collected from rankings in www.indexmundi.com/facts/indicators. Data on mangroves is obtained from the little green data book 13 from World Development Indicators, the World Bank (https://issuu.com/world.bank.publications/docs/9780821398142).

The summary statistics (Table 7.3) provide the structure of the data used to estimate the model. The average fishery production in the IOR is 1,532.4 thousand tons. The average MPA in the IOR is 6.1% of the territorial area. The standard deviation and coefficient of variation reveal a wide disparity among the countries with regard to the fishery production, the MPA, and the mangrove areas. The distribution of the variable under study is also highly skewed.

The following model is used to test this hypothesis of whether the area under MPA, coral reefs, and mangroves determine the size of fish production. A linear regression model is used to test the hypothesis (Table 7.4).

TABLE 7.3
Summary Statistics of the Variables Used in the Model

Variables	Mean	Median	Std. Dev.	C.V.	Skewness	Ex. kurtosis
Fishery	1,532.4	148.50	4,239.7	2.7667	4.0522	16.427
MPA %	6.1228	2.4000	9.7196	1.5874	2.8895	9.4449
Coral Reef	4,869.1	1,030.0	12,024.0	2.4695	3.4396	10.258
Mangroves	2,350.4	78.500	5,913.9	2.5161	4.1379	17.861

Notes: Fishery = Fishery production in thousand tons, MPA = share of MPAs to the territorial area in percentage, Coral Reef = Coral Reef area in square kilometers, Mangroves = Mangroves area in square kilometers, SD = Standard Deviation, C.V. = Coefficient of Variation, Ex. Kurtosis = Excess Kurtosis, for detail definition see footnotes of Table 7.1.

TABLE 7.4
Results of the Regression Model

Dependent Variable: Fishery Production

	Coefficient	Std. Error	t-Ratio	P-Value
Constant	484.117	377.185	1.283	0.2098
MPAs %	− 76.9717	43.9502	− 1.751	0.0908*
Coral Reef	− 0.0537239	0.0624928	− 0.8597	0.3973
Mangroves	0.757802	0.108710	6.971	< 0.0001***
R-squared	0.858685	Adjusted R-squared		0.843544
F(3, 28)	56.71289	P-value(F)		0000***
N = 32				

Notes: *** Significant at 1 % level, * significant at 10% level. For details see footnotes of Tables 7.1 and 7.2.

$$Y = \beta_1 + \beta_2 X_1 + \beta_2 X_2 + \beta_3 X_3 + u \ldots$$

where, Y = Total Fishery Production in IOR countries
      X1 = Area under MPA
      X2 = Area under coral reefs.
      X3 = Area under mangroves
      u = error term

The model shows that the MPA and mangroves do determine the amount of fish production in the IOR. The MPA and fish production are however negatively related as the coefficient pertaining to the MPA is statistically significant (at 10%). It appears that more MPA lowers the fish production, e.g., even a one-unit increase in MPA could bring down the fish catch by 76,000 metric tons. On the other hand, higher

MPA in the long run might control overfishing and, in turn, encourage sustainable development of fisheries.

However, the coefficient pertaining to mangroves is highly significant (at 1% level), indicating a positive relationship between mangroves and fish production. If the area under mangroves increases by one square kilometer, fish production is expected to increase by 750 tons. The model reiterates the importance of mangroves in maintaining sustainable fishery in the IOR, and suggests protection of mangroves. The model has a good fit with an adjusted R^2 of 0.84% and the estimated model is highly relevant as indicated by distinct significant F value.

Therefore, our model suggests that (a) the higher the area under mangrove the higher will be fish production, and (b) the more area under MPA the lower will be the fish yield. Furthermore, in contrast to the popular belief that coral reefs support fishery, our model does not show a significant relationship between fishery production and coral reef.

7.2 PROJECTED ECONOMIC BENEFITS

The growth of the blue economy in the IOR is comparable to any other regions in the world that have experienced galloping progress in three major sectors: fishery, shipping, and tourism. Before using a statistical method to estimate approximate economic returns from these three sectors for the IOR countries, the performance of the top ten countries in these three sectors are listed.

7.2.1 PRESENT SCENARIO

7.2.1.1 Fisheries

Fisheries play a crucial role in the blue economy in providing much essential food and protein to hundreds of millions of people in the IOR. Fisheries contribute immensely to food security, poverty alleviation, income generation, and job creation. The last few years have witnessed a reduction in the fish stock at a global level. It was estimated that of 441 stocks worldwide, 47% were fully exploited, 18% overexploited, and 9% depleted (www.iora.int/en/priorities-focus-areas). There are reasons to believe that the climate change and overfishing have resulted in reduction of fish stock in the IOR too.

In the IOR, some countries dominate in fishery production, both in aquaculture and capture fishery. For example, Indonesia, India, and Bangladesh are the leading countries in the IOR region engaged in aquaculture. These countries together make up around 85% of total aquaculture production in the region. Indonesia's aquaculture production (16.6 million tons, MT, in 2016) was three times higher than India's (5.7 MT in 2016), the second-highest producer. Bangladesh generated a little more than 2.2 million tons in the same year. The rest of the countries produced less than a million tons. This indicates that aquaculture has not made progressive inroads in most of the IOR countries.

The contribution of the IOR to global landings in terms of marine capture fisheries is third after the Pacific and Atlantic oceans. In this case also the leading producers are Indonesia (>6.58 MT) and India (>5.08 MT), followed by Myanmar (>2.07 MT) and Bangladesh (>1.67 MT). These four countries together constitute 65% of total production of capture fisheries. The other two countries that are dominant in capture fishery production are Malaysia (>1.58 MT) and Thailand (>1.53 MT). Although the rest of the countries have production of less than a million tons, it is to be noted that almost all countries in the region have their presence in capture fisheries.

7.2.1.2 Shipping

By transporting 127.88 MT of cargo in 2016, Singapore has had the major share of shipping activity in the IOR, registering a whopping 57.8% of total merchant fleet negotiated through this ocean. In fact, three countries—Singapore, India, and Indonesia—together share 76% of the total shipping activity in the IOR. In the same year, Indonesia transported 22.313 MT, India carried 18.481 MT, Saudi Arabia passaged 13.523 MT, and Malaysia transported 10.230 MT of cargo. The remaining countries have only 10% of the shipping share in terms of deadweight tons carried.

7.2.1.3 Tourism

Coastal tourism and related services are key components of the blue economy. The IOR is also seeking ways to generate greater economic benefit from these marine sectors. The tourism sector in Seychelles contributes approximately 25% to its GDP. If we add to it the indirect (especially services like leisure, cultural heritage) contribution, it would amount to about 60%. The direct and indirect contribution of tourism to the Seychelles' workforce is also more or less 25% and 60%, respectively.

The arrival of tourists in various countries of the IOR region in 2016 has been interesting. Thailand remained at the top with a footfall of 35.592 million (M) tourists, followed by Malaysia (25.948 M), Saudi Arabia (16.109 M), India (15.543 M), Indonesia (14.040 M), Singapore (13.903 M), Bahrain (11.370 M), South Africa (10.285 M), Australia (8.815 M), and Egypt (8.517 M). Though the distribution of tourist arrivals is somewhat skewed, most countries in the region have the presence of tourism activities. What is important as regards to the tourism sector is not just the number of tourist arrivals but the receipts from tourism. The receipts from tourism depend on the type of tourists arrived, types of services provided, and also the quality of services.

The pattern that appeared in the case of tourism receipts is different from the tourism arrivals. Thailand continue to occupy the first position both in arrivals and receipts. Australia, India, UAE, and Singapore occupied other positions in order. However, Malaysia, which occupied second position in terms of tourist arrivals, was pushed to sixth position in terms of tourist receipts. The differences are due to, perhaps, the quality of tourist services provided and the types of tourist arrivals. Maintenance of coastline in terms of ecology and environment is very important in attracting quality tourists. Moreover, local cost of living, available comfort facilities, and value-added additional attraction could influence tourist arrival as well as receipts.

7.2.2 THE FORECAST

We have made a case study of 11 countries in the IOR and estimated the projected economic returns from three major sectors. Our study has resorted to the ARIMA-based forecasting technique to project the growth of these sectors for the next ten years.

ARIMA is an autoregressive integrated moving average process, and is one of the most popular forecasting models that completely ignores the independent variables. Detailed methodology of ARIMA is given at the end of this chapter. For the present study, the following equation was used:

$$Yt = \phi + \alpha 1 Yt{-}1 + \beta 0ut + \beta 1ut{-}1$$

7.2.2.1   Fish Production in Selected IOR Countries

Two types of fish productions are considered for analysis: production by capture fisheries and that through aquaculture. The data used for the study are from World Bank database on fishery output for the period from 1990 to 2018. In some cases the data available are from 1990 to 2016.

Aquaculture: The aquaculture output is on an increase in most of the IOR countries. The graph below shows the trend of aquaculture production between 1990 and 2016, and beyond (Fig. 7.1, Table 7.5). Indonesia and India were the major producers of aquaculture fish in this region between 1990 and 2016. Indonesia has showed higher growth rate in aquaculture production as compared to other countries. If we look at the trend we find that there is a high growth in aquaculture production since 2005. The following ARIMA (p, d, q) models are being estimated and the forecasts have been made for each selected country as follows.

AU	BD	IN	ID	KE	MU	OM	SC	ZA	LK	TH
1,0,1	2,1,1	2,1,2	2,1,2	1,1,1	1,1,1	2,1,1	1,1,1	2,1,1	1,1,1	2,1,1

Values represent p, d, q, where p = number of autoregressive terms, d = number of differencing needed for stationarity, q= nymber of moving average terms in the equation (see end for detailed description of ARIMA). AU = Australia, BD = Bangladesh, IN = India, ID = Indonesia, KE = Kenya, MU = Mauritius, OM= Oman, SC = Seychelles, ZA = Republic of South Africa, LK = Sri Lanka, TH = Thailand.

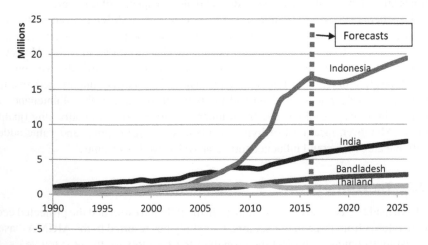

FIGURE 7.1   Aquaculture Production: Forecasts for 2017–26 (Million Tons)

TABLE 7.5
Forecasts of Aquaculture Production for 2016–26

Year	AU	BD	IN	ID	KE	MU	OM	SC	ZA	LK	TH
2016	0.0968	2.2035	5.7030	16.6000	0.0154	0.0010	0.0001	0	0.0080	0.0310	0.9626
2017	0.0933	2.3033	5.9497	16.3691	0.01503	0.0010	0.0001	0.00003	0.0077	0.0343	0.9725
2018	0.1020	2.3919	6.0994	16.0154	0.0156	0.0011	0.0002	0.00006	0.0072	0.0336	0.9863
2019	0.1028	2.4503	6.2901	15.9390	0.0161	0.0010	0.0002	0.00009	0.0071	0.0365	1.0030
2020	0.1056	2.5095	6.4804	16.1886	0.0165	0.0012	0.0002	0.0001	0.0073	0.0362	1.0219
2021	0.1100	2.5489	6.6505	16.6595	0.0171	0.0011	0.0002	0.0001	0.0076	0.0387	1.0424
2022	0.1121	2.5979	6.8450	17.2277	0.1760	0.0012	0.0002	0.0001	0.0078	0.0387	1.0641
2023	0.1157	2.6334	7.0204	17.8090	0.1810	0.0012	0.0002	0.0001	0.0080	0.0410	1.0867
2024	0.1189	2.6834	7.2065	18.3658	0.1861	0.0013	0.0002	0.0002	0.0082	0.0412	1.1100
2025	0.1218	2.7242	7.3893	18.8933	0.1911	0.0013	0.0002	0.0002	0.0083	0.0432	1.1338
2026	0.1252	2.7816	7.5707	19.4004	0.1962	0.0013	0.0002	0.0002	0.0084	0.0437	1.1581

Notes: Figures in million tons (MT), AU: Australia, BD: Bangladesh, IN: India, ID: Indonesia, KE: Kenya, MU: Mauritius, OM: Oman, SC: Seychelles, ZA: South Africa, LK: Sri Lanka, TH: Thailand.

The model specification indicates that the variable pertaining to aquaculture production for most countries are non-stationary in nature. However, it is becoming stationary at the first difference.

The forecasts reveal that in most of the countries, aquaculture fishery is showing a positive growth but in some cases growth is sluggish. The forecasted production shows the same trend as in the past, and the three countries, Indonesia, India, and Bangladesh are going to be the major producers in this region for the next ten years too. Indonesia's aquaculture production would reach close to 20 MT by 2026. While India's aquaculture fish production would be around 7.5 MT, the same of Bangladesh will be 2.78 MT. Thailand would be the fourth-largest producer in this region with 1.15 MT. The aquaculture production, thus, is going to be an important blue economy activity in the coming years (Fig. 7.1, Table 7.5).

Capture Fisheries: The share of capture fisheries in total fish landings, in general, is high as compared to aquaculture. Production of capture fisheries of select IOR countries during 1990–2016, and beyond are shown (Fig. 7.2, Table 7.6). The production structure follows more or less the same pattern as in the case of aquaculture. In the case of capture fishery, the outputs of India, Indonesia, and Thailand were much higher throughout 1990 to 2016. However, in 2015 and 2016 Bangladesh had overtaken Thailand. From 2000 onwards, the production of capture fisheries in Bangladesh showed an upward trend, while in the case of Thailand a downward trend was noticed. The longer coastline of Indonesia and India might have contributed to increased production of capture fisheries in these countries. The gap in production between India and Indonesia is increasing throughout the period.

ARIMA (p,d,q) for capture fishery forecasting is as follows.

AU	BD	IN	ID	KE	MU	OM	SC	ZA	LK	TH
2,1,2	2,1,2	1,2,1	3,2,2	1,1,1	2,0,1	1,2,1	2,0,2	2,1,2	2,1,2	1,2,1

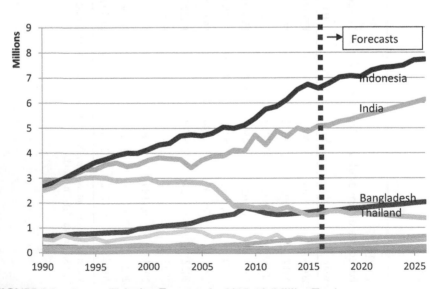

FIGURE 7.2   Capture Fisheries: Forecasts for 2017–18 (Million Tons)

TABLE 7.6
Countrywise Forecasts of Capture Fisheries, 2017–26

Yr.	AU	BD	IN	ID	KE	MU	OM	SC	ZA	LK	TH
2017	0.175	1.680	5.111	6.782	0.172	0.019	0.297	142,635	614,797	536,458	1.673
2018	0.169	1.718	5.263	7.023	0.169	0.019	0.316	155,162	660,886	547,911	1.669
2019	0.162	1.783	5.338	7.074	0.169	0.181	0.335	166,440	658,269	558,285	1.565
2020	0.155	1.800	5.465	7.041	0.167	0.170	0.356	177,293	639,957	570,359	1.595
2021	0.152	1.836	5.562	7.292	0.167	0.157	0.377	188,110	624,874	584,343	1.527
2022	0.149	1.893	5.679	7.401	0.165	0.141	0.399	199,075	618,067	598,378	1.514
2023	0.147	1.919	5.786	7.425	0.165	0.125	422618	210,278	617,368	611,074	1.476
2024	0.144	1.954	5.902	7.491	0.163	0.110	0.447	221,759	619,124	622,843	1.447
2025	0.142	2.005	6.014	7.698	0.163	0.009	0.471	233,539	620,901	634,953	1.416
2026	0.139	2.036	6.131	7.725	0.161	0.008	0.497	245,626	621,842	647,959	1.385

Notes: Figures in million tons, except for SC, ZA, and LK, Country legends same as Table 7.5.

The forecast shows that the production of capture fishery is going to increase in countries like India, Indonesia, and Bangladesh. However, the forecast also shows that it is going to decline in the case of Thailand. For other countries, the production is likely to remain more or less constant in the next ten years.

7.2.2.2 Shipping

Ocean represents a very cost-effective method of transportation for global trade and it is carbon friendly too. About 90% of the world's trade is seaborne. The number of merchant ships engaged in international trade has already crossed 50,000. The seaborne trade has shown an upward trend over the last several years except for a contraction in 2009. In 2009 sea-based cargo handling fell by 4.5% from 8,229 MT to 7,858 MT. The size of seaborne trade in 2013 was approximately 9,600 MT (Mohanty et al., 2015; UNCTAD, 2016) and in 2017 it reached 10,700 MT.

No wonder shipping is a major economic activity in the IOR's blue economy paradigm. India continued to be the major nation in the IOR with regard to seaborne trade up to 2014 (Fig. 7.3, Table 7.7), till Indonesia overtook India to became one of the leading nations involved in seaborne trade. The presence of other countries in seaborne trade is quite insignificant and was below 5,000 tons in 2018, except Oman.

The following ARIMA (p,d,q) models have been estimated to make forecasts for shipping.

AU	BD	IN	ID	KE	MU	OM	SC	ZA	LK	TH
1,1,1	2,2,2	1,2,1	1,2,1	1,0,1	2,1,2	1,1,1	2,1,2	2,2,2	1,1,1	1,1,1

The forecasts put Indonesia in a much better position in 2028. Though the shipping activity of India will increase during the next ten years, the gap between India and Indonesia is going to increase. Another highlight of the forecast is that Bangladesh

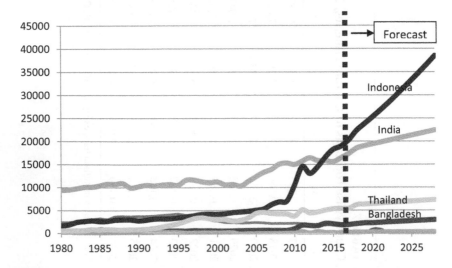

FIGURE 7.3 Shipping: Forecasts for 2017–28 (Dead Weight Tons in Thousands)

TABLE 7.7
Countrywise Forecasts of Shipping Activity, 2019–28

Year	AU	BD	IN	ID	KE	MU	OM	SC	ZA	LK	TH
2019	2,290	2,204	19,017	23,754	8.85	143	13.2	29	421	421	6,321
2020	2,360	2,217	19,424	25,207	848	168	13.5	126	454	408	6,429
2021	2,415	2,310	19,805	26,704	813	184	13.37	85	451	412	6,538
2022	2,460	2,410	20,182	28,247	7.8	182	13.48	109	481	411	6,647
2023	2,496	2,482	20,557	29,834	7.48	162	13.42	100	481	411	6,755
2024	2,526	2,565	20,932	31,467	7.12	138	13.45	106	509	411	6,864
2025	2,551	2,652	21,307	33,145	6.87	121	13.43	104	511	411	6,973
2026	2,572	2,733	21,682	34,868	6.59	121	13.45	105	537	411	7,081
2027	2,590	2,816	22,057	36,636	6.32	138	13.44	105	540	411	7,190
2028	2,607	2,899	22,432	38,449	6.06	162	13.45	105	565	411	7,299

Notes: Figures in Dead Weight tons in 000s, Country legends same as Table 7.5.

is going to make its presence felt among the IOR countries. The shipping activity of Oman is also projected to increase (Fig. 7.3, Table 7.7).

7.2.2.3 Tourism

The tourism sector has a great growth potential in the IOR. The United Nations World Tourism Organisation (UNWTO) in its Tourism Vision 2020 has forecasted that IOR would receive 179 million international tourists in 2020 and this would show an annual growth of 6.3% over the period 1995–2020. Coastal tourism is a major sector in the blue economy, as it has tremendous potential for job creation and economic growth. In 2015, the IOR as a whole (except Somalia) represented 8.5% of the world tourism industry.

Thailand is far ahead of other countries in the IOR with respect to tourist arrivals. India occupies second position, while Indonesia is ranked third in. Other countries that have made their presence felt in tourism are South Africa and Australia (Fig. 7.4, Table 7.8).

The following ARIMA (p,d,q) models have been estimated to make forecasts for the tourism arrivals in IOR.

AU	BD	IN	ID	KE	MU	OM	SC	ZA	LK	TH
1,1,1	1,2,1	1,2,1	2,2,2	1,0,1	1,1,1	1,1,1	1,2,1	2,1,2	1,2,1	1,1,1

Thailand is projected to continue its first position in the case of tourist arrivals in 2027 also. Indonesia is going to show its strength in the tourism sector by pushing India into third position by 2027. If the same trend continues by 2050, Indonesia will catch up with Thailand. Australia and South Africa are also expected to show a reasonable progress in tourist arrivals in the next ten years (Fig. 7.4, Table 7.8).

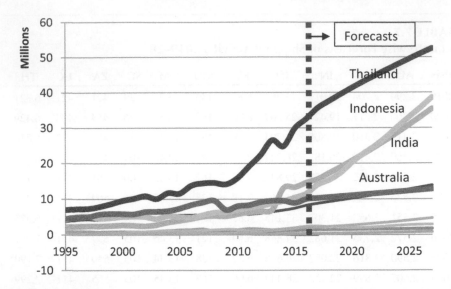

FIGURE 7.4   Tourist Arrivals: Forecasts for 2017–27

7.2.2.4   Tourist Receipts

What matters most is the tourist receipts rather than total number of tourist arrivals. In the IOR, though some countries are getting larger number of tourists their receipts from tourism are not commensurate with the number of tourist arrivals.

In the case of tourist receipts also Thailand is the leading country at present. However, Australia came up to second position and pushed India down to third spot and Indonesia to fourth place. As compared to number of tourists, the gap between the countries has come down with regard to tourist receipts. Until recently, Australia was the leading country in the region receiving maximum revenue from the tourist arrivals (Fig. 7.5, Table 7.9).

The forecasts show more or less the same trend with Thailand's dominance. One of the notable findings is that India is going to overtake Australia in terms of revenue from tourism. All selected countries show an increase in tourist arrivals in the next ten years (Fig. 7.5, Table 7.9).

What emerges from the above section is that the blue economy of the IOR has great potential to grow in particular sectors like fishing, transportation, and tourism that are booming in this region and as per the forecasts these segments would maintain the same importance in future. However, the benefits of the blue economy are not evenly distributed in the whole region. What we see is that in most sectors only three or four countries dominate.

7.3   SYNTHESIS

The activities under all the three sectors discussed here, namely, fisheries, shipping, and tourism are going to flourish in coming years. However, these projections are based on the assumptions that whatever conditions prevailed in the past will continue to prevail in the future also. The most likely factors that could offset projections

TABLE 7.8
Forecasts of Tourist Arrivals, 2018–27

Year	AU	IN	ID	KE	MU	OM	SC	ZA	LK	TH
2018	9.2887	17.2451	15.3123	1.3864	1.3793	2.2206	383,276	10.538	2.257	37.553
2019	9.7481	18.9011	17.2659	1.4054	1.4240	2.0519	420,264	10.792	2.444	39.431
2020	10.2131	20.6667	20.1210	1.4268	1.4652	1.9578	458,640	11.045	2.661	41.237
2021	10.6906	22.5108	22.6205	1.4465	1.5081	1.9847	498,713	11.298	2.899	42.982
2022	11.1828	24.4396	24.7284	1.4674	1.5502	2.1255	540,442	11.551	3.153	44.673
2023	11.6906	26.4517	27.2544	1,4875	1.5926	2.3328	583,833	11.804	3.423	46.317
2024	12.2142	28.5476	30.1540	1.5081	1.6349	2.5447	628,885	12.057	3.705	47.923
2025	12.7538	30.7270	32.9110	1.5284	1.6773	2.7100	675,597	12.310	3.999	49.494
2026	13.3093	32.9901	35.5974	1.5489	1.7196	2.8049	723,971	12.563	4.305	51.035
2027	13.8808	35.3368	38.5384	1.5693	1.7620	2.8360	774,005	12.816	4.623	52.551

Notes: Figures in million, except in SC, Country legends same as Table 7.5.

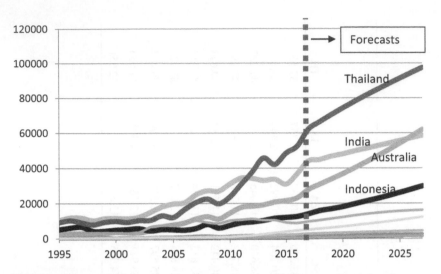

FIGURE 7.5   Receipts from the Tourism Sector: Forecasts for 2018–27 (US$ million)

are maritime security, the geopolitical scenario, economic stability of the country, and climate change, amongst others. The fish resources are also not going to remain as they are today, since overfishing in previous years could affect the future catch. Similarly, the coastal tourism that has flourished in the last few years may not necessarily be consistent due to various environmental reasons.

The top three IOR countries in order, and in terms of existing and projected performance until 2027–8 (Figs. 7.1 to 7.5, Tables 7.5 to 7.9) are: (a) Indonesia, India, and Bangladesh in fish production using aquaculture; (b) Indonesia, India, and Myanmar in the captured fish sector; (c) Singapore, Indonesia, and India in shipping trade volume; (d) Thailand, Malaysia, and Saudi Arabia in tourists arrival; and (e) Thailand, Australia, and India in receipts from tourism.

Fisheries is probably the oldest traditional activity in the world's oceans. Of these, captured fisheries are the oldest and requires limited sophistication in terms of using modern software and instrumentation. Bangladesh replacing Myanmar at the third spot in aquaculture sector could probably indicate the economic, scientific, and technological strength of the former over the latter. In the shipping sector Singapore could take the lead due to its large fleet of bulk carriers, sophisticated port terminals (including modern facilities of warehousing, shipbuilding and ship repairing), and customer-friendly cargo ecosystem. Tourism is considered as the most urbane and cultured area of operation where Thailand, Malaysia, and Australia have excelled, closely followed by India. Saudi Arabia coming third in tourist footfall is largely due to the annual Mecca-Medina Haz pilgrimage by millions of people of the Muslim (Islam) religion, and does not exactly represent the real picture. Interestingly, India is all set to replace Australia for the second spot with more high-end tourists visiting its ancient civilization in the future.

Due to ethical considerations, the size of the ethical blue economy will be still less than both from the blue and ocean economy. While the blue economy concept

TABLE 7.9
Forecasts of Revenue from Tourism, 2018–27

Year	AU	BD	IN	ID	KE	MU	OM	SC	ZA	LK	TH
2018	44,739	465	30,750	15,897	1526	2,065	2,979	683	5,625	10,437	66,323
2019	46,680	580	33,633	16,895	1458	2,128	3,161	779	6,209	11,285	70,298
2020	47,906	691	36,653	18,056	1405	2,192	3,336	886	6,834	12,180	74,106
2021	49,564	800	39,818	19,645	1364	2,256	3,506	997	7,494	13,059	77,766
2022	50,960	908	43,129	21,207	1331	2,322	3,672	1,115	8,189	13,870	81,295
2023	52,515	1,016	46,586	22,712	1305	2,387	3,833	1,239	8,917	14,574	84,709
2024	53,974	1,126	50,188	24,359	1284	2,453	3,990	1,370	9,678	15,151	88,021
2025	55,491	1,236	53,937	26,126	1268	2,519	4,143	1,506	10,470	15,596	91,242
2026	56,973	1,349	57,831	27,932	1256	2,584	4,294	1,648	11,295	15,920	94,385
2027	58,476	1,465	61,871	29,800	1246	2,650	4,442	1,797	12,150	16,147	97,457

Notes: Figures in US$ million, Country legends same as Table 7.5.

is based on sustainable development through legislations and government control (coercion), the ethical blue economy aims at the sustainable development through self-motivation and encouragement. One may think that the ethical blue economy is wishful thinking, but it can work wonders. When citizens of each country become aware of the significance of oceans in their day to day life and act accordingly to save the oceans from all kinds of threats (see Chapter 5), the goal of the ethical blue economy could be reached. Given the unique nature of the IOR where only three to four countries dominate, the need for an inclusive and ethical blue economy remains a necessity.

7.3.1 ARIMA Methodology

For the present study, the following equation was used: $Yt = \phi + \alpha 1 Yt{-}1 + \beta 0 ut + \beta 1 ut{-}1$, where $Yt{-}1$ is a first order autoregressive process and $ut{-}1$ is a first order moving average process. Autoregressive process is given as AR (P) and the moving average process is given by MA (q). In general, in an ARMA (p, q) process, there will be p autoregressive and q moving average terms. ARMA (1, 1) implies one autoregressive term and one moving average term. The equation given above is an ARMA (1,1). Hence, ARMA (1,1) implies one lag of Y variable $(Yt{-}1)$ and one lag of ut $(ut{-}1)$. Economic time series in general are non-stationary. But ARMA needs to be made stationary before we go for any kind of econometric analysis. If a time series is integrated of order $I(1)$, its first difference is $I(0)$, that is, stationary. Similarly, if a time series is $I(2)$, its second difference is $I(0)$. In general, if a time series is $I(d)$, after differencing it d times we obtain an $I(0)$ series. If we have to differentiate a time series d times to make it stationary and then apply the ARMA(p, q) model to it, we say that the original time series is ARIMA(p, d, q), that is, it is an autoregressive integrated moving average time series.

8 Ethical Blue Economy

We have seen in earlier chapters that understanding the complex dynamics of the ocean vis-à-vis the blue economy has been hugely challenging. The change in climate and its implications on every aspect of human endeavor appear to have resulted from the malfunctioning of the prevailing economy. Hence, there is a need to redefine the way success and profit are perceived today. For instance, richness does not necessarily mean earning more, or spending more or saving more, rather richness is when you probably need "no more."

Considering various resources (food, minerals, and energy), services (shipping, transport, tourism, and leisure), threats (climate change, geopolitics, and human mindset), and processes (carbon sequestration, bio-invasion, monsoon and natural disasters) that the ocean is famous for, it becomes necessary that a more sustainable and responsible mechanism governs the area of three-quarters of this planet. Such a mechanism calls upon ethics to be a part of the governance. Consequently, this chapter designs an ethical paradigm for the blue economy involving rational reforms in various sectors of governance. Modeling a possible transformation of the IOR into a hub of excellence in human values, ethics and sustainable economy following the principled doctrine are also examined and road-mapped. Although the concept of such governance could be country-specific, the approach may be global in scope and holistic in theme.

8.1 BTVG DOCTRINE

The concept of ethical blue economy (EBE) that this chapter is proposing is based essentially on the Buddha-Tagore-Vivekananda-Gandhi (BTVG) doctrine. Nature, forest, and humanity have been very special to all these philosophers of the modern world. It was under a tree that Lord Gautama Buddha, a prince born in Lumbini (Nepal), achieved enlightenment and attained nirvana in Bihar (India). The Noble Laureate author and poet Rabindranath Tagore embarked upon an open-sky under-tree education in his university in Santiniketan (India). Spiritual leader Swami Vivekananda, although he was an urban person, found peace in Belur Math near Calcutta, a seminary of his mentor Paramhansa Ramakrishna. Philosopher Mohandas Karamchand Gandhi spent a large part of his life in Sabarmati Ashram (Gujarat, India) amidst nature and miles away from the capitalist economy. The divine teachings of Lord Buddha, poet-philosopher Tagore, spiritual-leader Vivekananda, and ethical-activist Gandhi to cultivate a society devoid of greed, manipulations, deceptions, and shortcuts could form the foundation of an alternate ethical governance.

The EBE is suggested as an alternative to the seemingly meaningless madness of the current social and individual mindset. This concept works on "shared prosperity." The way a regulator controls the speed of any chemical (fission) reaction, ethical governance= could balance and normalize the direction, speed, and fulfillment of human desire and growth in GDP. Social dynamics of the IOR seems to suggest that rational aspirations and ethical economics could take the best care of the society by reining in aspirations to a responsible level. The present mindset of pointless consumption, irrational borrowing, and trusting that a high GDP can buy peace, happiness, and harmony probably needs a revisit. In this regard, the ethical governance model inspired by the BTVG could be an alternative administrative action plan for a successful and effective blue economy (Fig. 8.1; Mukhopadhyay et al., 2018b; Mukhopadhyay, 2019).

Advocating that man's growth must come from within, from the dictates of higher consciousness through yoga, the art of cosmic union, Gandhi, while speaking on austerity once, articulated that this "Earth has everything in abundance for the need of all persons, but not for their greed." Today, Gandhi's insight is being put to the test as never before. The BTVG doctrine can be used to work out an alternate responsible path of governance to usher in a sense of responsibility. The doctrine suggests ethical economy as a new system of governance, alternate to four other major market

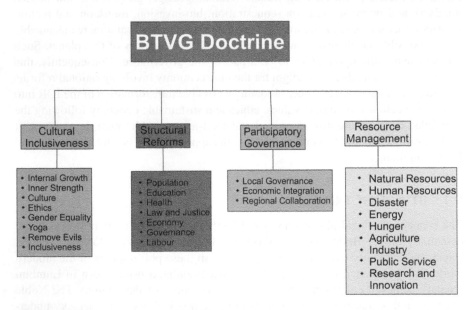

FIGURE 8.1 The Component of BTVG Doctrine

Note: BTVG= Lord Buddha – Poet Tagore – Spiritual Vivekananda – Philosopher Gandhi. The BTVG doctrine proposes for a Tangible Growth Index (TGI) to auto-correct GDP estimation and set right an responsive governance mechanism for the IOR countries under a liberal ethical economy (LEE) regime (Modified after Mukhopadhyay et al. 2018, Mukhopadhyay 2019)

economy theories: Capitalist Economy, Socialist Economy, Circular Economy, and Religious Economy. While the capitalist (free-market) economy lay stress on profit and requires perpetual growth, the socialist economy that is much more regimented and regulated largely by the government includes a series of do's and don'ts prompting large-scale corruption. The circular economy combines reduce, reuse, and recycle activities for economic prosperity through a systemic shift, but might get into trouble with low market demand against a consistent urge to earn more. The religious economy divides the world into narrow religious bigots/fraternities. The ethical economy on the other hand provides a value-based spiritually inspired balanced standard of living in complete harmony with the biosphere, hydrosphere, and atmosphere (Mukhopadhyay et al., 2018b; Mukhopadhyay, 2019).

Tagore found nature essentially related to human beings (Tagore, 1913). In the vastness of nature, humans are not unknown strangers; they are her kith and kin, her wards. He argued that the divine is not isolated from the world. There cannot be anything that cannot be subsumed by the human personality, which proves that the truth of the universe is human truth. The BTVG doctrine functions on this simple philosophy.

In implementing the blue economy paradigm, if public administration is considered as the hardware, then mindset becomes the software. To make public administration successful, visible, and relevant, the mindset of the individual and society needs moderation. The BTVG doctrine laments that what humanity is facing today is essentially the ecological consequences of its collective karma (their act) over the last few centuries (Fig. 8.1).

The doctrine would involve a need for re-engineering of the existing public administrative system through reforms. Such a revised and reformed public system would be ethics based and may have four major principles: cultural inclusiveness, structural alterations, participatory governance, and resource management. Dealing with a blue economy paradigm requires changing mindsets and attitudes, and developing a climate-literate citizenry, particularly youth, equipped with knowledge and skills, for building a culture of resilience and preparedness (Gijzen, 2013).

While integrating philosophy into the study of good governance, Satpathy et al. (2013) found that UNESCAP's (United Nations Economic and Social Commission for Asia and Pacific) norms of good governance are closely related to the Bhagavad-Gita's (an ancient Indian religious text) intrinsic perspective of governance. Such an approach starts with self-governance, then is elevated to corporate governance, and finally raised to global governance.

8.2 ETHICAL BLUE ECONOMY

In this section we put forward an ethical blue economy (EBE) model, a revised version of the blue economy approach proposed earlier by Pauli (2010). Considering the three major threats to the blue economy, such as climate change, geopolitics and militarization, and human mindset and unending aspirations of the average human being (see Chapter 5), the EBE model (formulated following the BTVG doctrine), campaigns for an overhaul of the existing system. The re-engineering of the model in turn calls for distinct alteration in the way IOR people presently perceive a series of aspects, for instance, population control, hunger management, economic growth, industrial progress, agricultural improvement, resource advancement, administrative apparatus, cultural inclusiveness, gender equality, and ethical values, etc. Later, a possible roadmap to achieve such an EBE is discussed. Hence, the EBE would involve large-scale reforms in thoughts, activities, and mechanisms of the IOR people and by their governments.

8.2.1 System Re-Engineering

As mentioned earlier, the system re-engineering involves large-scale restructuring in the areas of social system, community thinking, and public administration. The reforms could have as many as 16 components bolstered essentially on four philosophies (Fig. 8.1, Table 8.1).

The first philosophy of reforms in governance is the *cultural inclusiveness*, which insists that one has to change him/herself first from within before thinking out. This situation can be equated with the soul of a human being, and so also to the natural environment. Culture emerges from the interdependence of human and nature (both plants and animals). Accordingly, the IOR nations need to be made internally strong and disciplined. This is possible only when a majority people of IOR nations can have a mind ready for accepting new innovations emerging through laboratory-proven

TABLE 8.1
Components of Strategic Reforms in Governance of IOR Countries

Sl.	Philosophies	Components
01	CULTURAL INCLUSIVENESS	Ensure internal growth and stability, and enhance inner strength through Culture-Ethics-Gender equality-inclusiveness-Yoga combination
02	STRUCTURAL REFORMS	Bring in Structural Reforms for effective implementation with respect to – (a) Population and demography, (b) Education, (c) Health, (d) Law & Justice, (e) Economy, and (f) Democratic governance
03	PARTICIPATORY GOVERNANCE	Encourage Participatory governance to ensure sustainability-- (a) Local governance, (b) Economic integration, and (c) Regional collaboration
04	RESOURCE MANAGEMENT	Efficiently manage the Natural Resources, Human Resources, Disaster, Hunger, Agriculture, Energy, Industry, Public Services, including Communication (road/river/rail/air/internet/telephone) through Innovation and Research The resources would include all natural resources found in air, on land and at sea/river. The land resources would include forest & wetland cover, agricultural land, mines, while the resources from sea would include fisheries, coral, seagrass, monsoon, and mangroves. The human resource would include education, health, skill development, jobs and employment.

Source: Modified from Mukhopadhyay et al. (2018b).

scientific research, and the time-tested teaching that ensures mutual respect, gender equality (inclusiveness), ethical behavior, and social goodness.

The second philosophy of *structural transformations* involves undertaking necessary reforms by all IOR countries with respect to population growth, education, health, law and justice, economy, and governance. Such alterations will help these nations to efficiently manage their natural and human resources. Additionally, such reforms would help better manage energy need, hunger, disasters, and all public services, including communication (road/river/rail/air/internet/telephone).

The third philosophy of reforms would advocate a *participatory governance*. People's participation at the individual, local, national, and global levels would increase trust and confidence of people and stakeholders in the government, and may bring regional cooperation and economic integration among the IOR countries. In fact, both reforms and management of resources would need an outstanding level of cooperation and enthusiastic participation of as many countries as possible in the IOR. The local government should be empowered to carry forward the execution of such agreements. Even a minor mechanism of cooperation in trade, commerce, and economy among the IOR countries would be immensely beneficial in the long run. Such a measure would recover much-lost trust, and usher cooperation among the 50 IOR nations (Tables 8.1, Fig. 8.1). As stated in Chapter 5, the mutual trust and cooperation between India and Bangladesh, among the ASEAN countries, and among the IORA are prime examples of fruitful effective regional collaborations.

Emboldened by the actions and teachings of these philosophies, the IOR countries could now delve into managing their resources. The challenges under this principle are to efficiently manage the natural resources (lands, oceans, fisheries, corals, seagrasses, mangroves, rivers, mines, forests amongst others), human resources (education, health, jobs, employment, skills), climatic disasters (earthquake, pollution, cyclone, flood, fire, drought, hunger and so forth) and day-to-day services (energy production and distribution, and public services, including communications road/river/rail/air/internet/telephone). Although the next four sub-sections (8.2.2 to 8.2.5) are technically and fundamentally a part of system re-engineering, we have accorded separate headings to reflect their importance.

8.2.2   CULTURAL INCLUSIVENESS

A strong dose of principled culture ensures growth, stability, and strengthening of inner character in a human being or in an organization, or in a country, as the case may be. The analogy of Dr. Seltarman's cruel ethical choice, in which the greed of 10% students denied grace marks to all students in the class (see Chapter 5) appears relevant in help remove greed (Table 8.1, Fig. 8.1). Such efforts should be led by ethics, inclusiveness, gender neutrality, and yoga.

8.2.2.1   Ethics

Nature and culture, and science and ethics are inextricably linked (Northcott, 2007). It is increasingly being felt that the blue economy cannot be dealt with appropriately if its ethical dimensions are not highlighted. Culture is often described as a comprehensive knowledge that fosters human interactions in a certain situation, such as knowledge of language, habits, rituals, opinions, values, and norms (Hofstede and McCrae, 2004). It is further concluded that cultures are not static and do not exist in isolation, rather they evolve with time in response to changes in surroundings, wars, and new inventions (Tagore, 1913; Strauss, 2012).

Many a time artificial needs are deliberately created to make an economy appear strong and prosperous, ignoring the future implications. Wars are sometime "fashioned" to increase market demand (through sales of arms and ammunitions, food, clothes, etc.) and to strengthen the economy of some particular countries. Demand surges productivity, but in the process, it also augments the rich-poor gap, increases anomaly in resource distribution, and causes destruction of ecology. Such progress negatively impacts ecology and environment. Many of the IOR cities are now bloating with materialistic demand at the cost of humane development. The culture of a human being or an organization under the EBE must make sure to craft an ecosystem where "greed" is replaced by the legitimate aspirations of the people.

The blue economy is a challenge, as well as an opportunity for the entire IOR populace. Therefore, enhancing the resilience and capacity of the society to respond positively to any threat to the blue economy will be an essential feature of ethical responses to such changes. The ethical objectives do not necessarily need to be pursued by policies and regulations, but by voluntary commitment based on certain principles (faith/belief) unilaterally, without waiting for any instruction or incentive.

8.2.2.2 Gender Neutrality

The IOR society is immensely gender-biased and violently masculinized. This is reflected by numerous types of crimes happening against women, such as dowry deaths in India, honor killings in Pakistan, acid-throwing incidents in Bangladesh, female genital mutilations in some African countries, and religious skirmishes in Afghanistan and Sri Lanka (Visweswaran, 2004). Therefore, women need to be treated carefully and with much empathy and respect to neutralize patriarchal or hierarchical structures prevailing in the IOR.

The saying goes that "Men pollute, and Women suffer." In short, women often face societal, financial, and political barriers, and it is vital to find gender-sensitive strategies to respond to the environmental and humanitarian crises arising out of any developmental campaign (Doherty et al., 2014). As the livelihoods of women are more dependent on natural resources, women will be more benefitted by human rights issues and gender equality in an EBE paradigm.

The distribution of money and power since time immemorial always puts men on a higher pedestal, in terms of possessing greater wealth, freedom of movement, and subsequently in emitting greenhouse gases, and polluting the surroundings (Sousounis, 2016). For example, in many IOR countries, women walk for long distances to fetch water and firewood owing to the negative intervention of climate change (drought/lack of rainfall/drying of the nearby river), wherein a sensible dispensation of resources under the EBE could come as a tool to ease such burdens.

The EBE could also act as a secular and inclusive instrument, as it could motivate people of various nations, sects, religions, and economies to work together. However, such networking will remain toothless, unless the world starts treating women as an equal partner. Unlike capitalism, which is alleged to have been founded by the subordination of women and nature, the blue economy could, in fact, act as an anti-racial and de-patriarchal issue. In fact, every activity of women (choice of voting, choice in purchasing goods, choice of sex, choice to conceive, choice of work) can constitute a political statement. The endless economic and technological growth will remain environmentally unsustainable and socially imperfect until women have healthy living conditions (Awadalla et al., 2015). Hence the success of EBE depends to a great extent how the world (and the IOR region, in particular) starts treating its women.

It is reported that countries that resist implementation of blue economy measures have a predominating patriarchal structure. The entire debate might get derailed with a vicious lobby conveniently coupling the blue economy with nationalism and religion, knowing well that the blue economy cannot be done at a national level. Instead, the success of the blue economy requires gender equality, after bidding goodbye to consumer culture, masculinity, distorted religious rituals, and the narrow definition of nationalism. Hence, a women-centric pattern for socioeconomic development is therefore necessary (Sikri, 2015) for EBE to succeed.

8.2.2.3 Yoga

The biggest problem the world is facing today, after religious terrorism and climate change, has been human behavior (i.e., unreasonable aspirations, greed, manipulations, corruptions, and unethical lifestyle). In fact, the solutions to a majority of these problems could be found in Yoga.

One of the fundamental requisites of ethical behavior is self-realization through self-inquiry. For this, the mind needs to be refined and rendered pure to be able to get connected with the supreme self. However, until the person is ready for a direct path, one can take indirect methods, such as paths of bhakti (devotion), pranayama (breath regulation), and karma (action). Yoga brings unison of bhakti, karma, and Gyana (Taneja, 2019). This self-inquiry, inquisitiveness, and passion are even the pillars of good scientific capacity-building exercises, which in many of the IOR countries are missing. School education systems need a major revamping. For example, the IOR would better be served by removing focus from exams to asking questions, from memorizing to creating a congenial environment where curiosity, appetite, and hunger for knowledge rule. IOR countries must realize that no society has grown with a consistent rate without continuous and systematic investment in science and technology.

Yoga integrates various forcing parameters that influence the mind, body, and soul of a human being. It gives a complete and comprehensive knowledge, without compartmentalizing various aspects (Matilal, 1986; White, 2012). Following this philosophy, a person could bring about a change in his/her priority, style of living, and way of looking at their surroundings.

The lessons of Yoga are (a) be the change yourself to recognize and embody your unity with the environment, and through this recognition, change your personal practices, and (b) relate individual smallness to the vastness of nature and the natural environment, as human beings are not separated from nature. Yoga offers people a sense of "harmony with self, society, and nature" and could create a "social consciousness." Through Yoga, the urge to gather unnecessary resources at the expense of others' legitimate share will vanish. Through Yoga, the need will dominate the greed, and if that can be achieved this world will be a different place to live in.

Yoga balances one's existence, relates one to nature and offers one the ability to appreciate all the dynamic forces in this world. The teachings of Yoga could make a person strong from the inside so much so that any undeserving aspiration for material benefit will not even emerge in her/his mind. A higher level of Yoga could help remove bad practices and ill-manners from human beings. If people could even eliminate a few trivial evil practices, the IOR society and surroundings will change for the better. For example, violating traffic rules, spitting in public, grabbing public places for personal use, telling lies, manipulating, and encouraging a shortcut attitude to get any work done may be stopped quickly. Additionally, discouraging mobocracy, encouraging plurality, opposing social intolerance, being punctual, ensuring hygiene and cleanliness at home, at the workplace, and on the streets, respecting dignity of labor, and abhorring corruption in any form, could go a long way to make the IOR society ethically disciplined.

The above list could be enlarged and modalities finalized through debate, and discussion among various stakeholders. All these issues may be taught to toddlers under "moral study" at the kindergarten and primary sections. It would also be interesting to trace the moral, philosophical, and psychological reasons for the continuation of such vices for so long in the IOR region.

8.2.3 STRUCTURAL TRANSFORMATIONS

After proposing ethical correctness required for EBE in the earlier section, the second philosophy of the "system re-engineering" involves structural reforming of many of the ongoing policies, activities, and programs in the IOR countries. The most significant areas for the IOR countries where reforms are overdue and must be taken up quickly are population and demography, education, health, law and justice, economy, and democratic governance. All these are briefly discussed below.

8.2.3.1 Population and Demography

Although the growth rate of the world's population decreased marginally from 1.49% in 1950 to 1.0% in 2019 (and is expected to fall further to 0.75% by 2030 and 0.5% in 2050; USGHGIR, 2011), the overall global population is increasing at a rate faster than the carrying capacity of the Earth. The growth is however not a universal challenge, as it is specific to the developing world including the IOR, and particularly to the least developed countries. The developed world (shown in blue in Fig. 8.2) has witnessed a marginal increase in population between the years 1950 and 2015, whereas for the developing world (in red) the population has escalated from ~1 billion to ~6 billion during the same period (see also Table 8.2).

Population growth is a product of a vicious circle of poverty leading to malnutrition > to diseases > to illiteracy > to gender discrimination > to women's disempowerment, > and back to population growth. The world's population is expected

Global Population: 1950-2015

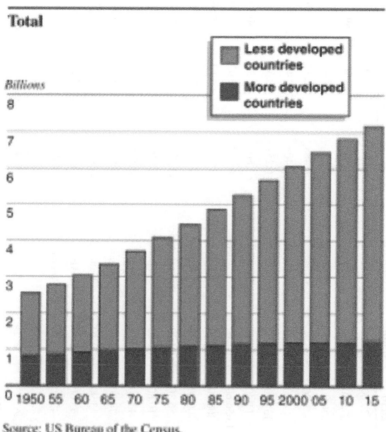

Total

Source: US Bureau of the Census.

Growth Rates

FIGURE 8.2 Growth of Population in the World and IOR WWI-World Bank 2016. Note: More the population, less has been the growth propelling poverty. Also note that while there is only marginal increase in population in the developed world (<1 billion in 1950 to about 1.2 billion in 2015, the same in developing world increased from about 1.8 billion in 1950 to about 6 billion in 2015

to reach close to 9 billion by the mid-twenty-first century from 7 billion at present, against a sustainable carrying capacity of only 3–4 billion by this Earth (WWI, 2013).

Drawing an example of population explosion from India will not be out of context here, where virtue of small families seems to have been long forgotten. The recent census data of the nationwide Family Health Survey of the UN Population Division

TABLE 8.2
Trend of Population in IOR Countries

Countries	1950	2000	2050	Youth	Countries	1950	2000	2050	Youth
Australia	8.30	19.10	29.00	67.70	Mauritius	0.48	1.20	1.40	70.90
Bahrain	0.114	0.655	1.80	77.30	Mozambique	6.30	18.00	59.00	51.40
Bangladesh	45.60	132.20	250.20	61.60	Myanmar	19.50	47.40	70.70	68.70
Comoros	0.148	0.545	1.20	----	Oman	0.488	2.40	5.40	66.10
Djibouti	0.079	0.669	1.40	62.30	Pakistan	40.40	152.40	290.80	61.00
Egypt	21.20	65.20	137.90	62.80	Qatar	0.251	0.64	2.60	86.60
Eritrea	1.40	4.20	11.40	54.80	Saudi Arabia	3.80	21.30	40.30	68.20
India	369.90	1,006.30	1,656.60	65.10	Seychelles	0.033	0.079	0.10	71.50
Indonesia	83.00	214.10	300.20	66.30	Singapore	1.02	4.10	8.60	78.40
Iran	16.40	68.60	100.00	71.00	Somalia	2.40	7.50	22.60	53.00
Iraq	5.20	22.70	56.30	59.20	South Africa	13.60	45.10	49.40	65.80
Israel	1.30	6.10	10.80	62.20	Sri Lanka	7.50	19.10	25.20	67.20
Jordon	0.56	4.70	11.20	60.10	Sudan	6.50	27.10	59.11	54.80
Kenya	6.10	30.60	70.80	54.70	Tanzania	7.90	33.20	118.60	52.00
Kuwait	0.14	2.00	3.90	72.30	Thailand	20.00	62.90	66.10	71.00
Madagascar	4.60	15.70	45.80	55.60	UAE	0.072	3.20	8.00	78.60
Malaysia	6.40	23.20	42.90	65.50	Yemen	4.80	17.20	46.10	54.90
	1	2	3			1	2	3	
Total IOR	2,538.41	8,796,145	3,465		World	6,973.74	70,000,000	10,038	
% of World	36.40	12.57	34.00						

Notes: Population in millions in three years (1950, 2000, 2050). Youth (%) in age bracket of 15–64 years, 1 = Population in million, 2 = GDP in million, 2 = GDP in million US$, 3 = Per-capita GDP in US$. Data from US Census Bureau, www.census.gov/population/international/.

suggest that if the present uncontrolled growth of population continues, India will overtake China as the most populous nation by 2026. Globally, a fertility rate of 2.1 is considered as replacement rate, i.e., an average 2.1 babies per woman will keep the present numbers unchanged. India's fertility rate has been 2.2, compared to China's 1.7, the world's 2.4, and Africa's 4.3. By 2060, India is expected to reach the population peak with 1.65 billion (and Africa with 3 billion), but both may start dipping after that. China however will cross her peak in 2030 with 1.4 billion. Between 2006 and 2016, while the fertility rate among Muslims and Buddhists in India saw a fast dip by 22.9% and 22.7%, respectively, the same rate of fall has been unfortunately low for Christians (15%) and Hindus (17%).

The population growth in the IOR from the 2000 as base year is projected (Table 8.2) and can be divided into three clusters. In Cluster 1 the population of Qatar is increasing by 4.06 times, Tanzania by 3.57 times, Mozambique by 3.28 times, and Somalia by 3.01 times. Increase in population will be somewhat high in Cluster 2 that includes Madagascar (growing by 2.92 times), Yemen (2.69), UAE (2.50), Iraq (2.48), Oman (2.25), Sudan (2.18), Egypt (2.11), and Singapore (2.10). The rest of the IOR countries form Cluster 3 that will show a low growth in population. In this category, Pakistan would record a growth of 1.91 times by 2050, followed by Bangladesh and Saudi Arabia (1.89 a-piece), Malaysia (1.85), Israel (1.77), India (1.64), Myanmar (1.49), Iran (1.45), Indonesia (1.40), Sri Lanka (1.32), South Africa (1.09), and Thailand, wherein population is growing at the least speed (increasing by 1.05 times only).

With regard to the availability of youth for employment (Table 8.2), in Cluster 1 (growth above 3 times), about half of the population will fall in the 'youth' bracket and will be potentially available for jobs. For instance, except Qatar (>86%), the percentage of youth available for employment for all the fast populating nations would range between just around 51% and 53%. In Cluster 2 (growth between 2 and 3 times), again except for UAE and Singapore, the youth available for job would hover around between 55% and 66%. However, a little more than 78% of the population would be potentially asking for jobs in UAE and Singapore. The reason for these two countries behaving anomalously could be attributed to the fact that these two countries are largely active in providing service (in fact both these two countries are expected to register a moderate growth in population). In Cluster 3 (growth below 2 times), except an aging Malaysia with 42.9%, all other countries would have to face a challenge to arrange jobs for 61–71% of their population who will be potentially available for employment.

Hence, the IOR nations must take immediate step(s) to reduce the rate of growth of population (see the Structural Reforms section for suggestions) and bring the growth down to only one time (i.e., replacing the existing population through death and birth). Similarly, the IOR is expected to provide employment to as many as 50–70% of its population as job aspirants will pose a huge challenge to the economy of the region. The evolving situation would certainly insist for a radical change in the existing economic philosophy being pursued by the IOR countries.

Hence, the world, especially the IOR countries, must design and quickly carry out a well-conceived, clearly articulated, flexible, fair, and internationally coordinated population control program to bring about at least 20% reduction in the population of 2015 level by 2050. Achieving such an aspiring target would demand a major orientation of human thoughts, values, expectations, lifestyles, and willpower of the governments and by individuals.

Among the steps that could be rationalized in the IOR countries in this regard could include (a) stopping early marriages through strict legislation, (b) encouraging parents to adopt the second (and the last) child from an orphanage, (c) institution-alizing such norms by offering cash allowance/incentive in health and education for the entire family, (d) making both contraception and abortion legal and cheap, and (e) facilitating social groups and NGOs to help cut the vicious circle of pov-erty ↔ population growth through awareness and education.

8.2.3.2 Education

The overall level of education in the IOR region has not been very exciting. Moreover, the culture prevailing in schools, colleges, and universities of this region largely discourages students from asking questions to their teachers. This has single-handedly done the greatest harm in developing a culture of rational thoughts and healthy debate. More than innovation and curiosity, memorizing the syllabus is often given priority. Professional degrees are valued more, presently, than acquiring wisdom. As a result, the educational institutions in the IOR regions largely produce a host of trained and skilled workers, but very few innovators and thinkers.

The very purpose of education should be made clear in the beginning. It is not the degree but the knowledge that matters. Hence, a thorough overhauling of curriculum of pre-primary, primary, secondary, and tertiary levels is needed on a war footing in IOR countries. The courses need to be purposeful and practical that would encourage independent thinking. Students should be able to express their thoughts in their own way and language. For example, mathematics is an outdoor subject and should be taught in the open through geometrical interpretation. A near-uniformity in courses under various educational boards, including those run by religious institutions, is desirable. The syllabus taught in the religious schools also may be monitored by the government authorities at regular intervals. Innovation, inquisitiveness, passion, fearless transparent school and college education, with special emphasis on skill development, may perhaps make the mainstay of the reformed IOR education under the blue economy. Each country may be advised to spend not less than 5–7% of its GDP for education and research.

8.2.3.3 Health

Health is an important component of the prescribed EBE governance model. Many of the cities and towns in the IOR region have a polluted atmosphere. The higher organic carbon concentration in air seen over this region indicates the significant contribution of biomass burning (Ramanathan et al., 2007; Babu et al., 2011). Increase in such contribution would cause climate-sensitive diseases. Setting up databases and early warning system on climate-sensitive vector-borne and water-borne diseases may be required to track their geographic distribution. The government must prioritize research and education on climate-related diseases, and train health professionals and educate communities. Since a majority of the IOR population can be bracketed under the "very poor to poor" category, instead of tertiary health investment (super-specialty and expensive) these countries must place emphasis on primary health care (at the village/town level) in terms of budget and facilities. These primary healthcare units must be specialized enough, accessible, and affordable even for the poorest of the poor. A good primary health facility would reduce pressure on secondary and tertiary health care. Each IOR country may be encouraged to spend not less than 8–10% of her GDP on the health sector.

8.2.3.4 Judiciary

Besides education and health, maintaining law and order, and justice delivery system are fundamental priority issues for any country. Unfortunately, in all these areas the overall performance of IOR countries has been below average. Some of the laws, although they have lost their relevance, are still continuing in several countries. A few others were formed by colonial powers to destroy any freedom of movement, however, those have still lasted for several decades after independence. The justice delivery system in IOR countries, in general, is painfully slow and horrendously cumbersome.

The ratio of judge/lawyer/public in most of the IOR countries has been abysmal, and needs a quick improvement. To make this situation further worse, all cases are constitutionally eligible to be heard by the Supreme Courts or High Courts, which is impossible, and has been one of the main reasons for the accumulation of cases and delay in the delivery justice system. Some of the following suggestions for reforms could be worked out for a speedy and quality judgment.

Police departments in most of the IOR countries are awfully undermanned and inadequately trained to carry out both the jobs of law enforcement and investigation of crimes. It is suggested that a department of scientifically trained investigating professionals (armed with modern technological knowledge of cyber-security and armaments) may be formed to do the investigation of all crimes in a professional time-bound manner. By this, a substantial employment opportunity for people with very special investigative skill could be created.

The reform in the judiciary may include clear demarcation of type of cases a court would entertain. For example, the Supreme Court of a country could address anomalies related only to constitutional provisions, public interest litigations, disputes between Federal government and Provincial government, and cases of life imprisonment or death sentence. Again no cases may be examined by more than two levels of courts, and judgment for all cases may be delivered preferably within 200 days from its filing, but no way later than 365 days.

It is suggested that the governments of the IOR countries may institutionalize setting up hundreds of "Grassroots Consultancy Courts (GCCs)" in all parts of the country. GCCs will be manned by professionals (retired judges, helped by young fresh pass-out lawyers from Universities). Litigants must first approach the GCC with their complaints (particularly disputes of civil, social, and property types). The GCC will be mandated, empowered, and encouraged by the government and higher judiciary to find a low-cost quick balanced solution for a rapid settlement of cases/disputes. In fact, the GCC is supposed to help clear about 70% of the disputes at their levels within 60 days. If not reasonably satisfied, the aggrieved person can file his/her case to a higher court, but the same can be heard only by one more level of the court.

The measure as outlined above is expected to make the justice delivery system rapid, low-cost, and least cumbersome. Also, the stupendous requirement of trained investigators and law professionals would create a tsunami of employment opportunities for thousands of highly skilled professionals.

8.2.3.5 Economy

The major issues that ail the economy of most of the IOR countries are the absence of efficient and fair public services, and the right incentives for private initiatives. In addition, there is an absence of rationality in choosing suitable major public investments (projects/activities), no inducement for use of science and technology in cultivation, and inadequate development of skill for productive use. Europe and Japan got over these shortcomings in the nineteenth century, while Korea and China recovered in the twentieth century (Matilal, 2017). In fact, an equitable affordable and inclusive outlook in policymaking and implementation are what the IOR needs most now.

Economic priorities for the IOR nations should be for a strong, sustainable, job-creating economic growth. The large population of this region desires decent employment. In contrast, the growth in the IOR during the last three decades has largely been capital-intensive, and not labor-intensive. Technological progress has improved capital productivity and reduced the relative price of machinery, but the inordinate delay in labor reforms has kept the effective cost of labor too high. This partially explains the increase in unemployment (Chinoy, 2019).

Again, over the decades, job creation has been the most among the least educated. For example, in India between 2016 and 2018, of the total jobs created, about 53% were created for those who could not even complete 5th standard in school, and 25% jobs were made available for those who left studies before they could reach 10th standard. While only 18% matric-pass (junior college/12th standard) people obtained jobs, a paltry 4% of opportunities were created for graduates and master's degree holders (Vyas, 2019). It raises the very question of why one should pursue higher education, when jobs are mostly created at the lower strata.

Hence, the question is how to create good jobs? The answer lies in increasing the pace of growth in labor-intensive areas (or job-creating sectors), which is extremely important. For this to happen it is required to quickly reform the economy and labor laws. For example, in the construction sector (housing, roads, bridges, dams, seaports, airports, railways, commercial real estate, interlinking of rivers, amongst others) many moderately skilled personnel (more than 50% of the unemployed youth) could be accommodated. But what holds back the construction sector from booming is the cumbersome process of land acquisition, environmental clearance, and lack of funding from public sector banks (as their non-performing assets are only increasing), or private houses (Rajan and Banerjee, 2019). The reforms have to work out for a transparent, speedy, and fair process of acquiring land through land-bank after necessary environmental clearance. A good degree of reform in this area will also elevate IOR standing in "ease of doing business" globally.

Two other problems the economy is facing today are the public sector borrowing and the current account deficit (CAD is the gap between investment and savings). While the former can be controlled by strict fiscal discipline, the latter could be managed through higher household savings. Continuing lowering of interest rates on individual savings in banks might boost investment but will increase CAD, which is harmful for long-term economic growth (Chinoy, 2019). Again, political will and state capacity must be on the same page in ushering in reforms. If capacity utilization rates keep rising, and the non-performing assets problem is resolved in the near future, the situation will be ideal for large-scale private sector investment. Additionally, improved bank governance, and bank recapitalization would encourage increased private investment for a sustained economic growth (Table 8.3).

Another aspect the IOR nations must work out is to frame a strong mechanism to export their products within the IOR and outside. For example, in the year 2000, two-thirds of India's exports included labor-intensive material (agriculture, textile, gems, jewelry, and leather). Presently, such exports have come down below the 50% mark. Again, data between 2005 and 2008 suggest that the middle-income group of 500–600 million people contributed in economic growth. It was possible because of substantial earning by the farmers and other middle-class people. However, presently this layer is shrinking and only the top 100 million people do the purchasing. It defeats the very philosophy of inclusive growth (National Institute of Public Finance & Policy, 2019).

To support the ever-increasing aspirations of youth, a minimum income amount may be worked out which gives a decent living (Banerjee, 2019; Table 8.3). Credit Suisse in their Global Wealth Report of 2018 found that the richest 10% of Indians own 77.4% of the country's wealth, while the richest 1% own 51.5% of the resources. In contrast, the bottom 60%, the majority of the population, own only 4.7% of the resources in India (Figs. 8.3, 8.4). It means that while wealth has been rising in India, not everyone has share in this growth. There is still considerable wealth poverty, reflected in the fact that 91% of the adult population has wealth below US$10,000. At the other extreme, a small fraction of the population (0.6% of adults) has a net worth over US$100,000 (Chakraborty et al., 2019). The new EBE hence must reform labor regulations, cut the effective cost of labor, incentivize the job-contract, make health and education robust for all citizens, and avoid doling out fiscal subsidies.

The existing neoliberal capitalism in the world today has its own pros and cons. This philosophy has struck well with the masses as it not only allows people to dream and aspire, but also lay out a fairly detailed mechanism to fulfil such dreams and achieve the goals, however selfish that might be. This philosophy encouraged private investments, and caused an increase in competitiveness between people, between peers, and among companies. However, this capitalist economy suffers from several diseases, a few being chronic in nature. The foremost cons of this philosophy has been the fact that it failed to distribute the wealth (see Fig. 8.3). Thailand, Russia, and India are the top three offenders. India ranks third in the inequality index, preceded only by Russia in the second rank (top 1% people controlling 60% of resources and top 10% owning 85% of resources) and Thailand as the top unequal nation (70% and 90%, respectively) in the world. The richest 1% in India control more than 51.5% of the country's wealth, and the top 10% rein in 80% of the resources, while the poorest 60% of people had to share just 4.7% of the country's resources. This increasing inequality is a potential tool for social implosion, which seem to be appearing in the form of attrition, violence, ultra-nationalism, leading to terrorism.

TABLE 8.3

Economic, Trade, and Commerce Scenario in the IOR

IOR Nations	01	02	03	04	05	06	07	08			09		
Australia	1,417	55,421	2.1	2.0	4.8	33	0.939	CHN	JPN	SKR	CHN	USA	JPN
Bangladesh	314.7	1,889	7.3	5.4	NA	32.4	0.608	GER	USA	UK	CHN	IND	SNG
Comoros	0.7	833	2.8	2.0	NA	45	0.503	FRN	IND	GER	UAE	FRN	CHN
Egypt	299.6	3,020	5.5	14.5	9.8	31.8	0.696	ITA	TUR	UAE	CHN	SAB	USA
India	2,972	2,198	7.0	3.9	NA	33.9	0.640	USA	UAE	CHN	CHN	USA	SAB
Indonesia	1,101	4,123	5.2	3.3	5.2	39.5	0.694	CHN	JPN	USA	CHN	SNG	JPN
Iran	484.7	5,820	-6.0	37.2	15.4	40	0.798	CHN	IRQ	UAE	CHN	UAE	SKR
Iraq	225.9	5,759	2.8	2.0	NA	29.5	0.685	IND	CHN	USA	TUR	CHN	SKR
Israel	381.6	42,144	3.3	0.9	4.0	42.8	0.903	USA	CHN	UK	CHN	USA	TUR
Kenya	99.2	2010	5.8	4.4	NA	42.5	0.590	PAK	UGN	USA	CHN	IND	UAE
Kuwait	136.9	29,128	2.5	2.5	1.3	–	0.803	CHN	SKR	IND	CHN	UAE	JPN
Madagascar	12.7	471	5.2	6.7	NA	44.1	0.519	FRN	USA	GER	CHN	UAE	IND
Malaysia	373.4	11,385	4.7	2.0	3.3	41	0.802	SNG	CHN	USA	CHN	SNG	USA
Mauritius	14.8	11,693	3.9	2.1	6.9	35.8	0.790	FRN	USA	UK	IND	CHN	SAF
Mozambique	15.4	493	4.0	4.2	NA	45.7	0.437	IND	NTL	SAF	SAF	CHN	UAE
Myanmar	65.7	1,238	6.4	3.9	4.0	38.1	0.578	CHN	THL	JPN	CHN	SNG	THL
Oman	79.5	18,081	1.1	1.5	NA	–	0.821	UAE	QTR	SAB	UAE	CHN	IND

Country	01	02	03	04	05	06	07	08			09		
Pakistan	278	1,358	2.9	7.6	6.1	33.5	0.562	USA	CHN	UK	CHN	UAE	SAB
Qatar	193.5	70,288	2.6	0.1	NA	41.1	0.856	JPN	SRK	IND	USA	CHN	IND
Saudi Arabia	752.3	22,507	1.9	-0.7	NA	45.9	0.853	CHN	JPN	IND	CHN	USA	UAE
Seychelles	1.7	17,154	3.4	3.0	NA	46.8	0.797	UAE	FRN	UK	UAE	SPN	FRN
Somalia	7.9	NA	3.5	NA	NA	--	--	OMN	UAE	SAB	CHN	IND	ETP
South Africa	371	6,331	0.7	5.0	27.5	63	0.699	CHN	GER	USA	CHN	GER	USA
Sri Lanka	84.2	3,837	3.5	4.5	4.4	39.8	0.770	USA	UK	IND	IND	CHN	UAE
Tanzania	61	1,172	4.0	3.5	NA	37.8	0.538	IND	SAF	VTN	CHN	IND	UAE
Timor-Leste	3.1	2,422	5.0	2.5	NA	--	0.625	IDN	USA	CHN	CHN	SNG	HK
UAE	427.9	39,806	2.8	2.1	NA	36	0.863	IND	JPN	CHN	CHN	IND	USA
Vietnam	260	2,725	6.5	3.1	2.2	37.6	0.694	USA	CHN	JPN	CHN	SRK	JPN
Yemen	29.1	919	2.1	20	NA	36.7	0.452	OMN	THL	BLR	CHN	UAE	TUR

Major Global Players

	01	02		06	07	
USA	20.580	62,869	Export 1,545 B$	41.5	0.924	Import 2,407 B$, Trade Balance (−) 861 B$
China	27.331	19,520	Export 2,263 B$	46.2	0.752	Import 1,843 B$, Trade Balance (+) 419 B$
Japan	05.747	45,546	Export 698 B$	37.9	0.909	Import 671 B$, Trade Balance (+) 26 B$

Notes: Most data of 2019. **01** = GDP (US$ billion), **02** = Per-capita GDP (US $), **03** = Real growth in GDP (US $), **04** = Inflation (%), **05** = Unemployment % of labor force, NA = Not available, **06** = Gini—an indicator of wealth distribution, 0 is the best, **07** = Human Development Index, 1 is the best, **08** = top three countries wherefrom goods are imported, **09** = top three countries where goods are exported to. CHN = China, JPN = Japan, IND = India, SKR = South Korea, SNG = Singapore, GER = Germany, FRN = France, ITA = Italy. TUR = Turkey, SAB = Saudi Arabia, IRQ = Iraq, PAK = Pakistan, UGN = Uganda, IDN = Indonesia, HK = Hong Kong, SAF = South Africa, NTL = The Netherlands, THL = Thailand, SPN = Spain, OMN = Oman, ETP = Ethiopia, BLR = Belarus. GDP of USA, China and Japan is in trillion US dollar, B$ = billion US dollar.

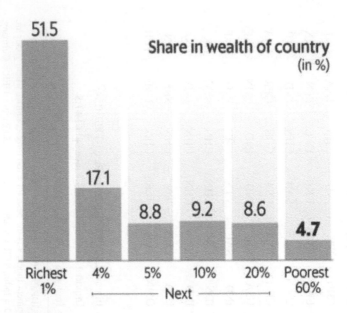

FIGURE 8.3   Wealth Distribution in India

The job-less growth has been another gift of the existing neoliberal capitalist economy. Many IOR nations blindly followed this trend in their country, which has been somewhat reasonably popular in the Europe and North America. The jobless growth in the IOR increased inflation and unemployment, unevenly distributed wealth, and inadequately made money available to the lower strata of the society. Another issue which this book has insisted for is the trade and commerce among the IOR nations. This intra-regional trade can boost the economy of every nation, establish new industrial set-ups, create job opportunities, and alleviate property. However, as of today, only 35.6% of trade emanating from this region is consumed by the IOR countries, and more so by only three nations: UAE, India, and Australia (Table 8.3). The trade needs to be more broad-based that must include many other countries of the region.

8.2.3.6   Governance

The necessity for overhauling the governance mechanism in the majority of the IOR countries emerges from the inability of these governments to answer many of the questions even after several decades of independence. Why have these countries failed to remove poverty and create large economies? Who is responsible, is it the extremely high rate of population growth **or** the cultural dispensation that sabotaged competition, entrepreneurship, and productivity, **or** the corrupt and religion-caste-based politics, **or** the lack of professionalism, **or** the hiding of inefficiency under the garb of inclusiveness, **or** the weak policing and judicial system, **or** the unique attitude of the IOR population never to focus on real issues?

Taking an example from India, it is found that too many Indians are involved in low productivity sectors. For instance, about 50% of India's labor force works in the

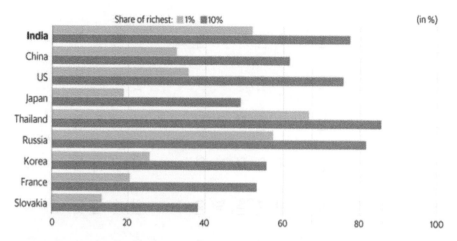

FIGURE 8.4   High Degree of Wealth Inequality in India

agriculture sector that contributes only 14% to her GDP, while only 0.7% of Indians working in the IT sector contributes 8% to the GDP. Similarly, demography has an interesting relationship with low productivity geographies (the state of Karnataka even with one-third population of UP produces the same GDP), low productivity firms (63 million enterprises account for only 19,500 companies with a paid-up capital of US$1.4 million), and low productivity skills (bottom 10% of engineers will make less salaries than 1% of IT graduates; Sabharwal, 2019, Tables 8.4, 8.5). A comparable situation prevails in several other IOR countries, indicating a lack of strategy to integrate labor force, geography, skills, and entrepreneurship, vis-à-vis GDP.

In the IOR in general, the ease of doing business (EDB) is very poor with hundreds of compliance rules and tens of filings. The late Jawaharlal Nehru (the first prime minister of India) said in 1955 that "India cannot become a welfare state unless its national income goes up. India has no existing wealth to divide among its people, there is only poverty to divide." Hence, the Indian economy must aim for plenty. It is said that poverty is not like cancer that needs only to be removed. Rather it is a system obesity and requires a long process to control through diet and exercise.

The IOR must value the future over the past, and must stop being seen as a region that is more "interesting than important" to the world. It must have an infrastructure that offers opportunity. Joint ventures with the world's leading firms must be encouraged. This will bring technology, skill, investment, and change the IOR's economic landscapes in the next few years. In tune with this philosophy, governments of several IOR nations must revisit their policy of continuing with the "monthly unemployment allowance" (a type of social security benefit) to the unemployed youth, as besides reaching harm to the accountability/sustainability of financial system, an income without work will not boost anybody's confidence (Sabharwal, 2019), and will fail the dream of the poet Tagore—"mind is without fear and head is held high."

Freedom of speech, thoughts, movement, expression, free media, and fair election (Table 8.4) in IOR countries has not been very profound. However, Australia, followed

TABLE 8.4
Governance Factors and Ranking of IOR Nations

Country	01	02	03	04	05	06
Persian / Arabian Gulf Region						
Bahrain	188	178	65	81	55	67
Kuwait	150	100	103	79	101	99
Oman	169	81	82	70	69	82
Qatar	157	17	46	38	53	35
Saudi Arabia	205	142	90	83	95	90
UAE	171	57	36	70	57	36
West Asia						
Iran	199	190	134	169	195	159
Iraq	178	201	182	205	189	193
Israel	72	176	30	47	32	46
Jordon	154	148	97	79	90	82

Country	01	02	03	04	05	06
Australia and Southeast Asia						
Australia	09	42	13	11	07	09
Indonesia	104	154	117	140	120	149
Malaysia	131	116	42	72	63	72
Singapore	98	07	01	09	01	07
Thailand	133	184	82	106	88	111
Horn of Africa						
Djibouti	195	102	180	154	136	115
Egypt	154	195	157	127	140	138
Eritrea	209	163	201	199	207	146
Sudan	203	207	197	192	199	207

	01	02	03	04	05	06
Yemen	192	209	191	195	153	193
South Asia						
Bangladesh	140	192	163	171	**168**	**166**
India	89	186	111	100	138	136
Myanmar	**203**	**173**	201	199	205	**186**
Pakistan	161	209	161	171	157	180
Sri Lanka	148	163	113	102	**109**	101
East Africa and South Africa						
Comoros	146	**138**	203	178	**193**	**155**
Kenya	129	**190**	136	165	122	184
Madagascar	159	152	178	169	145	145
Mauritius	57	45	49	47	**44**	69
Mozambique	121	87	147	140	136	140
Seychelles	95	62	76	98	**122**	**69**
South Africa	74	119	76	**89**	78	97
Tanzania	123	**112**	151	138	**132**	163

Source: Modified after World Bank (2013), Cordesman and Toukan (2014).

Notes: Governance (with 1 as the best situation indicator) is described as a combination of six factors: **01** = *Freedom* (apparent freedom of speech, thought, expression of citizens, free media, free & fair election), **02** = *Stability* (of government against terrorism/violence/unconstitutional ways/military coup), **03** = *Government Effectiveness* (represents professed quality of policy formulation, quality of public & civil services, credibility of government), **04** = *Rule of Law* (represents sanctity of rules, regulations, property agreements; freedom & effectiveness of judiciary, police and security agencies, human rights), **05** = *Regulatory Mechanism* (ability of the government to frame and implement rules that promote a level playing field for all stakeholders to contribute for development), **06** = *Controlling Corruption* (supposed authority and commitment of the government to root out big or petty corruption). **Bold digits** = Improved from proceeding years, rest deteriorated from earlier years or register no-change.

TABLE 8.5
Relationship between Governance and Economy in the IOR Countries

A: Governance and Ease of Doing Business

Range	Countries
000–020	Singapore, Australia
020–040	No country
040–060	UAE, Mauritius, Malaysia, Qatar, Oman
060–080	Israel, Bahrain, South Africa, Saudi Arabia
080–100	Thailand, Seychelles
100–120	Kuwait
120–140	Sri Lanka, Jordan
140–160	No country
160–180	Indonesia, Mozambique, India, Egypt, Tanzania
180–200	Kenya, Pakistan, Djibouti, Madagascar, Bangladesh
200–220	Iran, Yemen, Comoros, Iraq, Sudan, Eritrea, Myanmar

Source: Data from World Bank Country Data Report (2013), Cordesman and Toukan (2014), www.doingbusiness.org, http://info.worldbank.org/governance.
Notes: Combined indicator range of Governance and Ease of Doing business (EDB). 000 is the best governed country, 220 is poorly governed. A positive relation exists between good governance and EDB.

B: Governance and Security Risk

Risk Intensity	Countries
Low-Moderate	Singapore, Australia, Qatar, UAE, Mauritius, Malaysia, Oman, Seychelles, Bahrain, South Africa, Kuwait
Moderate	Israel, Jordan, Saudi Arabia, Thailand, Mozambique, Madagascar, Tanzania, Sri Lanka
High	India, Kenya, Indonesia, Djibouti, Egypt
Very High	Pakistan, Iran, Bangladesh, Comoros, Eritrea, Yemen. Myanmar, Iraq, Sudan, Somalia

Notes: Security of a nation comprise her strength in economy, robust finance, good governance, compassionate society, sustainable environment, and technological finesse. Low risk indicates good governance. Countries are listed in order of best secured (Singapore) to poorly secured (Somalia).

TABLE 8.5 (Continued)
Relationship between Governance and Economy in the IOR Countries

C: Governance and Macroeconomic Stability

Range	Countries
000–020	Singapore
020–040	Qatar, UAE
040-060	Malaysia, Australia, Oman, Israel
060–080	Bahrain, Mauritius, Saudi Arabia, Kuwait
080–100	Seychelles, Thailand
100–120	Indonesia
120–140	India
140–160	South Africa, Sri Lanka
160–180	Tanzania, Mozambique, Iran, Jordan, Bangladesh
180–200	Myanmar, Iraq, Madagascar, Egypt
200–220	Djibouti, Pakistan, Kenya, Comoros, Eritrea, Yemen, Sudan

Notes: Range is the combined strength of economy and governance. 000 is the best governed country with strong macroeconomy (Singapore), 220 is poorly governed State with weak macroeconomy (Sudan). Data source same as 8.4 (A).

by Mauritius, Israel, South Africa, and India, are amongst the top five governments respecting and committing to freedom. The worst five countries, on the other hand, are Eritrea, Saudi Arabia, Myanmar, Sudan, and Iran. With regard to ensuring political stability, Singapore, Qatar, Australia, Mauritius, and Seychelles occupy the top five positions in the IOR, while Iran, Iraq, Sudan, Pakistan, and Yemen are the worst performers. On the question of how effective the governance is in terms of professed quality of policy formulation, quality of public & civil services, credibility of government, Singapore, Australia, Israel, UAE, and Qatar rank top five in order, while Yemen, Sudan, Eritrea, Myanmar, and Comoros occupy the lowest five positions. In terms of Rule of Law, encompassing sanctity of rules, regulations, property agreements; freedom & effectiveness of judiciary, police, and security agencies; and maintaining human rights, the top five countries in the IOR have been Singapore, Australia, Qatar, Israel, and Mauritius, while the most lowly placed five nations have been Yemen, Sudan, Eritrea, Myanmar, and Iraq. With regard to supposed authority and commitment of the government to root out big or petty corruption, Singapore, Australia, Qatar, UAE, and Israel have been at the forefront, while Kenya, Myanmar, Iraq, Yemen, and Sudan are failing consistently (Table 8.4).

A quick glance suggest that Singapore, Australia, and Qatar are the leading countries as far as the good governance in the IOR region is concerned. Lack of freedom of expression in Singapore is however remaining a concern amidst its splendid performance in almost all other performing sectors. While critics would give credit to the small territory of Singapore and Qatar for their good governance, the lower population must have certainly helped Australia to excel in an outstanding manner. On the other hand, Eritrea, Iraq, Myanmar, Sudan, and Yemen are the worst performers in all aspects of governance. However, if the IOR has to win the battle of poverty, lack of human development, violence, and lawlessness, the entire region has to start respecting life, humanity, and dignity of work. The developed countries in the region must help these disturbed countries quickly in kind, and with expertise. A good country cannot live happily with bad neighbors. This anomalous distribution of humanity must be addressed on a war footing.

Governance has a telling impact on economy. Good governance encourages EBD and attracts foreign direct investment (FDI) into the country. And the six factors that constitute good governance (Table 8.5) conclude that Singapore and Australia are the best nations in the entire IOR. These two countries are followed in the good governance chart by UAE, Mauritius, Malaysia, Qatar, Oman, Israel, Bahrain, South Africa, Saudi Arabia, Thailand, and Seychelles. The positioning of Indonesia and India, the two most populous nations in the IOR, beyond 100 marks is however unfortunate, reflecting on some hard facts that there is no alternative to good governance and macroeconomic stability. Both these countries are huge in area and storehouses of manpower, and cannot be ruled from one place through popular politics and centralization of power.

All these suggest for a quick and comprehensive overhauling of the economy of the IOR, so also the reforms in farming, skill development, and education sectors. The reforms should be done in an integrated manner to make the EBE work optimally and ensure its sustainability in the IOR region.

8.2.4 Participatory Governance

One of the foremost requirements of the EBE is its sustainability. That is possible when the government-initiated blue economy campaign transforms into a people-run movement. To make this happen, one needs to encourage participatory governance in every sphere. The movement must involve local stakeholders, public and private sector investment, and ease regional collaboration to encourage the best integration of economy, trade, and commerce.

8.2.4.1 Regional Cooperation and Economic Integration

Reports of the Asian Development Bank (ADB) and the World Bank cautioned the over-reliance of the IOR on exports, and highlighted the need to rebalance its sources of growth, putting more emphasis on domestic demand (ADB, 2014; World Bank, 2014a). In fact, with more than two and a half billion people, the IOR countries constitute a huge market. Consolidation of this market in terms of consumption of goods created locally and regionally is required. The trades and services among the IOR countries may work to facilitate this possibility. Such trade is certainly possible by inducting a dose of ethics, when sky-rocketed prices of products are reduced to a realistic level, and the purchasing power of the market (general public) is increased. Such growth and increasing domestic demand would encourage greater intra-regional trade in services, and would eventually attract more local and foreign investment, enhance employment, and reduce socioeconomic anomalies. Integration of trade and commerce thus may lead to the IOR having a common economic growth policy. An economically strong IOR may force reforms in various political, economic, and strategic decision-making institutions in the world.

The IOR economy has the potential to boost the socioeconomic resilience in the entire Afro-Asian region and could ensure that it contributes to global economic realignment, only if it integrates (Tables 8.4 and 8.5). Such an integration will render the IOR economy with strength, stability, and predictability. The IOR should not only be a producer and exporter of its goods and services, but also a consumer. The philosophy is to cooperate to prosper, and prosper to contribute to a more inclusive and environmentally sustainable brighter future for all (ADB, 2010). The possibility of a similar collaboration finding success is high, and such prospect lies in the fact that the borders of these countries are largely contiguous, and traditionally they have similar values, ethics, and perceptions (Agranoff and McGuire, 2003; Carlson, 2007).

We argue that the growth of the IOR cannot be remarkable and sustainable till the growth runs in complete harmony with nature. To win the battle, one need not only be innovative, but need to have a change in mindset, that encourages gender equality, and ethical behavior. The IOR can still remain afloat with regard to constraints in its economy, food, shelter, health, education, and employment, as long as ethics, integrity, trust, goodwill, regional collaboration, and trade/economic integrations among the member nations grow.

8.2.4.2 Local Administration

Local Administration (LA) should play an important role in the ethics-based system re-engineering. For this to happen, *first*, LA may be given a prime place in the planning and implementation of the EBE concept. *Second*, LA must be made well-informed, responsible, responsive, well-trained, and efficient. *Third*, LA may make efforts not to treat EBE as a special localized SEZ type (special economic zone) measure, but making it a tool to change the mindset of people and that of society, and *four*, reduce the multiplicity of authorities working at the grassroot level.

To make the blue economy model work, the existing discourse of multiple authorities working in the same area for the same or similar project must be dispensed with. On many occasions, ambiguous overlapping of authorities proved a disaster in speedy tackling of developmental projects (Satpathy et al., 2013; Mukhopadhyay et al., 2018b). A SOP (Standard Operating Procedure) may be prepared by the LA to take appropriate measures to mitigate any threat (Tanner et al., 2009).

8.2.5 Resource Management

To ensure a strong blue economy architecture, ethically managing natural resources (NR) and human resources (HR) will be challenging for all IOR governments. We discuss a few such issues briefly.

8.2.5.1 NR: Water Management

Scarcity of water for human consumption and for cultivation is posing a severe problem, not only for the IOR but also for the entire world. Perhaps water scarcity could be the main reason for a third world war. For example, the per-capita availability of water in India has declined considerably from 5,177 m³ in 1951 to only 1,820 m³ in 2001, and is expected to decrease further to 1,341 m³ by 2025 and 1,140 m³ in 2050 (SAARC Technical Report, 2010). The city of Chennai is running out of groundwater reserve by this year, and same will be the fate of another 21 cities in the near future. It seems that 40% of Indians will have no access to drinking water by 2030 and almost the entire Indian populace will have no drinking water by 2040. The groundwater depletion rate has been 10–25 mm/year between 2002 and 2016, as the rainfall has declined from 1,050mm in 1970 to <1,000mm in 2015 (Chandra, 2019).

To counter such a threat, the existing strategy is to expand rainwater harvesting, improve storage capacity, put up several desalination plants, install conservation techniques, and make irrigation efficient, in close coordination with other sectors. This strategy would need framing of integrated water resource and hazard management policy for the IOR countries.

Some other measures that could play vital roles in water conservation may include recharging natural aquifers. These aquifers may be identified preferably in upstream areas of major rivers with favorable porous geology to capture and store flood water. During a dry spell, the same water may be pumped out for farmers to irrigate their land. This technology will be particularly useful for the countries of West Asia and South Asia.

Linking of regional rivers could be another major contribution. For example, the entire region of South Asia may be considered as one unit while planning and implementing interlinking of rivers (Mukhopadhyay, 2008). To start with, all the rivers of this region may be mapped in detail with regards to their catchment area, the flow rate in different seasons, gradient, and other parameters. River interlinking could reduce floods during the monsoon by discharging excess waters to relatively dry places, and avoid drought by supplying excess waters from such areas. In addition, river interlinking will initiate the least expensive transport network of goods and passengers, contribute to tourism, facilitate the generation of hydroelectricity, and help increase afforestation and greenery all around.

8.2.5.2   NR: Managing Agriculture

As the world's population is expected to grow from 7 billion now to 9 billion by 2050, pressure on staple crops will increase. However, production of wheat—one of the world's most important staple crops—is set to fall by 6% for every 1°C rise in temperature (Asseng et al., 2014). Again, about 54% of methane, 80% of nitrous oxide, and almost 100% of terrestrial carbon dioxide are contributed by agriculture alone (IPCC, 2014). Hence, there is an urgent need to change the present agricultural practices.

The reformed strategy could include: (a) recognizing agriculture as an industry and apply all rules and provisions used in industry; (b) introducing cooperative farming with several square kilometers of land being cultivated collectively in the form of industry; (c) modifying pattern, variety, place and date/time of crop sowing; and (d) encouraging genetically modified (GM) crop.

As a system re-engineering measure, the contiguous cultivable land of a reasonably large area (several square kilometers) could be "leased" (not sold) by the land-owning farmers to a Public-Private Consortium (PPC) on payment of existing/ agreeable market price. Farming over such a vast area in a scientific manner with the latest technology and collective dispensation would expect to be both economically and commercially viable. Besides, one able-bodied person from each family whose land has been leased may be given a job to work in the Agricultural Farm against a fixed monthly salary. This mechanism will offer financial stability to farmers, would insulate them from crop loss or financial bankruptcy and stop them from committing suicide, while the land will continue to be theirs and being effectively used to benefit humankind. Sowing, cultivation, storing, and marketing of the crop in the local, regional, or global market will be decided, and implemented by the PPC. The government will only facilitate the entire process and will act as a distant-regulatory body to provide a level playing field to stakeholders and neutralize exploitation of farmers, if any. The fact is that the land deserves to be used more scientifically and rationally rather than keeping hostage to rituals, customs, and hollow emotions.

With change in climate, the farmers of IOR countries must change their historic mindset and approach to food and agriculture. The people must accept that all agricultural fields are not meant for all types of crops. The farmer awareness program being conducted by Indian Meteorological Department and by BOBLME (Bay of Bengal Large Marine Ecosystem) in this direction may be adapted by the other IOR nations with appropriate ground-level changes, as needed. Hence, focused nationwide testing of the soil, climate, weather, rainfall, and availability of groundwater are required to be undertaken. The aim would be to first, find and earmark each agricultural farm with the most suitable crop. The scientific way of cultivation over a larger area by forming Agricultural Farms (community farming) with special attention on the type of crop to be grown and giving financial security to all farmers would probably be the most sustainable ways.

It is generally accepted that higher agricultural productivity holds the key in the fight against rural poverty. Again, with enhanced demand for land for infrastructural purposes, and increasing intrusion of saline water in the coastal land-spaces, the overall shortage of land in IOR countries has been quite serious. In contrast, the agricultural yield is predicted to reduce by 10–35% by 2050 because of the change in climate. In this regard, biotechnology promises to boost productivity and suitable GM crops could increase rural incomes, much in the same way that the green revolution did in large parts of Asia during the 1960s to 1980s. Toward this, cultivation of GM crops is one option through which good, healthy, quality, and a substantial amount of food grains could be produced most optimally in the land made available (Fig. 8.5). The GM crops are those seeds whose DNA (deoxyribonucleic acid) has been modified through genetic engineering to introduce a new trait to an existing naturally produced crop. Such crop would have increased resistance to pests, diseases, and spoilage, but with increased nutritional value and improved shelf life. Cultivating GM would mean higher crop and livestock yields, lower pesticide and fertilizer applications, less demanding production techniques, higher product quality, better storage, and easier processing, or enhanced methods to check the health of crop and yield.

The GM crops are sustainable, comparatively cheap, nutritious, and probably can take care of increasing mouths in the IOR region, as well as in the world. Realizing its benefit, the countries like the USA, Canada, Argentina, and Chile are involved in GM crop cultivation in a major way. Among the rest, India has more than 9 million hectares under GM cultivation, followed by China (3–9 million hectares), Pakistan and South Africa (1–3 million hectares; Fig. 8.5). However, greater research is needed on GM crops followed by an effective awareness campaign to remove the doubts from the minds of people with regard to adaptability, if any. IOR must catch up to the world in such research and production of GM crops.

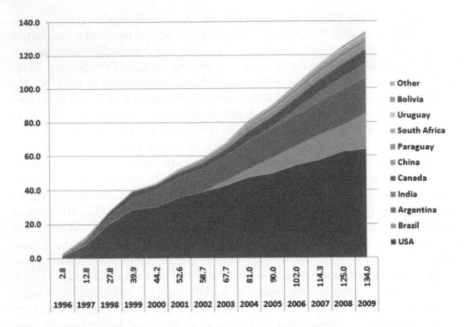

FIGURE 8.5 Land Area (in Million Hectares) used for Genetically Modified Crops Figures in million hectare

Note: 160 million hectares of land used for GM crop cultivation in 2011. India figures prominently in GM crop cultivation (ADB 2014)

8.2.5.3   NR: Coastal Zone Resources

Coasts play a vital role in the blue economy paradigm. Billions of people worldwide—especially the world's poorest—rely on healthy oceans to provide jobs and food. In fact, the ocean contributes US$1.5 trillion annually in value-addition to the overall economy (OECD, 2016), with millions of people being employed in fisheries and aquaculture. Healthy oceans, coasts, and freshwater ecosystems are not only crucial for economic growth and food production (Fig. 8.6) but are also fundamental to global efforts to mitigate climate change.

Tagore once said, "The roots below the earth claim no rewards for making the branches fruitful." It is time IOR governments introduce a "Resource Formation Cost" (RFC). Except for a nominal license fee or royalty paid by the user, nobody really compensates the cost of making of the resources, e.g., fishermen do not pay for the making of fishes (although pay royalty for fishing), miners not for the making of ores (only pay for mining), and farmers not for the making of soil (pay only for seeds and irrigation). However, the concerned persons spend money on the fishing/mining/farming process, and on the mechanism to extract or collect these resources. These entire public and private entrepreneurs take the mother Earth for granted. Nobody pays for the formation and growth of the resource that nature painstakingly did/does over several millions of years through optimum integration of several physical and chemicals parameters. One must at least respect the grueling scientific processes that resource formation involves. Once this RFC starts, respect to nature may probably increase.

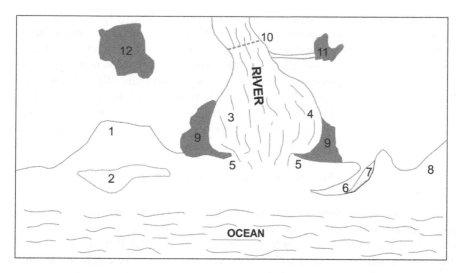

FIGURE 8.6 Coastal Dynamics and Economy (After ADB (2016)). 1= Lagoon, 2= Barrier Reef, 3= River, 4= Bay, 5= Spits, 6= Tied Island, 7= Tombolo, 8= Cuspate Foreland, 9= Tidal saltwater coastal wetlands, 10= Limit of Tidal incursion, 11= Tidal freshwater coastal wetland, 12= Non-tidal freshwater wetland. (Modified after ADB 2016)

Protection of the coast is important for the coastal economy. The best way to deal with coastal erosion, inundation, loss of land, and coastal pollution that will impact the EBE, is to go for soft adaptation using natural mechanisms, such as dune reinforcement, creation and management of wetlands, and protection of existing natural barriers. It is prudent to avoid any short-term solutions, such as by the construction of seawalls, dikes, or other hard artificial structures that may result in long-term problems. Strict implementation of rational geomorphology specific coastal regulation measures, restoring mangrove, and applying judicious land-use policies are required to protect coastlines.

In the IOR, the integrated coastal zone management program of India (ICZM) could be an initiative for many other countries to follow. The ICZM finances national and state-level capacity building, land-use planning, and pilot investments in pollution management, resource conservation, and livelihood improvements. The program is pioneering "Hazard Line" mapping for the entire coastline of India, to better manage coastal space and minimize vulnerabilities through shoreline protection and land use plans. So far, 1.7 million people have benefitted from the program (nearly half of them women), and 16,500 hectares of mangroves have been planted along the coast, making it one of the World Bank's biggest habitat restoration projects. Many sewage treatment plants have been set up that help to prevent the daily flow of more than 80 million liters of untreated sewage into the ocean to protect over 400 km of coastline in India. Very few of the IOR countries have completed Marine Spatial Planning (MSP), which is essential to regulate the development along the coast.

Under an EBE regime overfishing, use of explosives, and cyanide for fishing, coral mining and poor tourism practices must not find a place. Over the past four decades about 35% of global mangrove forests have been destroyed for firewood, coastal developments, aquaculture, and infrastructure build-up (Fig. 8.6). These have a telling effect on some of the most endangered species, including the Royal Bengal tiger of the Sundarbans. Sundarbans is a 10,000 km^2 marshy mangrove land at the confluence of three rivers—Ganga, Brahmaputra, and Meghna—emptying into the Bay of Bengal, and spread over Bangladesh (60%) and India (40%). It comprises closed and open mangrove forests, agriculturally used land, mudflats, and barren land, and is intersected by multiple tidal streams and channels. Hence preserving these "rainforests of the sea" is essential for coastal risk reduction and mitigating climate change besides preserving the aquatic organisms.

8.2.5.4   HR: Ethical Blue Economy Technology (EBET) and Employment Generation

Technological innovations are fundamental to future growth (and vis-à-vis creating employment), and research is the key tool to cope with such an existential change-over to meet the ever-increasing human aspiration. The EBET, therefore, insists to broaden the skill, education, and human capacity to make research and teaching meaningful to the society. The IOR society must remain ready to challenge its past achievements, and transform with high-quality teaching and globally visible techno-logical innovations (see also Tables 8.3, 8.4, 8.5).

Besides creating jobs in innovation and research, EBET may design and develop a new set of field equipment and laboratory instrument to generate renewable energy from solar insolation, river current, ocean-tide, ocean-wave, wind, river (hydroelec-tric), biomass, and hydrogen. All these will create local jobs and would strengthen rural economies as these interventions would use cheap, renewable, and environmen-tally friendly raw material available locally. A whole lot of possible new technology and related opportunities is discussed in Chapter 6.

8.2.5.5   HR: Disaster Management

A successful EBE has to deal with a number of natural threats. More than the anthropogenic disasters that include oil spills, leakage of poisonous and destructive substances, illegal dumping, and bio-invasion, the IOR is also identified as one of the most hazardous areas in terms of natural disasters. The loss of lives as well as the damage to property and the natural environment had been incalculable. The domino effect of these disasters on poverty, famine, societal imbalance, and other resultant human tragedies had been enormous.

In response to these, the IORA has geared up by forming Disaster Risk Management (DRM), particularly for the smaller island countries. Following the Sendai Framework to reduce vulnerabilities to natural hazards, the DRM of IORA, besides giving a major thrust on capacity-building initiative, also works on forecasting, preparation, mitigating, and recovering from hazards and emergency situations. DRM is a multi-disciplinary concept involving the participation of several stakeholders, including national governments, non-governmental organizations, regional and international partners, donors, civil society, and the private sectors. The aim is to facilitate and enhance regional cooperation on preparedness and response strategies to fragile and unpredictable situations.

The IORA Action Plan (2017–21) for disaster management aims to enhance cooperation and develop resiliency in the IOR. All national data, reports, digital maps, and information collected can be hosted in a national institution identified by the respective country. An electronic information exchange platform may be developed to allow access for continuous updating and interactions. Collaboration with universities and national institutes are bearing fruits. Regional simulation exercises for preparedness to disasters in partnership with non-States are also conducted with regional partners.

8.2.5.6   HR: Hunger Management

Nearly one out of every seven people on Earth now suffers from chronic hunger or food insecurity (Russell, 2014). Climate change could further affect food production causing widespread hunger. In addition to fossil fuel, globally livestock contributes about 15% of GHG emission (=~7 billion tons of CO_2) through consumption of meat and milk. This contribution includes 39% methane from enteric fermentation in the digestive system of the livestock, 21% nitrogen oxides through fertilizers used to grow feed for livestock, another 26% nitrogen oxide from the manure produced by these animals, and the rest 14% caused by transportation, processing, and deforestation to offer land for growing feed (TOI, 2015).

It is found that among all meats, beef has the highest carbon-emitting potential. In fact, a kilogram of beef emits 22.6 kg CO_2 (equivalent to running a car for 160 km), compared to 2.5 kg from pork, 1.6 from poultry, and 1.3 kg from milk. The consumption of beef worldwide has increased from 70 million tons (MT) in 1960s to 278 MT in 2009 and is projected to be about 460 MT by 2050. Compared to meat, plant cultivation is responsible for much fewer emissions. For example, while a kg of beef emits 16 kg CO_2, one kg of wheat was found to emit just 0.8 kg CO_2. The average daily per person meat consumption is a little more than 300 gm in the USA, Australia, and New Zealand; about 200 gm in Europe, Brazil, Argentina, and Venezuela; about 160 gm in China; and only 12 gm in India. Hence, reducing consumption of meat could be a great battle win by a blue economy over climate change impact (IPCC, 2014; TOI, 2015). Hence, the amount of water used and CO_2 released during the processing of 1 kg of beef may make people turn to vegetarianism.

Average annual global GHG emission between 2007 and 2016 has been 52 Giga-tons CO_2 equivalent (or 52 billion tons), of which agriculture and deforestation contribute 12 Gt CO_2 and rest 40 Gt CO_2 is supplied by energy, industry, and waste burning. About 23% of the world's land area is facing degradation due to deforestation, and would directly impact agricultural yield and food security, as it uses 70% of global fresh water. Prices of cereal may increase by 23% by 2050, as about 25–30% of total food production is presently lost or wasted. It appears that cutting GHG emission alone may not be sufficient to tackle the global warming effect, till people adopt sustainable farm practices and, possibly, plant-based vegetarian food habits (IPCC, 2019).

8.2.5.7   HR: Tourism

Culture and heritage play major roles in tourism. Among the several contributions that the tourism industry provide, the two most significant are: (a) it remains an engine of economic growth, and (b) it opens up a new vista of understanding, goodwill, and friendship among people of various nations. The domino effect of tourism on creating jobs, alleviating poverty, and sustainable development has been tremendous (see Tables 7.8 and 7.9).

Special attention should be given to creating a new flavor to the entertainment and tourism industries by diversifying attraction. Adventure sports, such as skiing and glacier treks, paragliding, rope-way traveling can boost the tourism industry. The plan must include estimating the carrying capacity of any tourism industry and ensuring integrated linkage to other sectors. Improvement of the road and rail networks could also spur new livelihood opportunities and boost sustainable growth.

A number of initiatives in this direction may be made. The regional cooperation among IOR members to develop regional projects, capacity building, and workforce development could be the first step. This may be followed by promoting large-scale private investment and Public-Private Partnerships (PPPs) to encourage state-of-the-art tourism business networks in the region and in the globe. Next, there is a need to create an IORA culture and heritage database, so as to be able to grasp the challenges and opportunities connected with this area. Research in the cultural field in the IOR needs to be intensified. Ultimately, visa-on-arrival and e-visa are priorities to help increase tourist arrivals from the member states and from other parts of the world. Acknowledging the diverse cultures and rich heritage of the IOR nations, the EBE concept must explore how these can be useful for the growth and development of the region. As cultures and heritage mold the daily lives, identities, and values of individuals and societies, these can contribute much for environmental protection, social capital, and economic growth.

8.2.5.8   HR: Communication Network

Managing the physical and virtual linkages among the IOR countries are extremely important for the success of the blue economy. More than 100,000 shipping routes connect the Mediterranean and South China Sea and Pacific and Atlantic oceans through the Indian Ocean. These shipping lanes form the umbilical cord and the arteries of the IOR's economic health. These lanes serve as commercial trade lines during times of peace, but during the war, these routes are considered strategic pathways to keep the war machine fully oiled. Recently, these routes have been threatened by piracy, drug trafficking, gun-running, human smuggling, pollution, accidents, interstate conflicts, and territorial disputes. None is independent of the other and failure in one often leads to failure in others. This calls for greater cooperation among states to enhance security and safety of the maritime enterprise (Sakhuia, 2010).

The well-secured road and rail links from the ports to the hinterland are essential to make sea-trade meaningful. So also high-speed internet, email, and real-time data sharing facility among the IOR countries and their commercial establishments would be hugely beneficial. Optimum use of the present genre of communication facilities and that due in the future would be desirable.

8.3   ROAD MAP FOR EBE

In summary, the Indian Ocean Region (IOR) with 32 countries and 18 island-nations (total 50 entities) shelters more than 2.53 billion people or 36.40% of the world's population. Yet, the cumulative GDP of the IOR nations has been a paltry 12.57% of that of the global GDP. Similarly, the per capita GDP of the IOR nations has been only 34% of the global GDP-PPP. This is despite the fact that the IOR is strategically significant from resource (as the world's most important energy and trade transmission belt) and security (military) points of view. In fact, the IOR has eight nuclear powers active in its waters: the USA, Russia, China, France, the UK, India, Pakistan, and Israel.

A well-performing economy must have a high labor force (HLF) as a percent of population and a low unemployment as a percent of labor force. On this account, the five best-performing IOR nations (i.e., low-risk countries in terms of providing employment to >60% of its population) have been Qatar, UAE, Bahrain, Thailand, and Singapore. On the same account, the least-performing nations have been South Africa, Yemen, Iraq, Sudan, and Iran.

The discussions held over the last seven chapters and in the last two sections of this chapter indicate that to save this world and the IOR from economic and environmental disasters, a major change with regard to dealing, understanding, and perspective of the prevailing discourse is required. It appears that the free-market economy, so also the unbridled greed, and undeserving aspirations need to be controlled, if this globe is to be saved. Considering several facets of threats to the blue economy (from carbon emission to ocean acidification, from extreme events to human mindset, from geopolitics to gluttony), this section will try to roll out a possible generic roadmap to overcome the ensuing challenges for the entire IOR region.

Although it looks formidable, one could find shortcomings in the existing legal and technical framework followed by the IOR countries, which have been found wanting during the conservation of ocean resources, and at attracting large-scale private sector investments in the marine sector. These inadequacies have been particularly visible in terms of duplication of projects, overlapping authorities, and multiple coordinating agencies. These problems are in addition to a financial resource crunch and at times too ambiguous target that at the end very few would know where they were supposed to reach.

However, the major problem seems to lie somewhere else. The increasing unmerited aspiration quotient of an individual and nations, and strategic manipulation by some countries are the fundamental problems of the blue economy that needs ethical treatment. Social dynamics of IOR countries suggest that a rational aspiration and ethical economics could take the best care of their societies by adjusting the aspiration to a responsible level. As ethics could effectively balance the direction, speed, and fulfillment of human desire, the present mindset of irresponsible indulgence, irrational borrowing, and having faith that a high GDP can buy contentment, harmony, and prosperity needs reconsideration. Accordingly, this chapter proposes to include ethics as an important component to revise the blue economy concept and model.

Money can buy a comfortable bed, not sleep, the way it can buy a house, but not a home. EBE must make sure of an environment where "greed" is replaced by the legitimate aspirations of the people. Also, such an environment would discourage the creation of artificial needs. Hence, from the governance angle, the first step for the IOR countries could be to undertake focused quantitative research to collect data on specific activities, such as potentials of maritime trade, investment and security, economic returns from fisheries/minerals/energy, coastal zone economy, pollution and health, tourism and culture, disaster risk management, scientific research, and capacity building. Next, the IOR nations may prepare unbiased status reports for each sector, find out loopholes in the existing system, and plan for a revised policy based on scientific data. It would be prudent for each country to formulate a mechanism to coordinate optimally and avoid duplication of efforts. Countries may as well consider bringing all industrial and environmental affairs with regard to the marine environment under one umbrella authority, since it will minimize the governance, institutional, and coordination issues and reduce spending by every IOR country. The people's rights to live in a safe and secured environment need to be incorporated in such a governance model.

The next step for the IOR countries would be to ratify international agreements. This will close the loopholes and increase collaboration and cooperation. A strong political will could help make this happen. Essentially the IOR must continue to try to transform threats from climate change, geopolitics, and unethical human attitude to ethical opportunities. Such a transformation could bring sustained and inclusive growth not in the economy alone, but also in society. As prevention (mitigation) is considered better than cure (engineering) all technological innovations may be ensured to follow moral practices under EBE. This would possibly transform the mindset, and a change in values of the people, government, and entrepreneurs. The mechanism to arrest the anomalous increase in population and swelling of selfish greed of people should be in focus. Reforms in agricultural and water management sectors and in food habit must also be undertaken (Mukhopadhyay et al., 2018b).

Regional collaboration among the IOR countries would be essential particularly in areas of scientific research on climate change, hydrography, and the early warning mechanism for natural disasters, patrolling sea-lanes, humanitarian assistance, and emergency evacuation. Helping the small islands developing states (SIDS) should take center-stage of such collaborations. Deepening economic and security cooperation with maritime neighbors and island states, and promoting cooperation in low-carbon shipping, fishing and food processing, ocean surveillance, information-sharing, marine biotechnology, and IT services, would strengthen EBE.

An inclusive society may also catalyze market integration, and probably deeper economic partnership among the IOR nations. The IOR economy has the potential to boost the region's economic resilience and can contribute to global economic realignment, only if it cooperates. The region should see itself as not only a producer and exporter of its goods and services but also as a consumer. Such well-protected structured cooperation would remove uncertainty from the market and strengthen stability and sustainable economic growth. As a fallout, the mutual trust and confidence among the not-so-friendly or even hostile nations of the IOR will be expected to increase (Mukhopadhyay, 2019).

For accelerated growth, it will be prudent for EBE to encourage PPP as a force multiplier. EBE must also ensure just and fair environment for seizing business opportunities in the IOR for private investments, by facilitating coordination between mature sectors (e.g., fishery, shipping, ports, maritime logistics) and emerging sectors (e.g., mineral exploitation, renewable energy). Also, it would be ideal to integrate elements of national and human security with marine safety and ecological integrity, both in planning and operations stages. Particularly for SIDS, adaptation measures to counter the climate change impact could generate larger benefit when combined with other developmental activities, such as disaster risk reduction and community-based approaches to development.

The EBE must also address current social, economic, and environmental issues, raise awareness through outreach and capacity-building/training programs, and communicate future risks to local communities. The adaptation and mitigation measures in the IOR countries could be made complementary, particularly in areas of energy supply, tourism infrastructure, and coastal wetland services. The rich nations, however, prefer to pay money for mitigation, rather than adaptation and refugee rehabilitation. This is simply because mitigation keeps the door open for commercial interest, and adaptation, it is alleged, kills the market demand (NOAA, 2014; USGCRP, 2014; USEPA, 2016).

Frustrated with lame excuses of the politicians, the IOR people are increasingly asking for good governance. For example, it seems necessary to explore new ways to strengthen food security and improve nutrition in a region like South Asia where 60% of the world's undernourished people live. Rapid population growth, stagnating crop productivity, water scarcity and pollution, climate change, and other constraints are making it increasingly difficult to substantially and sustainably improve food output and security in the region. The steep rise in global prices of rice and other food staples in 2007–8, which caused widespread distress among poor and vulnerable groups, have highlighted the need for such action (ADB, 2010). Referring to the reforms required in the agricultural sector listed in Section 8.2, the IOR with 2.5 billion people accounts for one-third of the world population and with an average age of IOR people of 30 years compared to 38 years for the USA and 46 years for Japan, this region represents the world's best hope for achieving the United Nations Millennium Development Goals (MDGs). To begin with, each country in the IOR may first adopt the Adaptation Policy Framework and integrate the same to its national strategic policy and developmental planning.

In this regard, social protection, in terms of finance, health, food, and shelter by the government could bring inclusive economic growth. For example, social assistance like guaranteed work programs, contributory pensions, old-age pensions, and post-retirement incomes could limit the social impacts of economic crises, serious illness, natural disasters, degrading environments, and the migratory labor movement. In fact, such protection will stimulate domestic consumption, and develop healthier, more educated, and better-skilled workers, and overall build up the region's physical infrastructure (ADB, 2010). In fact, India has taken some significant steps to build up her social protection systems (National Rural Employment Guarantee Scheme, Atal Pension Yojana, Prime Minister Jan Dhan Yojana, etc.), but much is desired to be done in IOR nations.

Another factor that probably cannot be denied is that poverty is the biggest pollution maker. The Western countries have proved this saying correct by cleaning up their air, water, and environment within 100 years of industrial revolution. A clear vision followed by notable achievements in scientific research and technological innovations accelerated their stupendous economic growth and took these countries away from atmospheric pollution. However, it is unfortunate that most of the IOR countries have failed to do so. Loose and inefficient governance has costed the IOR region dearly, resulting in these countries still being amidst the vicious circle of poverty with a least to moderate degree of economic growth. A summary of the roadmap to establish a successful EBE in the IOR is laid out (Table 8.6).

TABLE 8.6
Standard Operating Procedure (Roadmap) for Establishing and Sustaining EBE in the IOR

Bare Facts
- Threats to ocean health, maritime trade, and regional peace are increasing every day. Philosophy of existing economic order is under tremendous stress. So also is human nature—of trust deficit, ruthless manipulation, and lack of honesty.
- Do-not-do-anything will not solve the problem. Economic socialism has to come to save this Earth, while guaranteeing humanity a decent living. The costs of such repairing must be across the globe and over generations.

Spiritual fulcrum
- People may adopt ethical doctrine to move slowly from a consumer mindset to a conserving mindset, and from using society to giving back to society. Such moves can only sustain the Earth and humanity.
- Similar shifts have occurred earlier also, from whale oil to petroleum, from horse to automobile, from lamp light to electrification, from typewriter to computer, from telephone land lines to cell phones, and from Test cricket to One-day to T-20.

Lifetime Opportunity
- The threats (see Chapter 5) can be seen as the best instruments to bring in the BTVG doctrine (economic socialism and ethical governance). The doctrine may be used for distributing wealth of the developed countries among the poor nations legally.
- Fighting against a big threat can make humanity disciplined and innovative. In this regard, BET (Blue Economy Technology) could bring revolution in ocean management in terms of optimum use of ocean, ethically and responsibly. BET will open up new types of jobs in manufacturing and service sector scenarios, so also in the areas of infrastructure, transportation, agriculture, health, living style, and urban development.

Strategy of Implementation
To promote the blue economy concept and to maximize financial, social, and environmental returns in an ethical and sustainable manner, the IOR countries may do the following:

- POPULATION CONTROL: Stop early marriages, encourage adopting for a second child, offer incentive, ensure easy and cheap access to contraception tools, decriminalize abortion, break the vicious circle of population ↔ poverty ↔ illiteracy ↔ diseases ↔ malnutrition ↔ women empowerment ↔ population.
- WATER MANAGEMENT: Explore linking the rivers of neighboring countries scientifically for irrigation, power generation, and transport via rivers. Recharge aquifers.
- AGRICULTURAL REFORMS: Recognize agriculture as industry. Initiate collective farming through PIP/PE intervention. Take farmers out of poverty and suicide net. Change crop pattern appropriate to the soil condition. Research, cultivate, and use Genetically Modified food.
- CLEAN ENERGY: Make it a regional campaign at all levels to opt for solar and wind energy. Use biofuel and hydrogen for vehicles.
- RESOURCE COST: Introduce appropriate cost to the natural resources.
- HUNGER MANAGEMENT: Change food habits, manage livestock scientifically, and manage meat consumption.
- REGIONAL COOPERATION: Remove uncertainty from market and strengthen economic stability. Integrate IOR market and encourage trade within IOR countries. Encourage building of trust even among hostile neighboring nations.

TABLE 8.6 (Continued)
Standard Operating Procedure (Roadmap) for Establishing and Sustaining EBE in the IOR

- LOCAL ADMINISTRATION: LA should play an important role in the ethics-based system re-engineering. It must be professional in its approach, must reduce the multiplicity of authorities working at the grassroot level.
  The approach that can be followed by LA may include: (a) appropriate planning to regulate optimum land use, (b) manage forest conservation that can reduce carbon emissions, (c) create healthy ecosystems for livelihoods and industries, (d) regulate fisheries that can provide fallback options during periods of drought or shortfalls in food production.
- CHANGE IN MINDSET: Change in mindset, Ethics, Culture, Feminism, and Yoga-Way-Of-Life are desirable. Make Yoga as way of life—irrespective of creed, caste, nationality, religion. Live for need not for greed. Yoga offers people a sense of harmony with self, society, and nature and could create a social consciousness.
- REVISED GOVERNANCE POLICY UNDER ETHICAL BLUE ECONOMY (EBE): Time has come to adopt alternate strategy for sustainable IOR countries. The measures may include:
- Accurately value potentials of blue economy resources within their respective EEZ and ECS areas.
- Develop a mechanism to measure the contribution of blue economy to national GDP.
- Prioritize blue economy resources based on accurate valuation of its capital investment, natural occurrence, human involvement, and productivity potential.
- Make policy with regard to trade-offs amongst different sectors of the blue economy.
- Use the best available science, data, and technology to harness already demarcated BE resources.
- Anticipate and take cognizance of various threats including climate change, pollution, geopolitics, and adaptation measures.
- Ensure ocean health by investing financially and technologically through blue bonds, insurance, and debt-for-adaptation swaps. Following UNFCCC, the SDG and NDC commitments, especially economic diversification, job creation, food security, poverty reduction, and economic development, could be achieved.
- Ensure effective inclusion and active participation of all societal groups, especially women, young people, local communities, indigenous peoples, and marginalized or underrepresented groups. In this context, traditional knowledge and practices can also provide culturally appropriate approaches for supporting improved governance.
- Government must enhance social protection to bring inclusive economic growth in the society by carrying all sections together.
- Developing coastal and marine spatial plans (CMSP) is an important step to guide decision making for the blue economy, and for resolving conflicts over ocean space. CMSP brings a spatial dimension to the regulation of marine activities by helping to establish geographical patterns of sea uses within a given area.
- The private sector can and must play a key role in the blue economy. Business is the engine for trade, economic growth, and jobs, which are critical to poverty reduction.
- In view of the challenges facing SIDS and coastal LDCs, partnerships can be looked at as a way to enhance capacity building. Such partnerships already exist in more established sectors, such as fisheries, maritime transport, and tourism, but they are less evident in newer and emerging sectors. There is thus an opportunity to develop additional partnerships to support national, regional, and international efforts in emerging industries, such as deep-sea mining, marine biotechnology, and renewable ocean energy. The goal of such partnerships is to agree on common goals, build government and workforce capacity in the SIDS and coastal LDCs, and to leverage actions beyond the scope of individual national governments and companies.

Note: **PIP** = Public-Individual-Private entrepreneurship.

9 Epilogue

9.1 WHAT WE HAVE LEARNT?

An integrated development of an ocean economy that is socially inclusive, environmentally responsible, and has a novel business model was suggested first in the Sustainable Development Goal (SDG 14) of the UN Conference on Climate Change in 1992. This thought matured over the next couple of decades into the concept of the blue economy introduced in 2010 by Gunter Pauli while writing his book *The Blue Economy: 10 Years, 100 Innovations, 100 Million Jobs*. Since then the idea of the blue economy has been gaining ground, insisting for a drastic change in perception in the way the activities in this world have been carried out so far. More so, Pauli's thesis did not concentrate on the ocean alone, and emphasized more on revamping the existing course of technological development. For example, Pauli spoke at length on how using bamboo for house construction in South America is generating potable water, and how in China yields of paddy using saline water have been increasing every year.

The present book deviates from Pauli's one, and focuses entirely on the ocean alone, which occupies about two-thirds of the area of this planet. Moreover, overfishing, climate change, and plastic pollution have been considered by Pauli and others (more vocally by the World Ocean Summit in March 2018 in Mexico) as making up the main threats to the blue economy architecture. This book again swerves from the prevailing popular assessment as above, and includes geopolitics and underserved human aspirations as two other equally important threats to the blue economy.

In addition to offering ways to bring about technological revolution to counter threats to the blue economy, the present book prescribes for ethical governance to repair the prevailing uncertainty in human relationship, extreme individualism, slow-down in economy, rampant corruption, and menacing erosion of moral values. At the end, this book proposes an ethical governance model for the sustainable use of ocean resources for economic growth, improved livelihoods, jobs, and health of the ecosystem. Hence the notion and perception put forward by this book (Ethical Blue Economy, or EBE) is a much revised and value-added form of the concept advocated by Pauli (2010).

The blue economy, discussed in detail in earlier chapters of this book, encompasses harnessing various resources from the ocean for social and economic development in a sustainable manner. The resources include renewable energy (from wind, tide, wave, OTEC), minerals (like hydrocarbon, polymetallic nodules, hydrothermal sulfides, seamount cobalt crust, beach placers), protein-food (fisheries, aquaculture, mariculture), waterways (shipping, trading, port development), and tourism (cruise liners, yachts, harbor development).

In addition, as they contain 97% of the Earth's water, and represents 99% of the living space on the planet, the ocean and associated wetlands are important carbon sinks (blue carbon) that help mitigate the impact of climate change. The ocean protects biodiversity, keeps the planet cool, absorbs about 30% of global CO_2 emissions, and accounts for at least 3–5% of global GDP. The European Commission has estimated the world ocean economy to be worth around US$1.6 billion and this could be more than doubled by 2030. The blue economy, a much-sanitized humane version of the ocean economy, is a sunrise (bountiful, emerging, expanding rapidly) sector that should be kept safe, secured, and ethically sustainable. By doing so, the blue economy would help generate at least 10–15% of global GDP by the year 2030, and 20–25% by 2050 (Mukhopadhyay, 2019).

Given this background, the growth prospects of the blue economy, its capacity for future employment creation and innovation, and its role in addressing global challenges for the Indian Ocean Region (IOR) need an in-depth assessment. The innovation potential of emerging ocean-based industries and their likely contribution to address challenges such as energy security, food security, environment, and mitigating climate change deserve special attention. Finally, the future of the blue economy and its long-term development prospects in a responsible and sustainable manner should be ascertained. This book has discussed all these aspects in the last eight chapters. We accumulate a crux of those discussions.

After introducing the concept and rationale of the rapidly emerging blue economy, the first chapter hints at the shortcomings of the existing system, and the need for a change to a different public administration mechanism. An approach for sustained economic growth, enhanced social inclusion, improved human welfare, and creating opportunities for employment, while maintaining a healthy functioning of the Earth's ecosystems, was conversed. As commerce is interlinked to culture, the geographical significance of the Indian Ocean is traced, so also the culture, civilization, and human migration in the IOR is discussed. A possible blue economy vision for the entire Indian Ocean is also emphasized.

The next three chapters assess the blue economy potential of India, and ten major countries each from the IOR, and from the global ocean. The resources discussed are living resources, non-living resources, energy resources, blue carbon, and services. The IOR countries included in the discussion are: Australia, Bangladesh, Indonesia, Kenya, Mauritius, Oman, Seychelles, Sri Lanka, South Africa, and Thailand. The countries beyond the IOR whose blue economy prospect has also been discussed are: Brazil, Canada, China, Japan, Mauritania, New Zealand, Nigeria, Pacific Islands, the United Kingdom, and the United States of America.

The three major threats to the blue economy are discussed in the fifth chapter. The first threat comes from climate change—largely a natural phenomenon, but contributed enormously since the 1850s by human beings through unbridled carbon emission in the guise of industrial development. The threat is displayed by the rise in sea temperature and sea level, increased ocean acidification, and augmented frequency of natural calamities (Table 9.1). The second major threat comes from safety and security deficits in the region, with unprecedented militarization of the Indian Ocean by major powers. The seemingly unnecessary flexing of muscles under the pretext of trade, heritage, and development are emerging as real threats. The third threat comes from within, the mindset. This is manifested by greed, underserving aspiration of people and nation, unethical attitudes, and unsustainable development. In fact, the erosion of values, balance, and ethics had influenced the first two threats—climate change and militarization—and can be blamed for unequal distribution of wealth and resources in the world.

A series of scientific and technological innovations that could be a part of any future blue economy revolution is discussed in the sixth chapter. Also examined is how such innovations could end up in creating wealth, jobs, and a balanced society. Chapter 7 pertains to economic projections in which an exercise is carried out to quantify the possible contribution of the blue economy to the GDP. How much each

TABLE 9.1
Crossing Any of the Following Boundaries Will Be Dangerous to Sustainability

Sl	Process	Events	Current Value	Boundary Value	Status	Preindustrial Level
01	Climate Change	CO_2 concentration	400	350	Crossed	280
02	--- do ----	Radiative Forcing	1.5	1.0	Crossed	0
03	Biodiversity Loss	Extinction Rate	>100	10	Crossed	0.1–1
04	Biogeochemical	Atmospheric N_2	121	35	Crossed	0
05	--- do ----	Anthropogenic P	8.5–9.5	11	UC	−1
06	Acidification	Aragonite saturation	2.90	2.75	Crossed	3.44
07	Land Use	Converted to Cropland	11.7	15	UC	Low
08	Freshwater	Global Consumption	2600	4000	UC	415
09	Stratospheric Ozone	Global Concentration	283	276	Crossed	290

Source: Rockström et al. (2009); Diamond (2011).

Notes: CO_2 concentration in ppm in volume; Radiative Forcing in Watts per square meter, Extinction rate (number of species per million per year), anthropogenic nitrogen (N_2) removed from the atmosphere (millions of tons per year), anthropogenic phosphorus (P) going into the oceans (millions of tons per year), Global mean saturation state of aragonite in surface seawater (omega units), Land surface converted to cropland (%), Global human consumption of water ($km^3/yr.$), Stratospheric ozone concentration (Dobson units). Crossed = crossing the boundary value, UC = Under Control.

of the three selected sectors—fisheries, shipping and trade, and tourism—are going to account for a nation's economy in 2030 and 2050 are examined.

After the global economic recession in 2008, two Nobel economists, Amartya Sen and Joseph Stiglitz, showed concern that GDP cannot alone indicate the real progress of a country, and probably needs a revised real indicator to measure development at the grassroot level. Similar concern was put forward by Stiffler (2014, Fig. 5.16). It is seen that an increase in GDP is not necessarily linked to the enhancement in purchasing power of the people, or employment generation, or overall economic development of the country. Rather, GDP growth is often found to be related to growth of a few individuals. For instance, in India 73% (up from 58% in 2018) of the country's resources are locked up in the vaults of only 1% of people (OXFAM International, 2019). That GDP increase is not analogous to poverty alleviation is also proved by the fall of India in the Human Development Index (HDI) rating to an unenviable 130th position (UN-HDI, 2019).

This alarming situation necessitates the introduction of ethics into the economy (ABP, 2019). A recent report suggests that poverty in the USA has decreased from 14% in 1967 and 15.1% in 2009 to 11.1% in 2018 (US Census Bureau Shares, 2019). However, the inequality continues to grow in the USA as seen by the fact that the top 5% earners recorded a nearly 16% growth in income, while the bottom 10% in fact showed a decrease in their income. Also, the household average income has shown marginal growth in five decades—from US$47,085 in 1967 to US$63,179 in 2018. The economic inequality at the global level shows that although the income share by the top 10% of the population has fallen from 35% in 1990 to 32% in 2015, the trend shows a return to 35% by 2030, whereas the best situation allows only 25% of income to be shared by the top 10% of people (World Bank, 2017).

India would be better off following the well-worn track of a few countries that have shown tremendous progress in the blue economy sectors. In Australia, offshore gas, oil, and aquaculture led the blue economy contribution of US$47 billion to her GDP for 2011–12. Similarly, in Mauritius, the coastal tourism and port development as major blue economy activities accounted for US$90 million. Likewise, fisheries, tourism, and transport led China's marine economy sector to contributing 4.03% to its GDP in 2010 and employing 9.25 million people. India could focus on marine ICTs, and transport (shipping) and communication services, and the creation of a knowledge hub for marine research and development, alongside the more traditional sectors like fisheries and coastal tourism (Mittra, 2017).

As discussed in Chapter 5, India is lagging behind in outreach activities to IOR countries compared to China. This needs to be quickly attended to and India must pay particular attention to strengthening connectivity with IOR nations in terms of exploration, identification, and making use of offshore resources, developing infrastructure facilities in alignment with sustainability, increasing trade and commerce, and spreading ethics-based liberal democracy in the region. However, lack of coordination and overlapping of authorities among the 17 ministries of India who are associated with blue economy is a sad commentary. India probably needs one ministry to work through as a nodal agency—the Integrated Maritime Authority. To start with India must have satellite-based detailed Marine Spatial Planning of its entire coastal zone. The environmental data from the ongoing ICMAM project of her Ministry of Earth Science could help enormously in this endeavor.

The concept of port-oriented industrialization is premised on the liberal-economic model of the newly initiated Sagarmala project of India (Fig. 2.5). The project aims to integrate India with the global economy model, global finance capital, markets, and technology. It would possibly synergize India's land-sea domains into a single development model through multimodal connectivity through inland and coastal waterways. India presently transports only 6% of its cargo through inland water, against 47% by China and 34% by Japan.

In 2015, India redefined its maritime strategy (*Ensuring Secure Seas: Indian Maritime Security Strategy*) to develop its role as a "net security provider" in the IOR. Its vision for a blue economy development in the IOR could propel India to its aspiration to be a credible leader in the Indo-Pacific security paradigm. The crux of the three threats to blue economy from climate change, military geopolitics, and human mindset (discussed in detail in Chapter 5) and the resultant roadmap and SOP for the path of recovery is tabulated in Table 8.6.

9.2   LESSONS FROM AND FOR OTHER OCEANS

Contrary to the popular perception, there appears to be a hazy and uncertain relation between poverty, unemployment, and contribution to national GDP. This indeterminate relation is found in many IOR countries and even with the USA and UK (Table 8.3). In the case of India this undefined relation is quite prominent (Table 9.2). For example, in India, the states with high poverty like Odisha (32.59%), Karnataka (20.91%), and West Bengal (19.98%) record a moderate degree of unemployment (5%, 1.5%, and 4.9%, respectively) and contribute variably to the national GDP (2.6%, 5.9%, and 6.8%) In comparison, states like Goa, Kerala, and Andhra Pradesh that registered least poverty (5.9%, 7.05%, and 9.20%, respectively) show a high rate of unemployment (9.6%, 12.5%, and 3.9%) and contribute only peripherally to the country's GDP (0.5%, 3.8%, and 4.4%). The result suggests that socioeconomics is a complex area and no straight linear equation can explain the multifaceted dynamics.

The bottom line of the proposed ethical blue economy (EBE) is sustainability that could be achieved by taking integrated action primarily in areas like social, economic, environmental, geopolitical, and ethical (Fig. 9.1), as this world has not performed well with regard to maintaining sustainability. In this regard, this book has tried to put up a series of facts vis-à-vis major challenges to the blue economy (climate change impact, geopolitics and militarization, and human mindset) and a possible way out (Table 9.3).

The assessment of the progress of the blue economy in the IOR and elsewhere must start with the fact that resources available in the oceans are not limitless. Against this background, the ever-increasing demand for resources by human beings, inefficient governance, inadequate economic incentives, insufficient technological advances, scarce human capacities, and adoption of unethical approach have often led to poorly regulated activities in several countries of the world (UNDESA, 2017).

For example, it seems 57% of fish stocks are fully exploited and another 30% are over-exploited (FAO, 2016). Illegal, unreported, and unregulated (IUU) fishing

TABLE 9.2
Development Indicators for Coastal States in India

States	Unemployment Rate (2015–16)	Poverty Rate (2010–11)	2013–14 GDP Contribution	Service Sector Rate of Growth (2013–14)
Gujarat	0.9%	16.63	7.31	12.7%
Maharashtra	2.1%	17.35	14.4%	9.3%
Goa	9.6%	05.09	0.5%	10.3%
Karnataka	1.5%	20.91	5.9%	8.0%
Kerala	12.5%	07.05	3.8%	5.5%
Daman & Diu	0.3%	09.86	–	–
Tamil Nadu	4.2%	11.28	8.2%	9.3%
Andhra Pradesh	3.9%	09.20	4.4%	7.2%
Odisha	5.0%	32.59	2.6%	8.7%
West Bengal	4.9%	19.98	6.8%	9.4%
Pondicherry	4.9%	09.69	0.2%	–
All India Rate	5.0%	21.29	–	9.0%

Source: Labour Bureau (2017), Mittra (2017), MOSPI (2019).

FIGURE 9.1 The Mosaic of Social Economic and Environmental Network

TABLE 9.3
Major Challenges to Blue Economy and a Possible Way Out

FACTS AS OF TODAY	ACTIONS PROPOSED

THREAT I: CHALLENGES AGAINST CLIMATE CHANGE IMPACT

I-1: Restructuring Food, Agriculture and Energy Sectors

Since 1980 the global temperature has increased by 0.94°C, and sea level rose by 20–23 cm; Although CO_2 emission is down now, the accumulative effect would continue to warm the Earth.	• Collaboration to cut GHG emission among USA-BRICS-EU required; • Make soil-crop compatibility test a must; Boost community farming; Encourage seawater agriculture; Appropriately use GM crop; • Save energy, and transit from fossil fuel to renewable energy; • Reduce GHG on food (meat/milk) processing; Manage Hunger • Convert Smart cities to Eco-smart; Use S&T to improve human conditions;

I-2: Arranging Water for All

People's access to clean water increased from 76% in 1990 to 90% in 2018, but present consumption of water is almost twice than what nature can generate; About 70% of clean water is used for agriculture, 20% for industry (of which 70% by developed country), while 10% is used for domestic purposes.	• Increase R&D for lower cost of desalination; • Develop wastewater products such as fertilizer, algae for biofuel and feeding shrimp, and recover nitrogen and phosphorus; • Implement integrated water management to ensure universal water and sanitation access; Mass-produce electrochemical wastewater treated solar power toilets; • Interlink rivers at regional level (South Asia, West Asia, Africa, etc.)

I-3: Balancing Population Growth and Resources

Current world population is 7.6 billion, which is expected to grow to 8.6 billion in 2030 and 9.8 billion in 2050; The 2.2 billion more people in just 23 years will create unprecedented demand for food, water, energy, and employment; Population growth is expected to be most rapid in the 49 least developed countries, doubling in size from about 900 million today to 1.8 billion in 2050; There were only 1 billion humans in 1804; 2 billion in 1927; 6 billion in 1999; and 7.3 billion today, and 9.8 billion in 2050.	• Incentivize people to adopt a two-child policy to reduce population growth, the second child being adopted (see Section 8.2), Support policies to cut mortality rate for child at birth, inspire family planning, • Impart training in resilience, disaster forecasting, and management; • Develop eco-smart-cities and villages with modern facilities- ICT, hygiene, hospital and educational institutions. HDI should be a prime focus; • Encourage saltwater agriculture (halophytes) on coastlines to produce food for humans and animals; Also improve rain-fed agriculture and irrigation management; Invest in precision agriculture, aquaculture and mariculture; • Produce pure meat without growing animals (demonstrated in 2013); Use genetic engineering for higher-yielding and drought-tolerant crops;

(continued)

TABLE 9.3 (Continued)
Major Challenges to Blue Economy and a Possible Way Out

FACTS AS OF TODAY	ACTIONS PROPOSED
	• Build floating cities for ocean wind & solar energy, agriculture, and fish farms; Accelerate R&D for safe nanotechnology to help reduce material use per unit of output while increasing quality; Encourage vegetarianism; Examine expanding insect production for animal feed and human diets (remember insects have low environmental impact per nutritional unit)

I-4: Innovative Technology to Improve Human Condition

| Artificial intelligence and quantum computing are changing the nature and speed of new scientific insights and technological applications;

 In addition, advancement is made in synthetic biology, 3D and 4D printing, robotics, nanotechnology, ICT, drone- and renewable energy technologies,

 Collective intelligence systems will take civilization to an unimaginable world. | • Invest more in R&D to link research agendas to human needs and threats, and connect it throughout the world by ICT;
 • Prevent future artificial super intelligence evolving against human interests;
 • Prosecute "patent trolls" (firms that do not produce anything but simply file patent lawsuits for extortion);
 • Do intense research and public awareness campaign in connection with health facility, Strengthen primary healthcare system rather than tertiary one,
 • Encourage research and promote computational chemistry, computational biology, and computational physics in appropriate fields |

THREAT II: CHALLENGES OF GEOPOLITICS AND SECURITY ISSUES

II-1: Use Shared Values, and Modern Security Regime to Reduce Terrorism, Piracy, WMD

| Incidents of armed conflicts have increased from a mere 4,000 in 1990 to 16,000 in 2018 and projected to be 35,000 in 2030;

 About 65.6 million people have been forcibly displaced of which 22.5 million are now refugees;

 Although Global Peace Index has bettered slightly, cyber-attacks from government and organized crime groups on general population have only increased;

 Violence cost the world $13.6 T in 2015. | • Conflict prevention strategy must include transparency, accountability, reachability of government services to all people; Stop giving patronage to one group and corruption for any excuse whatsoever;
 • Modify the syllabus particularly of religious schools to stop making possible terrorists in future;
 • Bottom line of teaching would be that our education, research, teaching and learning (from school onwards) must make sense that superiority of human being does not rest in power of possession but in power of unison;
 • Encourage peaceful resolution of conflict, accepting institutionalize system to track the cross-border movement of WMD and human-trafficking. |

TABLE 9.3 (Continued)
Major Challenges to Blue Economy and a Possible Way Out

FACTS AS OF TODAY	ACTIONS PROPOSED

II-2: Controlling Transnational Organized Crime

Annual cost of organized crime is $3 trillion, this is now a growing enterprise; Business may lose $2 trillion in cybercrime; It is possible that 50% of medicine online is fake contributing $200 billion to crime industry;

Remember Columbia sustained armed insurgency for decades by money coming from illicit drug trafficking, illegal mining, and extortion.

- Formulate a global strategy including a financial prosecution system to counter organized crime;
- Increase military surveillance in the entire Indian Ocean through IORA or some such regional bodies, wherein all IOR countries would contribute by providing security staff, logistics, and finance.
- Engage farmers in high-income agricultural alternatives instead of producing drugs;
- Make laws to consider organized crime / terrorist activities as crime against humanity.

II-3: Emergence of Democracy from Authoritarianism

Democracy is threatened as antidemocratic forces are using cyber world;

Net freedom for people in the world is decreasing. Of the 195 countries, 44% are free, 36% not free, while 20% are partly free;

Internet and social media are carrying freedom and transparency to all sections of people, and these must be fortified;

Liberal and plural democracy must be spread, as powerful democrats tend to become autocrat

- Secure tamper-proof electoral systems and stop high election spending;
- Invest in R&D that could counter fake news and strict guidelines for the digital world;
- Establish and enforce measures to reduce corruption, promote transparency, participation, inclusion, and accountability in decision making;
- Promote new forms of e-governance, and introduce civics in all forms of education, implement UN treaties on minorities, migrants, and refugees;
- Create and protect spaces for dialogue and peaceful conflict resolution;
- Deal with military, security, and intelligence services—"apolitically"

THREAT III: CHALLENGES FOR HUMAN MINDSET AND ETHICS

III-1: Reduce Rich-Poor Gap through Ethical Market Economy

Extreme poverty reduced from 51% in 1981 to <10% in 2018 due mostly to income growth in India and China; However, inequality in Africa remains a threat to global stability;
The wealth of top eight rich persons equals that of 3.6 billion people in the world;

- Instead of exporting raw products, poorer regions must use innovative technology to develop finished product for exports—gaining in the process skill, money and employment; Tax-structure may be made friendly for this to happen;
- Ensure a friendly tax system for wealth creators and job producers;
- Promote decentralized autonomous institutions; Establish several community skill development centers; Take help of ICT in a big way;

(continued)

TABLE 9.3 (Continued)
Major Challenges to Blue Economy and a Possible Way Out

FACTS AS OF TODAY	ACTIONS PROPOSED
New S&T innovations must lower the cost of living; World economy is growing at 3.6%, and with population growth of 1.11%, the global income per capita is growing at 2.39% annually.	• Stark unequal distribution of wealth and job-less growth are the two major shortcomings of the existing capitalist market economy. This anomalous growth is unsustainable. Moreover, the GDP as indicator to measure growth is not real and does not respect ecology, humanity, and anything other than profit. Instead GPI may be considered as real measure of progress (Fig. 5.18)

III-2: Changing Women Status to Improve Human Condition

About 84% of world nations have guaranteed for gender equality; women as corporate board members increased by 54% since 2010 and in Parliaments by 23.5% since 1997; Yet barbarism (FGM, honor killing, dowry deaths) in some parts continues and about 600 million women from 15 countries are still not protected from domestic violence by law; Although women constitute 41% of global workforce, they earn 35% less than men.	• Parents should nurture gender equality at home among their male and female child. Charity begins at home; • Make policies to help women in their career and family responsibilities; • Legalise equal remuneration for equal value of work; • Increase participation of women in peace-building negotiation; • Provide maximum security and safety for women at workplace (use mobile app / siren to draw attention for any violation) and strict compliance of the law to deal with all types of violence against women (mental, physical, religious, and traditional).

III-3: Use Education to Make People Aware of Global Challenges

Artificial Intelligence is being used increasingly to know the best way to learn, as cognitive/ neuroscience has shown that brain performance can be improved; Internet and social networking could spread education fast; AI could help in curriculum formulating.	• Make individual wisdom and collective intelligence national objectives of education; Increase R&D funding of AI-human symbiotic evolution; • Begin shifting from mastering a profession to mastering a combination of skills; Teacher training schools should show how different teaching strategies affect neural activity of students' brains; • Explore alternative models of education and learning (both Finland and South Korea score top in the world but have quite different systems). For example, Mathematics is an outdoor subject and should be taught in the lap of nature with 3-D models and geometrical interpretations.

TABLE 9.3 (Continued)
Major Challenges to Blue Economy and a Possible Way Out

FACTS AS OF TODAY	ACTIONS PROPOSED

III-4: Integration of Ethics in Decision Making

Presently most of the decisions are made by artificial intelligence (data computed & processed by computer) and such process is not ethically neutral and remains unaccountable; Narrow interests of profit and strategic positioning takes precedence; The universal declaration of human rights insists for global ethics and justice; Ethics even do not receive much consideration in S&T innovations	• Introduce civics and ethics in all forms of education—ones' behavior must match the values s/he believes in; • Government and citizen must be on the same page of development; • Enforce ethics to reduce corruption; • Stop taking decisions based on artificial intelligence models and not before subjecting the model to pass through ethical guidelines; • Love for humanity and global consciousness must guide the formulation of policy decisions.

III-5: Ethical Business Leadership in the Global Economy

CORE ETHICAL VALUES:	SUPPORTING ETHICAL PRINCIPLES:
Trustworthiness, Respect, Responsibility, Fairness, Caring, Citizenship	Sincerity, Integrity, Promise Keeping, Loyalty, Honesty; Autonomy, Courtesy, Self-Determination; Diligence, Continuous improvement, Self-restraint; Justice, Impartiality, Equity; Kindness, Compassion; Philanthropy, Fairness

Source: Modified after Millennium-project.org (2018), and Mukhopadhyay et al. (2018b), Mukhopadhyay (2019). sustainability.asu.edu/sustainabilitysolutions/ece-cert/
Notes: ICT = Information & Communication Technology, T = Trillion, WMD = Weapons of Mass Destruction, FGM = Female genital mutilation.

accounts for roughly 11–26 million tons of fish catch annually, equivalent to US$10–22 billion, and in fact is considered as unlawful or undocumented revenue. Again, unplanned and unregulated development in the narrow coastal interface and nearshore areas has caused destruction of marine and coastal habitats and landscapes, often irreparably. Pollution in the form of excess nutrients from untreated sewerage, agricultural runoff, and marine debris such as plastics is also causing havoc and choking the sea.

The Ethical Blue Economy Technology (EBET) must focus on areas to turn the demand and threats into opportunities (see Chapter 6). In fact, any measure to ease the negative impact must be a new product or process or algorithm based on innovation. And any new such initiative will consequently open up new opportunities for employment, infrastructure expansion, and change in mindset (doing similar/same things, but differently). All these efforts must focus also on removing poverty in a time-bound manner, as poverty has been the biggest pollution maker. It is because poverty and those who are undeservingly ambitious, encourage short-cuts of the long procedure through corruption, and cutting corners of the law. This disturbs the social balance and pushes some people into the deeper hole of poverty.

Cooperation in developing EBET is extremely important, as the research is both cost intensive and time consuming. The patent data show that the USA and Europe are involved in a high number of cooperative projects in harnessing energy from solar insolation, followed by Belarus and Russia (in solar PV and thermal), India and the USA (solar PV, wind), South Africa and Europe (biofuels and wind), the USA and China (solar PV). Maximum co-invention has occurred in the field of CO_2 storage, followed by biofuels, CO_2 capture, solar PV, wind, hydro/marine, geothermal, and solar TH expansion (Hascic et al., 2012). For example, power generation potential from solar radiation in India is developing fast. India expanded its solar-generation capacity eight times from 2,650 MW in May 2014 to over 20 GW in January 2018.

The world saw the change in mindset, albeit subtly, during the climate conference in Paris 2015 (COP 21). Perhaps, the biggest gain for IOR is the way the "principle of common but differentiated responsibilities," which is part of the 1992 UN climate convention, has been elucidated in the Paris Agreement. In the end, it seems that none of the three worlds (developed, developing, and least developed countries) have won in Paris (COP 21). All have compromised their positions and rightly did so. But if somebody has really won in Paris it is the "possibility" that this blue planet could still be saved if all the countries in the world come together to see the reasons, voluntarily. The ethical blue economy (EBE) could take a cue from this understanding.

This world would be much safer and secure when EBE redefines the present-day neoliberal free-market capitalist economy. EBE will reoutline the way the things are perceived today. According to EBE philosophy, richness does not mean earning more, or spending more or saving more, rather richness is when you need "no more." The effort would result in developing a unique ethical socioeconomic model by re-engineering and reforming the existing system of governance (see Section 8.2). The re-engineered model will lead to a possible transformation of the IOR into a hub of excellence in human values.

The EBE doctrine, based on the BTVG doctrine (Buddha-Tagore-Vivekananda-Gandhi, see Sections 8.1 and 8.2), involves resolving the intrinsic relationship between humans and nature, encouraging in the process ethical innovation, gender equality, rational aspirations, and sustainable development. Moreover, unending greed, unfair trade, absolute selfishness, and seemingly unnecessary competitions tend to create disorder in the minds of people and entrepreneurs alike. To make that happen, the existing governance system of most of the countries including those in the Indian Ocean, needs a revamping and re-engineering.

Besides adopting EBE in letter and spirit, the global oceans (beyond the IOR) should look to embrace the BTVG doctrine of balancing economic benefits with sustainability for meeting the broader goals of security, growth, employment generation, equity, and protection of environment. As EBE aims to move beyond "business as usual" and to consider economic development and ethics as compatible propositions, it needs a long-term strategy. In order to become actionable, the EBE concept must be supported by innovative technology, trusted knowledge base, strong collaboration, informed citizens, and participatory governance (UNDESA, 2017). The world should not look at the oceans as just water bodies, but as a global stage for continued economic, social, ethical, and cultural dialogue. It must leave the oceans for future generations who would be grateful.

The world has deviated far from its expected course of sustainable use of resources. A greed-based economy is ruling the roost. Climate change is considered as the biggest manifestation of the failure of the market economy, as the inflated materialistic demands exceed the carrying capacity of the world. For some, climate change is seen as a social evil, a personification of greed. But the question arises of "how to deal with this catastrophe?"; the answer probably lies in "thinking differently" and to opt for a course correction—quickly and squarely. What is needed is to reinvent the relation of the individual to nature to make sense that superiority of human beings does not rest in power of possession but in power of unison. This could be achieved probably when EBE, as we propose here, replaces the present-day market economy.

9.2.1 SOME EXAMPLES OF ETHICAL GOVERNANCE

While many may think that an ethical governance model is simply a supreme-utopian idea, there are unbelievably few places in this world that do have a community governance that implements a policy for their people—"do not steal, do not lie, do not be lazy, do not do anything that may reach harm to others," and all of these voluntarily.

The first such example comes from Taquile Island in Lake Titicaca, 45 km offshore from the Peruvian city of Puno (Fig. 9.2). This non-uniformly extended island is 5.72 km² in area (5.5 × 1.6 km), wherein about 2,200 people live on the island. The highest point of the island is 4,050 meters above sea level and the main village is at a height of 3,950 meters. The inhabitants of Taquile Island speak Quechua, an indigenous language mostly spoken during the Inca Empire, and still spoken primarily

FIGURE 9.2   Taquile Island in Titicaca Lake, Peru

by about 8–10 million people living in the Andes Mountains of Peru and highlands of South America.

The Taquile islanders run their society based on community collectivism and on the Inca moral code *ama sua, ama llulla, ama qhilla'* (do not steal, do not lie, and do not be lazy). The first coming of tourists in 1970s was construed as a threat to the maintenance of the moral code, but the islanders confidently neutralized such danger by creating an innovative, community-controlled sustainable tourism model, offering home stays, transportation, lodging for groups, cultural activities, local guides and restaurants.

The second example comes from the village Shani Shingnapur in India's Maharashtra state, where homes have no front doors, shops are always left unlocked, banks remain unlocked, and yet locals never feel unsafe. The villagers shun security because of their undying faith in Lord Shani, the god of Saturn, who is considered the guardian of the village (Fig. 9.3). They leave their jewelry and money unsecured, firmly believing that their holy guardian will protect them from any mishap (Jain, 2016).

Shani Shingnapur is a village of 82 km² and attracts about 40,000 devotees each day due to its remarkable 300-year-old legend. The legend says that about three centuries ago, after a bout of rain and flooding, a heavy black slab of rock was found washed up on the shores of the Panasnala River, which once flowed through the village. When locals touched the 1.5m boulder with a stick, blood started oozing out of it. Later that night, Shani appeared in the dreams of the village head, revealing that the slab was his own idol. The deity ordered that the slab should be kept in the village, where he would reside from here on. But Shani had one condition: the rock and its colossal powers must not be sheltered as he needed to be able to oversee the village without hindrance. Shani then blessed the leader and promised to protect the village from danger.

After the villagers installed the huge slab on a roofless platform in the heart of town, they decided to discard all doors and locks. They didn't need them anymore, not with the Lord to watch over them. The police station, which only opened in September 2015, has not yet received a single complaint from the villagers. Locals

FIGURE 9.3   No Houses in the Village Have Doors or Locks. The Bank in the Village Is Also Not Locked

are so nonchalant that they don't even ask their neighbors to watch over their house while they are out of town (Jain, 2016). They believe that thieves will immediately be punished with blindness, and anyone dishonest will face seven-and-a-half years of bad luck.

9.3   THE LAST WORD

Although ocean governance has been perennially lacking in the IOR, the mode and mechanism of such governance has been evolving since the fifteenth century (Table 9.4). There is no one single medicine for all diseases. As the area of suffering differs the medicine may get changed or the illness may require the same medicine of different dose. For example, a field experiment on social science between the late 1990s and early 2000s amongst the poor populace of Africa and India showed that subsidies, which are considered as a demotivating factor under a capitalist economy (according to conventional wisdom), could in fact increase use of preventive health care among the poor (Duflo and Banerjee, 2019). This fact must encourage WHO to promote subsidy in health care for the poor and IOR countries to initiate focus welfare schemes.

In contrast to the conventional belief that high carbon emission reflects a high degree of development and superb health care facility in the country, it is now found that high life expectancy can be achieved even with low CO_2 emission (www.equalitytrust.org.uk). While high CO_2 emission (>15 tons/person) coupled with high life expectancy is recorded in the USA—75 years, Austrasia 79 years, and Saudi Arabia 71 years—some countries with the least CO_2 emission (<5 tons/person) also record high life expectancy (Costa Rica 75 years, Sri Lanka 72 years). Denmark, France, Portugal, Iceland, Japan, the UK, Norway, and Canada all have a high life expectancy rate (between 70 and 80 years) while carbon emission has been varying between 5 and 15 tons/person. India stands at 62 with a corresponding CO_2 emission at 2.5 tons/person. In contrast, Sierra Leone has the least life expectancy rate (<40 yr) coupled with least CO_2 emission (<1 tons/person, www.equalitytrust. org.uk).

The carbon footprint of an American is three times higher than that of an Indian. A recent report of The Emissions Database for Global Atmospheric Research (EDGAR, 2019) suggests that many highly developed nations have reached emission levels that are disproportionately high for their population. Despite its small population, the USA has been the top emitter of CO_2 up until few years ago. Positive development in Europe has meant that its emission share is gradually declining to match its population share. Presently China is the world's largest carbon emitter, while almost all oil-producing countries in the Middle East have the highest ratio of emission to population ratio. India's per capita emission is however significantly lower than its share in population. The emission increased from 2005 to 2017, by about 24% in the power industry, by 28% in other industries, by 3% in building construction, by 21% in transport, and by 40% in other sectors.

The global population has been recording a lower growth rate (%) for some time now as it fell from 1.49% in 1950 to 1.0% in 2019, and is expected to fall further to 0.75% by 2030 and 0.5% by 2050 (US Census Bureau Shares, 2019). In this

TABLE 9.4
History of Maritime Governance

Year	Events	Year	Events
2009	May 13, 2009 marks the end of the ten-year period for making submission to Commission on the Limits of the Continental Shelf. However, the proposal for extension of this deadline for the developing (Pacific region) countries considered	1964	Two conventions on Territorial Sea + Contiguous Zone, and Continental Shelf enter into force
2008	As ten-year submission period ending next year, detailed assessment of the status and intended submission date of applicant-countries on the Extension of Limits of the Continental Shelf made	1962	Convention on the High Seas enters into force
2001	Decision taken on May 29, 2001 with regard to the date of commencement of the ten-year deadline period for making submissions to the Commission on the Limits of the Continental Shelf in concurrence with article 4 of Annex II to the United Nations Convention on the Law of the Sea to consider May 13, 1999 as starting date	1960	UNCLOS II held, no new agreement, Developing nations participated … no voice
1999	Scientific and Technical guidelines on Limits of the Continental Shelf (beyond 200 nautical miles from shore) was adopted	1958–60	Four conventions under UNCLOS I held on Territorial sea + Contiguous Zone, Continental Shelf, High Seas + Fishing, Conservation of Non-Living resources in High Seas
1994	Agreement on UNCLOS III implementation reached and Law of the Sea comes into force on November 16, 1994, after ratification by 60 countries	1956	UNCLOS I: First Conference on Law of the Sea held
1990–4	Negotiations continued on the implementation of UNCLOS III	1946–50	Argentina (1946), Chile and Peru (1947), Ecuador (1950) claimed sovereign rights to 200 miles to cover Humboldt current fishing grounds. Other nations also claimed the same.

TABLE 9.4 (Continued)
History of Maritime Governance

Year	Events	Year	Events
1982	UNCLOS III kept open for signature. It sets limit for various maritime zones, shipping passages, protection of marine environment, scientific research, conservation & management of living & non-living resources, and settlement of disputes	1945	Truman Proclamation: USA unilaterally extends jurisdiction over its Continental Shelf, Other nations were quick to follow
1973–82	UNCLOS III: 160 countries participated. Emphasized on building consensus, rather than deciding on majority vote	1930	The League of Nations called a conference in Hague to decide on national claims, but the deliberation remained unsuccessful
1967	Differing Sovereignty: 25 countries used 3-mile rules, 66 countries implemented 12-mile rule, while 8 countries followed 200-mile to claim sovereignty	~ 1750	Adopted Cannon-shot-the-rule, i.e., a coastal state can enjoy sovereignty as far seawards as a cannon can fire a cannon ball (~3 miles)
1966	Conservation on High Seas + Fishing enters into force	~ 1650	Adopted Freedom-of-the-Seas-Doctrine, i.e., The seas are free to all nations but belong to none
		1494	Alexander VI divided the Atlantic Ocean between Spain and Portugal

Source: Modified after websites of UNCLOS = United Nations Convention on Law of the Sea; Mukhopadhyay et al. (2018b).

background, the threshold for high human development is 0.8, while the same for Earth's bio-capacity (ecological footprint) has been 0.021 km^2 per person. The bio-capacity is the biologically productive land and water area that a person needs to produce all the resources s/he consumes and to also absorb the waste it generates, using prevailing technology and resource management practices (Global Footprint Network Report, 2008; UN-HDI, 2019). Cuba appears to be the best country with high HDI (0.778 in 2018), yet well within the bio-capacity threshold ($<0.02 \text{ km}^2$), while the USA has recorded one of the best HDI (0.92) but with per capita utilization of 0.095 km^2. Norway and Canada probably record the highest HDI (0.954 and 0.922, respectively) with an ecological footprint covering 0.07 km^2. Most of the European and North American countries have crossed the thresholds both in terms of HDI and ecological footprints, while a majority of African and Middle East countries ironically are well under either of the thresholds.

Seeing the waves of criticism for bringing the world economy on the edge, capitalism is now busy to obtain a new look. Despite the creation of immense wealth, and subsequently taking out a large number of population from poverty, still the facts that millions of people still go to bed hungry each day, every six seconds a child dies of malnutrition, and a mere 5% people hold about 80% of the entire world's resources, suggest that capitalism is not working well. Hence, this book realizes somewhere down the line the traditional model of profit-driven conventional capitalism must be replaced with "compassionate capitalism." Under the new arrangement, corporations should have a social heart and business brain. It's the new business strategy of the twenty-first century. It's also about the means (methods), and not only the end (profit). Compassionate capitalism is increasingly finding support from top global business houses as they try to redefine companies' purpose statement, and undo the stakeholder's primacy. This economy advocates increased government expenditure and lower taxes as measures to stimulate all the three levels—the demand, market, and economy. The question however remains as to whether politicians and government could be trusted to spend astutely, efficiently, and prudently in an impartial way (Singh, 2019).

Most people are greedy, always ... to earn more money, materials, fame, publication, prestige, and sometime to receive divine grace. Greed, in one way or other, makes us what we are. All BTVG philosophers had to fight their entire life to win over these temptations of mind, power, undue fame, flesh, and money. They summarized that what we choose is often not what we want, at times we want what we do not want and for which we will regret wanting, and what we can have but cannot hold on to. This is because preferences of the human mind are not consistent (Duflo and Banerjee, 2019).

The first challenge of capitalism is to change the current greed-based economic system where GDP is given more prominence than the genuine progress indicator (GPI). The GPI is the difference between GDP – N (where N is the combination of environmental degradation + social anomaly + economic cost, all due to unsustainable extraction of resources, physical alterations and destruction of habitats and landscapes, climate change, and pollution). In fact, the average combined cost N is estimated to be little more than 40% of GDP between the year 1950 and 2007 in the USA (Stiffler, 2014). It means a GDP growth of 7% actually depicts a progress of only 4.2% (Fig. 5.16). In fact, an effort has now been initiated by some of the authors of this book to develop a Tangible Growth Index (TGI), to constrain Gross Domestic Product (GDP) estimation.

The second task is to develop capacity and human resource to appropriate the employment opportunities and other benefits resulting from innovative ethical blue economy sectors. The third test would be to correctly evaluate in dollar terms the economics of the blue carbon process (carbon sequestration by mangroves, seaweeds, wetlands, coral reef, sea-level rise, acidification, and higher temperature gradient for power generation, etc.; Mukhopadhyay et al., 2018b).

Poverty might play a major role in fueling greed at the beginning stage. For a beggar, a small tweak of moral turpitudes could be a necessity, but for the rich that is simply greed. Poverty is not just a lack of money, in fact it is not having the capability to realize one's full potential as a human being. Similarly, more impact on education will be when you have good teachers rather than offering free midday meals or free textbooks. Toward poverty alleviation, a little incentive (monetary or in kind) could make the healthcare system at the grassroot level work wonders (Duflo and Banerjee, 2019).

To neutralize any geopolitical aggression and aspiration of China, recently on September 27, 2019, Japan and the EU signed an infrastructure deal in Brussels to coordinate the transport, energy, and digital communication sectors between Asia and Europe. This will be a multi-billion-dollar (US$65.48 billion) link through the Indian Ocean and is sure to upset China's OBOR/BRI. This EU–Japan collaboration will build infrastructure in the above-mentioned areas, especially in the IOR countries, without putting them under mountain of debts, and without letting the loner nation (China in case of OBOR) impact the loaned IOR countries in their strategic, military, and internal matters.

The EBE that we put forward in this book following the BTVG doctrine, seeks to answer the questions that emerged out of climate change impact, geopolitical aggression, and human mindset. It promotes ethical economic growth, social inclusion, and improvement of livelihoods while at the same time ensuring environmental sustainability of the oceans and coastal areas. This contour of economy decouples socioeconomic development from degradation of environment and ecosystem.

The exact pathways for the blue economy in the IOR differ from country to country depending on their national and local priorities and goals, For instance, India's commitment for an inclusive blue economy regime for the IOR countries would emerge from regional collaboration, her credible past, and her age-old ethical leadership in the region. The high moral positioning of India must be strengthened further with the power of unison, empathy, and compassion.

Yet there are some common steps that might make this economy successful (UNDESA, 2017). By adding ethical dimensions to the existing blue economy paradigm, these steps are to:

- Ensure accurate valuation of blue economy resources
- Prioritize the specific resource area(s) for investment
- Invest substantially in best R&D program and in technology innovation
- Explore introducing blue bonds, insurance, and debt-for-adaptation swaps to raise funds
- Anticipate and adapt climate change impact, by strengthening regional and global cooperation
- Safeguard ocean health through new non-polluting techniques
- Follow UNCLOS regulations and build robust legal and institutional frameworks to achieve economic development
- Involve societal groups, especially women, youth, local communities, indigenous peoples, and marginalized or underrepresented groups to maximize participatory governance
- Use traditional knowledge and practices to follow culturally appropriate approaches for supporting improved governance
- Develop coastal and marine spatial plans (CMSP) to help regulate all blue economy activities and to resolve conflicts

- Make considerable investment in capacity building and human resource development, make room for private sector to play a larger role (UNDESA, 2017)
- Involve ethics in every step of planning and implementation to remove even the thought of quick gain compromising the ecology and human traits
- Undertake ethical measures for economic diversification, job creation, food security, poverty reduction
- Guarantee equal distribution of resources and accrued economic benefits
- Obtain a "social license to operate" and negotiate any inherent resistance from communities, workers, and the business community (ESG, 2018).

Although some information related to the blue economy potential for the Indian Ocean is available, albeit widely scattered and grossly inadequate, the following topic may be studied deeply in future. The list may include how different social scientific perspectives can contribute toward implementing a sustainable ethical blue economy. Or how a specific theoretical perspective can shed light on integrating social, environmental, and economic considerations at multiple scales (e.g., global, national, or community levels) and the development of governance arrangements and mechanisms that facilitate implementing sustainable blue growth? Or how social scientific issues emerge in the implementation of blue growth within particular regions or related to a specific sector? Or to what extent blue economy changes or reproduces dominant discourses and power relations? Before being awarded with a social license, the six major blue economy industries like shipping, oil and gas, renewable energy, deep-sea mining, fisheries, and tourism-leisure must satisfactorily answer concerns of any particular community against any particular industry, if any; and be able to align its activities if needed with the existing culture keeping the best possible balance (ESG, 2018).

IOR nations need not only be strong and wealthy, but also be democratic in a true sense and more equal in deliveries. This decentralized dispensation must include affordable health care and education for all their citizens. The middle class must aspire to hold 60–70% of national resources. Private sectors have to play a major role in this transformation in the IOR, while government spending will focus more on health and education (and certainly not in service sectors, such as, in running airlines, railways, buses, hotels, banks, etc.). However, government must have in place a strong, effective, and prompt regulatory mechanism to ensure nothing goes wrong and punish the exploiters, if any, in public forums. The stress on government to provide jobs to unemployed youth must be lessened with private ventures taking the lead in using the available human resource in the most appropriate way. Both private initiative and social well-being can and must go together (Switzerland is the best example). A rational country can have a business welcoming environment together with the social even-handedness, if one can maintain equilibrium. Accordingly the range of quality output an EBE can provide is visualized in Fig. 9.4 and the essential milestones to ensure a successful EBE regime is established in the IOR are given in Table 9.5.

It is time the world takes lessons from the old African saying—Ubuntu. The saying goes that when a basket of sweets was offered to children, instead of rushing they held each other's hands, ran together, and divided the sweets equally among

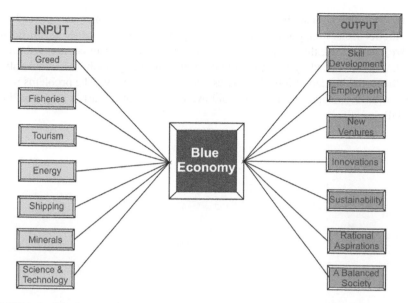

FIGURE 9.4   The Range of Quality Output When Raw Inputs Are Processed through Ethical Blue Economy

TABLE 9.5
Essential Milestones to Ensure a Successful Ethical Blue Economy in the IOR

External Dynamics	Threats	Internal Concerns
Follow Ethical Governance Framework		Encourage rational aspiration only
Innovate, and make Artificial Intelligence (AI) accountable to ethical standards	Climate Change Impact	Judge worker not by performance only, but by their conduct and integrity
Bring in Mutual Respect among Nations		Ensure impartiality, no corruption
Encourage Shared Prosperity	Geopolitics and Militarization	Take recourse of conscious spirituality
Reform Natural Resource Management		Guarantee quality and dignity in life
Reform Human Resource Management	Human Mindset	Encourage yoga, inclusiveness, ethics
Build Trust through Transparency		Reduce urge for material consumption
Use Labor-Intensive Technology		Pledge gender and social equality

Notes: Reform in natural resources includes that of trade, commerce, community agriculture, and minerals; reforming human resource includes that of population, education, health, capacity building, and labor.

themselves. The philosophy is "how can one be happy when others are sad." The EBE model seeks to apply Ubuntu philosophy in global governance (*I Am Because We Are*). Therefore, the existing tendency to see the growth in economy through a lens that is influenced by GDP, job creation, job market growth, the social value of jobs, the social burden of jobs, job losses, etc., needs a review. The problems may be assessed on a planetary scale, and avoid evaluations that are artificially based only on GDP growth.

Bibliography

ABP (2019) India aims for 5 trillion $ economy by 2024. *Ananda Bazar Patrika* Editorial, August 2019.

Abreu CT (2015) *Brazilian Coastal and Marine Protected Areas: Importance, Current Status and Recommendations*. New York: United Nations Division of Ocean Affairs and the Law of the Sea.

ADB (2010) Asian Development Bank, Research Agencies Target Food Security, Nutrition Gains in Asia.

ADB (2014) Asian Development Bank, Research Agencies Target Economy, Health and Social Dimensions in Asia.

AFD and IREDA (2014) A study on tidal and waves energy in India: Survey on the potential and proposition of a roadmap. www.ireda.gov.in.

AGR (2016–17) Annual Government Report, AGR, 2016–17.

Agranoff R and McGuire M (2003) *Collaborative Public Management: New Strategies for Local Governments*. Washington, DC: Georgetown University Press.

Ahmad T (2018) Integrating the GCC countries and Iran in a new Indian Ocean economic and security architecture: An Indian diplomatic initiative. In T Niblock, T Ahmad, D Sun (Eds.), *The Gulf States, Asia and the Indian Ocean: Ensuring the Security of the Sea Lanes*. Berlin: Gerlach Press, pp. 33–70. doi:10.2307/j.ctv4ncp9p.5.

Al Mamun MA, Raquib M, Tania TC, Rahman SMK (2014) Salt industry of Bangladesh: A study in the Cox's Bazar. *Banglavision* 14: 7–17.

Alam MK (2015) Ocean/Blue economy for Bangladesh. Proceedings of International Workshop on Blue Economy. Dhaka, Bangladesh: Bangladesh Ministry of Foreign Affairs. www.mofa.gov.bd/content/about-blue-economy. Accessed December 18, 2017.

Allen RB (2017) Ending the history of silence: Reconstructing European slave trading in the Indian Ocean. *Tempo* 23(2): 294–313.

Alongi DM (2012) Carbon sequestration in mangrove forests. *J. Carbon Management* 3: 313–322. https://doi.org/10.4155/cmt.12.20.

Annual Report 2016–17 of Department of Animal Husbandry, Dairying and Fisheries. Ministry of Agriculture and Farmers Welfare, Govt of India.

Anon. (2016). Blue economy interventions for food security in Bangladesh. Background paper of national seminar, Chittagong, Bangladesh. 6pp.

Arvis JF, Mustra MA, Ojala L, Shepherd B, Saslavsky D (2012) *Connecting to Compete 2012: Trade Logistics in the Global Economy—the Logistics Performance Index and Its Indicators (English)*. Washington, DC: World Bank Group. http://documents.worldbank.org/curated/en/ 567341468326992422/Connecting-to-compete-2012-trade-logistics-in-the-global-Economy-the-logistics-performance-index-and-its-indicators.

Asian Development Bank (2014) *State of Coral Triangle*. Papua New Guinea, Coral Triangle Initiative, ADB, The Philippines.

Asseng S, Ewert F, Martre P, Rötter RP, Lobell DB, and 48 others (2014) Rising temperature reduces global wheat production. *Nature Climate Change* 5: 143–147.

Atlee T (2016) *Mapping the Newly Emerging Alternative Economic Approaches*, P2P Foundation.

Attri VN and Bohler-Mulleris N (2018) *The Blue Economy Handbook of the Indian Ocean Region*. Africa Institute of South Africa, 496p.

Austen MC and Malcom SJ (2011) Marine Chapter 12, in *UK National Ecosystem Assessment: Technical Reports*, UK National Ecosystem Assessment, UNEP-WCMC, Cambridge.

Awadalla C, Coutinho-Sledge P, Criscitiello A, Gorecki J, Sapra S (2015) Climate Change and Feminist Environmentalisms: Closing Remarks, www.thefeministwire.com.

Azmy SAM (2011) *Sri Lanka Report on Coastal Pollution Loading and Water Quality Criteria*, s.l.: Bay of Bengal Large Marine Ecosystem Project.

Azmy SAM (2013) *Sri Lanka Report on Coastal Pollution Loading and Water Quality Criteria*, s.l.: Bay of Bengal Large Marine Ecosystem Project (BOBLME). Country Report on Pollution – Sri Lanka BOBLME-2011-Ecology-14, p. 89.

Babu SS, Chaubey JP, Moorthy KK, Gogoi MM, Kompalli SK and 6 others (2011) High altitude (~ 4520 m amsl) measurements of black carbon aerosols over western trans-Himalayas: Seasonal heterogeneity and source apportionment. *Journal Geophysical Research* 116, D24201.

Bailey M, Favaron B et al (2016) Canada at crossroads: Imperative for realigning ocean policy with ocean science. *Marine Policy* 63: 53–60.

Balboa EM, Conde E, Soto ML, Pérez-Armada L, Domínguez H (2015) Cosmetics from marine sources. In S.-K. Kim (Ed.), *Springer Handbook of Marine Biotechnology*. Berlin-Heidelberg: Springer, pp. 1015–1042. https://hdl.handle.net/10.1007/978-3-642-53971-8_44.

Balgos MC, Ricci N, Walker L, Didden J (2005) Compilation of Summaries of National and Regional Ocean Policies, www.researchgate.net/publication/237536645.

Banerjee A (2019) Time for minimum income: People's aspirations have grown, fewer want to sell "pakodas" for a living. Edit-Page, *The Times of India*, April 16.

Banerjee A and Duflo E (2011) *Poor Economics: A Radical Rethinking of the Way to Fight Global Poverty*. New York: Public Affairs, 320pp, ISBN 978-1-58648-7980-0.

Barone RSC, Lorenz EK, Sorioyoda DY, Cyrino JEP (2017) Fish and fishery products trade in Brazil, 2005 to 2017: A review of available data and trends. *Scientia Agricola* 74(5): 417–424.

Beejadhur YA, Kelleher K, Kelly T, Howells MI, Alfstad T, Farrell S, Smith J, Neumann JE, Strzepek KM, Emanuel KA, Willwerth J (2017) *The Ocean Economy in Mauritius: Making it Happen, Making it Last (English)*. Washington, DC: World Bank Group. http://documents.worldbank.org/curated/en/193931508851670744/The-Ocean-Economy-in-Mauritius-making-it-happen-making-it-last.

Benkenstein A (2018) Prospects for the Kenyan blue economy. *Policy Insights* 62. South African Institute of International Affairs, pp. 1–8.

Bergmann W and Feeney R (1951) Contribution to the study of marine sponges—The nucleosides of sponges. *J. Org. Chem.* 16: 981–987.

Bharatan V (2011) India needs a maritime governance authority. Gateway House, Indian Council of Global Relations. April 19, www.gatewayhouse.in

Bhatia R (2017) Impact of the Indian Ocean Connectivity on India-Kenya Relations. Gateway House, Indian Council on Global Relations.

BioMarine Organization (2012) About marine biotechnology. www.biomarine.org/about-us/biomarine-organization-ltd/.

Bird MI, Wurster CM, de Paula Silva PH, Bass AM, De Nys R (2011) Algal biochar—production and properties. *Bioresource Technology* 102: 1886–1891. doi:10.1016/j.biortech.2010.07.106.

Blue Economy bankable projects (2018) Sustainable Blue Economy conference proceedings, Kenya. www.blueeconomyconference.go.ke.

Boden TA, Marland G, Andres RJ (2012) Global, Regional, and National Fossil Fuel CO_2 Emissions (Oak Ridge, TN: Carbon Dioxide Information Analysis Center, Oak Ridge National Laboratory, US Department of Energy). http://cdiac.ornl.gov/trends/emis/overview 2010.html.

Boissevain J (1996) *Coping with Tourists: European Reactions to Mass Tourism.* UK: Oxford Press.

Boivin N, Crowther A, Prendergast M, Fuller DQ (2014) Indian Ocean food globalisation and Africa. *African Archaeological Review* 31(4): 547–581.

Bouillon S, Borges AV, Castañeda-Moya E, Diele K, Dittmar T, Duke NC, Kristensen E, Lee SY, Marchand C, Middelburg JJ, Rivera-Monroy VH, Smith III TJ, Twilley RR (2008) Mangrove production and carbon sinks: A revision of global budget estimates. *Global Biogeochemical Cycles* 22. doi:10.1029/2007gb003052.

Bramwell B (2004) *Coastal Mass Tourism: Diversification and Sustainable Development in Southern Europe.* Clevedon: Channel View.

Brewster D (2014) India's Ocean: The Story of India's Bid for Regional Leadership (Routledge Security in Asia Pacific Series). London: Routledge, 244p.

Brewster D (2015) Australia's roadmap for blue economy science for the next decade. *Journal of Indian Ocean Studies* 23(3): 360–374.

British Petroleum (2017) BP Statistical Review of World Energy 2017. pp. 12–34. www.bp.com/content/dam/bp/en/corporate/pdf/ energy-economics /statistical-review-2017/bp-statistical-review-of-world-energy-2017-full-report.pdf.

Brohan P, Kennedy JJ, Harris I, Tett SFB, Jones PD (2006) Uncertainty estimates in regional global observed temperatures: A new data set from 1850. *Journal of Geophysical Research* 111, D12106, doi:10.1029/2005JD006548.

Burke L, Reytar K, Spalding M, Perry A (2002) *Reefs at Risk: Revisited in the Coral Triangle.* World Resources Institute report. ISBN 978-1-56973-791-0.

Campbell G (2017) Africa, the Indian Ocean World, and the "Early Modern": Historiographical conventions and problems. *The Journal of Indian Ocean World Studies* 1(1): 24–37.

Carlson C (2007) A Practical Guide to Collaborative Governance. Policy Consensus Initiative, Portland, Oregon. webmaster@policyconsensus.org

Carnie P and QinetiQ (2011) A strategy for growth for the UK Marine Industries. Marine Industries Leadership Council, p. 38.

Castro de la Mata G (2012) Terminal Evaluation Partnerships for Marine Protected Areas in Mauritius and Rodrigues. GEF ID: 1246 – UNDP PIMS: 864, pp.1–64.

CCZM (2016) Centre for Coastal Zone Management and Coastal Shelter Belt. Database on Coastal States of India. Accessed December 2, 2016, http://iomenvis.nic.in/index2

Centre for Coastal Zone Management and Coastal Shelter Belt (2016) Database on Coastal States of India. Accessed December 2, 2016.

Cervigni R, Scandizzo PL (2017) *The Ocean Economy in Mauritius: Making It Happen, Making It Last.* Washington, DC: The World Bank, pp. 296.

Chakraborty M (2018) The Mint, October 23, 2018.

Chakraborty S, Chandrasekhar S, Naraparaju K (2019) Land distribution, income generation and inequality in India's agricultural sector. *The Review of Income and Wealth.* https://doi.org/10.1111/roiw.12434.

Chand Basha SK (2018) An overview on global mangroves distribution. *Indian J. Geo Mar. Sci.* 47: 766–772.

Chandra S (2019) Are parts of India becoming too hot for humans? CNN Report, July 4.

CHART (2014) *Asia Cruise Trends 2014: Analysis, Assessment, Appreciation.* November, for CLIA Southeast Asia.

Chatterjee S, Goswami A, Scotese CR (2013) The longest voyage: Tectonic, magmatic, and paleoclimatic evolution of the Indian plate during its northward flight from Gondwana to Asia. *Gondwana Research* 23(1): 238–267.

Chellaney B (2015) China's Indian Ocean strategy. *The Japan Times*, June 23.

Chi Z, Xie Y, Elloy F, Zheng Y, Hu Y, Chen S (2013) Bicarbonate-based integrated carbon capture and algae production system with alkalihalophilic yanobacterium. *Bioresource Technology* 133: 513–521. doi:10.1016/j.biortech.2013.01.150.

Chinonyerem NT, Ue MN, Chukwudi JC, Chinedum O (2017) Economic implications of marine oil spill to Nigeria: A case for improvement in coastal pipeline management and surveillance practices. *International Journal of Economy, Energy and Environment* 2(3): 40–47.

Chinoy SZ (2019) Elephant in the room: Next administration must have single-minded focus: Revive private investment. Edit-Page, *The Times of India*, May 3.

Chmura G and Anisfield S (2003) Global carbon sequestration in tidal, saline wetland soils. *Global Biogeochemical Cycles* 17(4). doi:10.1029/2002GB001917.

Chung IK, Beardall J, Mehta S, Sahoo D, Stojkovic S (2011) Using marine macro-algae for carbon sequestration: A critical appraisal. *Journal of Applied Phycology* 23: 877–886. doi:10.1007/s10811-010-9604-9.

Chung IK, Oak JH, Lee JA, Shin JA, Kim JG, Park KS (2013) Installing kelp forests/seaweed beds for mitigation and adaptation against global warming: Korean Project Overview. *ICES Journal of Marine Science* 70: 1038–1044. doi:10.1093/icesjms/fss206.

CLIA (2015) *State of the Cruise Industry Report.* Washington, DC: CLIA Europe.

CLIA (2016) *State of the Cruise Industry Report.* Washington, DC: CLIA Europe.

CLIA (2018) *Cruise Lines International Association.* Washington, DC. https://cliaasia.org/wp-content/uploads/2018/08/asia-cruise-trends-2018.pdf.

CMFRI (2016) Annual Report 2015–16. Technical Report of Central Marine Fisheries Research Institute, Kochi. ISSN 0972-2378, p294.

Colgan CS (2017) The blue economy of the Indian Ocean: Context and challenge. *Journal of Indian Ocean Rim Studies* 1: 7–37.

Colgan CS and Farnum MG (2016) *State of the U.S. Ocean and Coastal Economies 2016 Update.* http://midatlanticocean.org/wp content/uploads/2016/03/NOEP_National_Report_2016.pdf.

Commonwealth Foundation (2015) Small states and the green blue economy, Commonwealth insights discussion paper, CPF, 2015 series.

Commonwealth Marine Economies Programme (2018) Country View: Papua New Guinea, Commonwealth Heads of Government Meeting, London. https:// assets.publishing. service.gov.uk/government/uploads/system/uploads/attachment_data/file/769336/Commonwealth_Marine_Economies_Programme_-_Papua_New_Guinea_Country_review.pdf.

Conathan M and Moore S (2015) *Developing Blue Economy in China and the United States*, Centre for American Progress, www.americanprogress.org

Cordesman AH and Toukan A (2014) *The Indian Ocean Region: A Strategic Net Assessment.* Lanham, MD: Rowman & Littlefield.

Cordner L (2010) Rethinking maritime security in the Indian Ocean Region. *Journal of the Indian Ocean Region* 6(1).

Cruickshank MJ (1974) Model for assessment of benefit/cost ratios and environmental impacts of marine mining operation. Proceedings of the International Symposium on Minerals and the Environment. Institute of Mining and Metallurgy, London.

Curray JR, Emmel FJ, Moore DG (2002) The Bengal Fan: morphology, geometry, stratigraphy, history and processes. *Marine and Petroleum Geology* 19: 1191–1223. doi:10.1016/S0264-8172(03)00035-7.

Das VK and Ramaraju DV (1986) Estimation of salinity power potential in India. *Mahasagar* 19: 113–118.

De Silva SS (1998) *Tropical Mariculture*. New York: Academic Press, 487 p. doi https://doi.org/10.1016/B978-0-12-210845-7.X5000-4.

Defence White Paper (2016) Department of Defence, Australian Government. ISBN: 978-0-9941680-5-4. pp.1–188.

DeMers A and Kahui V (2012) An overview of Fiji's Fishery Development. *Marine Policy* 36: 174–179.

Demopoulos AWJ, Smith CR, Tyler PA (2003) The deep Indian Ocean floor. www.researchgate.net/publication/253047410.

Department of Fisheries (2015) *Marine Fisheries Management Plan of Thailand: A National Policy for Marine Fisheries Management 2015–2019*. Bangkok: Department of Fisheries (DoF) Ministry of Agriculture and Cooperatives.

Department of Marine and Coastal Resources (2014) Thailand. www.dmcr.go.th/ detailAll/2930/pc/3. ISBN: 978-616-3161-47-5.

Derrick B, Noranarttragoon P, Zeller D, Teh LCL, Pauly D (2017) Thailand's Missing Marine Fisheries Catch (1950–2014). *Front. Mar. Sci.* 4(402). doi:10.3389/fmars.2017.00402.

Diamond J (2011) *Collapse: How Societies Choose to Fail or Survive*. London: Penguin Books, ISBN 978-0-241-95868-1.

Dinarto D (2017) Indonesia's blue economy initiative: Rethinking maritime security challenges. RSIS Commentary, No. 206.

Dissanayake C (2015) Smugglers rely on Sri Lankan fishing captains and their boats to move migrants to Australia. *Global Press Journal*, November 8. https://globalpressjournal.com/asia/sri_lanka/smugglers-rely-on-sri-lankan-fishing-captains-and-their-boats-to-move-migrants-to-australia/.

Doherty B, Haugh H, Lyon F (2014) Social enterprises as hybrid organisations: A review and research agenda. *International Journal of Management Review* 16(4): 417–436.

Doloreux D and Shearmer R (2018) Moving maritime clusters to the next level: Canada's Ocean Super-cluster Initiative. *Marine Policy* 98: 33–36.

Douglass M and Miller MA (2018) Disaster justice in Asia's urbanizing Anthropocene. *Environment and Planning E: Nature and Space* 1(3): 271–287.

Duarte CM (2005) Major rule of marine vegetation on the oceanic carbon cycle. *Biogeosciences* 2: 1–8. doi:10.5194/bg-2-1-2005.

Duarte CM, Kennedy H, Marbà N, Hendriks I (2013) Assessing the capacity of seagrass meadows for carbon burial: current limitations and future strategies. *Ocean Coastal Management* 83: 32–38. doi:10.1016/j.ocecoaman.2011.09.001.

Duflo E and Banerjee A (2019) *Good Economics for Hard Times: Better Answers to Our Biggest Problems*. New Delhi: Juggernaut.

Eakins BW and Sharman GF (2010) *Volumes of the World's Oceans from ETOPO1*. Boulder, CO: NOAA National Geophysical Data Center.

Ebarvia M (2016) Economic assessment of oceans for sustainable blue economy development. *Journal of Ocean and Coastal Economics* 2(2). doi: https://doi.org/10.15351/2373–8456.1051.

Economist Intelligence Unit (2015) *The Blue Economy; Growth, Opportunity and a Sustainable Ocean Economy.* s.l.: Gordon and Betty Moore Foundation.

Ecorys (2013) Study in support of policy measures for maritime and coastal tourism at EU level Specific contract under FWC MARE/2012/06 - SC D1/2013/01-SI2.648530. Final Report. Client: DG Maritime Affairs & Fisheries, Rotterdam/Brussels.

EDGAR (2019) Emissions Database for Global Atmospheric Research, https://edgar.jrc.ec.europa.eu/.

Ekpo EI (2012) Impact of shipping on development: Implications for sustainable development. *Journal of Educational and Social Research* 2(7): 107–117.

Elizabeth C (2013) Indian Ocean islands tourism sector review: Seychelles (English). Washington, DC: World Bank. http://documents.worldbank.org/ curated/en/715191468004846420/Indian-ocean-islands-tourism-sector-review-Seychelles.

EPO (2016) *PATSTAT – the EPO Worldwide Patent Statistical Database: your backbone data set for statistical analysis.* Available at: http://documents.epo.org/projects/ babylon/eponet.nsf/0/15E2334DC50BE0AEC125767F005A91ED/$File/PATSTAT_flyer_en.pdf.

ESG (2018) Utrecht Conference on Earth System Governance, November 5–8, 2018. Utrecht University, Faculty of Geosciences and the Copernicus Institute of Sustainable Development, together with the Earth System Governance Project.

European Commission (2012) Blue Growth—Opportunities for Marine and Maritime Sustainable Growth. Publications Office of the European Union. 13 pp. Doi:10.2771/43949.

European Commission (2018) The Annual Economic Report on Blue Economy, Maritime Affairs and Fisheries.

Fahrudin A (2015) Indonesian Ocean Economy and Ocean Health. Inception Workshop on Blue Economy Assessment, Manila, July, 28–30 2015.

FAO (2014) *The State of the World Fisheries and Aqua Culture: Challenges and Opportunities.* Rome: Food and Agricultural Organization.

FAO (2016) *The State of World Fisheries and Aquaculture 2016: Contributing to Food Security and Nutrition for All.* Rome: FAO Publications. www.fao.org/fishery/fcap/CHN/en. www.fao.org/fishery/fcap/CHN/en

FAO (2018) *The State of World Fisheries and Aquaculture 2018: Meeting the Sustainable Development Goals.* Rome: Food and Agricultural Organization.

Folkersen MV, Fleming CM, Hasan C (2018) Deep sea mining future effects on Fiji's tourism industry: A contingent behaviour study. *Marine Policy* 96: 81–89.

Fourqurean JW (2012) Seagrass ecosystems as a globally significant carbon stock. *Nature Geoscience* 5: 505–509. doi:10.1038/ngeo1477.

FRAD-CMFRI (2018) Marine Fish Landings in India 2017. Technical Report, CMFRI, Kochi, 16pp.

Free CM, Thorson JT, Pinsky ML, Oken KL, Wiedenmann J, Jensen OP (2019) Impacts of historical warming on marine fisheries production. *Science.* doi: 10.1126/science.aau1758.

Friedlingstein P, Fairhead L, LeTreut H, Monfray P, Orr J (2001) Positive feedback between future climate change and the carbon cycle. *Geophysical Research Letters* 28: 1543–1546.

FSB (2017) Yearbook of Fisheries Statistics of Bangladesh 2016–17. Fisheries Resources Survey System Department of Fisheries Bangladesh Ministry of Fisheries and Livestock Government of the People's Republic of Bangladesh, p.128.

Gattuso J (2006) Light availability in the coastal ocean: Impact on the distribution of benthic photosynthetic organisms and their contribution to primary production. *Biogeosciences* 3: 489–513. doi:10.5194/bg-3-489-2006.

Ghosh M (2017) India's Strategic Convergence with Japan in the Changing Indo-Pacific Geopolitical Landscape. *Asia Pacific Bulletin*, No. 392, East-West Centre, Washington DC.

Gijzen H (2013) Moral and innovative leadership: A platform for sustainable peace and development. Global peace leadership conference Manila, Nov 20–22, UNESCO.

Global Foot Print Network Annual Report (2008) www.footprintnetwork.org/content/images/uploads/Global_Footprint_Network_2008_Annual_Report.pdf.

Global Information (2015) Market Research Report – 263147 Global Cosmeceuticals Market Outlook 2020. www.giiresearch.com/report/rnc263147-global-cosmeceuticals-market-outlook.html.

Global Volcanism Program (2013) Department of Mineral Sciences, National Museum of Natural History, Smithsonian Institution, www.volcano.si.edu.

GOI (2011) Government of India, Strategic Plan for New and Renewable Energy Sector for the Period 2011–17. Ministry of New and Renewable Energy, New Delhi, India.

Goonetilleke B and Colombage ADJ (2017) Indo-Sri Lanka Fishery Conflict: An Impediment to Development and Human Security. http://cimsec.org/indo-sri-lanka-fishery-conflict-impediment-development-human-security/30113.

Gopal R, Vijaykumaran M, Venkatesan R, Kathiroli S (2008) Marine organisms in Indian medicine and their future prospects. *Natural Products Radiance* 7: 139–145.

Government Office for Science (2018) *Foresight: Future of the Sea*. A Report from the Chief Scientific Advisor, Foresight, UK.

Government Portal (2018) Government of Seychelles, www.egov.sc/GovernmentAgencies/Sectors/BusinessIndustry.aspx.

Government Report (2018) www.mygov.go.ke/.

Greenpeace International (2007) Energy [R]evolution—a sustainable world energy outlook. GPI REF JN 035. Published by Greenpeace International and the European Renewable Energy Council (EREC), www.energyblueprint.info.

Greiner JT, McGlathery KJ, Gunnell J, McKee BA (2013) Seagrass restoration enhances "blue carbon" sequestration in coastal waters. *PLOS ONE* 8:e72469. doi:10.1371/journal.pone.0072469.

Gujar AR, Ambre NV, Mislankar PG, Iyer SD (2010a) Ilmenite, magnetite and chromite beach placers from South Maharashtra, Central West Coast of India. *Resource Geology* 60: 71–86. doi:10.1111/j.1751-3928.2010.00115.x.

Gujar AR, Ambre NV, Iyer SD, Loveson VJ, Mislankar PG (2010b) Placer chromite along South Maharashtra, Central West Coast of India. *Current Science* 99: 492–499.

Gupta M (2010) *Indian Ocean Region: Maritime Regimes for Regional Cooperation*. New York: Springer-Verlag.

Gusiakov V, Abbott DH, Bryant EA, Masse WB, Breger D (2009) Mega tsunami of the world oceans: Chevron dune formation, micro-ejecta, and rapid climate change as the evidence of recent oceanic bolide impacts. *Geophysical Hazards*. Dordrecht: Springer, pp. 197–227.

GWEC (2018). Global Wind Report. https://gwec.net/global-wind-report-.

Hamilton SE and Casey D (2016) Creation of a high spatio-temporal resolution global database of continuous mangrove forest cover for the 21st century (CGMFC-21).

Hamilton SE and Friess DA (2018) Global carbon stocks and potential emissions due to mangrove deforestation from 2000 to 2012. *Nature Climate Change* 8: 240–244. doi:10.1038/s41558-018-0090-4. ISSN 1758-678X.

Hardy JT (2003) *Climate Change: Causes, Effects and Solutions*. Chichester: John Wiley & Sons Limited.

Harris PT, MacMillan-Lawler M, Rupp J, Baker EK (2014) Geomorphology of the oceans. *Marine Geology* 352: 4–24.

Hascic L, Watson F, Johnstone N, Kaminker C (2012) Recent trends in innovation in climate change mitigation technologies. *Energy and Climate Policy*, pp. 17–53.

HDR (2006) *Human Development Report: Beyond Scarcity: Power, Poverty and the Global Water Crisis*. New York: United Nations Development Program.

Hiroshi T (2012) Japan's ocean policy making. *Coastal Management* 40(2): 172–182.

Hofstede G and McCrae RR (2004) Personality and culture revisited: Linking traits and dimensions of culture. *Cross-Cultural Research* 38: 52–88.

Hossain MAA (2012) Future of Bangladesh-India Relationship: A Critical Analysis. www.hsdl.org.

Hossain MS, Chowdhury SR, Navera UK, Hossain MAR, Imam B, Sharifuzzaman SM (2015) Opportunities and Strategies for Ocean and River Resources Management. Background paper for preparation of the 7th Five Year Plan. Food and Agriculture Organization of the United Nations, Dhaka, Bangladesh. pp.61.

Houghton JT, Ding Y, Griggs DJ, Noguer M, van der Linden PJ, Dai X et al. (2001) Climate Change 2001: The Scientific Basis. IPCC, Cambridge University Press, Cambridge, p 39.

Hurst D, Børresen T, Almesjö L, De Raedemaecker F, Bergseth S (2016) Marine biotechnology strategic research and innovation roadmap: Insights to the future direction of European marine biotechnology. Marine Biotechnology ERA-NET: Oostende, Belgium, 46pp.

Hussain GM, Failler P, Al Karim A, Khurshed Alam M (2017) Major opportunities of blue economy development in Bangladesh. *Journal of the Indian Ocean Region*. doi:10.1080/19480881.2017.1368250

IEA (2011) Guide to Reporting Energy RD&D Budget Statistics.

IEA (2014) Baseline case. Medium-term Renewable Energy Market Report 2014. Paris. https://doi.org/10.1787/renewmar-2014-en.

India State of Forest Report (2017) http://fsi.nic.in/isfr 2017/isfr-mangrove-cover-2017.pdf.

Indian Ocean Rim Association (2016) Mauritius Declaration on Blue Economy. Accessed December 3, 2016, www.iora.net/media/158070/mauritius_blue_economy_declaration.pdf.

Indian Trade Portal (2019) www.indiantradeportal.in, Department of Commerce, M/Commerce & Industries, Government of India.

Indonesian Maritime Council (2012) Analisis Input-Output Bidang Kelautan terhadap Pembangunan Nasional. [Input-Output Analysis on the Development of National Marine Sector]. Ministry of Marine and Fisheries, Republic of Indonesia. Jakarta [Bahasa Indonesia].

Indonesian Ministry of Marine Affairs and Fisheries (2018) https://kkp.go.id/.

Indonesian Statistical Council (2015). National Income of Indonesia 2010–2014. BPS Catalog: 9301501. Badan Pusat Statistik (Statistics Indonesia). Publications Number: 07130.1001. pp.1–204.

International Transport Forum (2015) Global trade: International freight transport to quadruple by 2050. January 27, 2015.

IORA (2015) Post-2015 Development Conference 2015, Indian Ocean Region's Blue Economy Conference 2015.

IORA (2017) Declaration of the Indian Ocean Rim Association on the Blue Economy in the Indian Ocean Region, Jakarta, Indonesia, May 8–10, 2017

IPCC (2014) Climate Change 2014: Impacts, Adaptation, and Vulnerability. Part A: Global and Sectoral Aspects. Contribution of Working Group II to the Fifth Assessment Report of the Intergovernmental Panel on Climate Change [Field CB, Barros VR, Dokken DJ, Mach KJ, Mastrandrea MD and 11 others (eds.)]. Cambridge and New York: Cambridge University Press, 1132 pp.

IPCC (2019) Choices made now are critical for the future of our ocean and cryosphere. The Intergovernmental Panel on Climate Change, IPCC Special Report, September 25, 2019.

IRENA (2016) *Remap: Roadmap for a Renewable Energy Future*. International Renewable Energy Agency, Abu Dhabi, 168pp (www.irena.org).

IRENA (2019) *Global Energy Transformation: A Roadmap to 2050*. International Renewable Energy Agency, Abu Dhabi, 51pp (www.irena.org).

IUCN (International Union for Conservation of Nature) (2007) Counting coastal ecosystems as development assets. *Coastal Ecosystems* (5). http://cmsdata.iucn.org/ downloads/ coastal_ecosystems_newsletter_july2007.pdf.

Ives JD, Shrestha B, Mool PK (2010) Formation of Glacial Lakes in the Hindu Kush-Himalayas and GLOF Risk Assessment. International Centre for Integrated Mountain Development, Kathmandu, May 2010, 66p.

Iyer SD, Amonkar AA, Das P (2018) Genesis of Central Indian Ocean basin seamounts: morphological, petrological, and geochemical evidence. *International Journal of Earth Sciences* 107: 2517–2538.

Iyer SS (2016) India's Indian Ocean policy: Challenges and opportunities. Unpublished Dissertation, Goa University, India, 86pp.

Jain S (2016) A village having no door and locks. The BBC World story on Shani Shingnapur, Maharashtra, India.

Jambeck JR, Geyer R, Wilcox C, Siegler TR, Perryman M, Andrady A, Narayan R, Law KL (2015) Plastic waste inputs from land into the ocean. *Science* 347(6223): 768–771. doi:10.1126/science.1260352.

Jessen S (2011) A review of Canada's implementation of the Oceans Act since 1997—from leader to follower. *Coast Management* 39(1): 20–56.

Johnstone P and Ketkar P (2017) Plastic Pollution and our Oceans: What Everyone Should Know. (https://camd.northeastern.edu /rugglesmedia/ 2017/02/08/ plastic- pollution-and-our-oceans-what-everyone-should-know/).

Kannangara P, Collins A, Waidyatilake B (2018) The Importance of the Indian Ocean: Trade, Security and Norms. Lakshman Kadirgamar Institute of International Relations and Strategic Studies (LKI), Sri Lanka, pp. 1–5.

Karim MS and Uddin MM (2019) Swatch-of-no-ground marine protected area for sharks, dolphins, porpoises and whales: Legal and institutional challenges. *Marine Pollution Bulletin* 139: 275–281. https://doi.org/10.1016/j.marpolbul.2018.12.037.

Katsarova (2013) www.europarl.europa.eu/RegData/bibliotheque/briefing/ 2013/130562/ LDM_BRI%282013%29130562_REV2_EN.pdf

Kaura V (2018) Securing India's economic and security interests in the Indo-Pacific. *Indian Journal of Asian Affairs* 31(1/2): 37–52.

Kedong Yi and Xuemei Li (2018) Recent advances in international marine econometrics. *Marine Economics and Management* 1(1): 20–42. https://doi.org/10.1108/ MAEM-07-2018-005.

Keesing J and Irvine T (2005) Coastal biodiversity in the Indian Ocean: The known, the unknown and the unknowable. *Indian Journal of Marine Sciences* 34(1): 11–26.

Keiji H (2011) Maritime policy in Japan. *Journal of Maritime Researches* 1(1): 65–84.

Kelkar U and Bhadwal S (2007) South Asian regional study on climate change impacts and adaptation: Implications for human development. Human Development Report 2007–2008, UNDP, pp. 47.

Khan A (2018) Muslim Cultures in the Indian Ocean: Diversity and Pluralism, Past and Present. Institute for the Study of Muslim Civilisations, September 12–14, 2018, The Aga Khan University, London.

Kidlow JR and Park KS (2014) Rebuilding the classification system of the ocean economy. *Journal of Ocean and Coastal Economies* 1 1–37.

Kiehl JT, Kevin E, Trenberth KE (1997) Earth's Annual Global Mean Energy Budget. *Bulletin American Meteorological Society* 78: 197–208.

Kirchherr J, Reike D, Hekkert M (2017) Conceptualizing the circular economy: An analysis of 114 definitions. *Resources, Conservation and Recycling* 127: 221–232.

Kopela S (2017) Historic titles and historic rights in the Law of the Sea in the light of South China Sea arbitration. *Ocean Development and International Law* 48(2): 181–207.

KPMG International (2015) Nutraceuticals: The future of intelligent food—Where food and pharmaceuticals converge. www.kpmg.com/ID/en/industry/CM/Documents/neutraceuticals-the-future-of-intelligent-food.pdf.

Kroeger KF, Funnell RH, Nicol A, Fohrmann M, Bland KJ, King PR (2013) 3D crustal-scale heat-flow regimes at a developing active margin (Taranaki Basin, New Zealand). *Tectonophysics* 591: 175–193.

Kumar K, Dasgupta CN, Nayak B, Lindblad P, Das D (2011) Development of suitable photo-bioreactors for CO_2 sequestration addressing global warming using green algae and cyanobacteria. *Bioresource Technology* 102: 4945–4953. doi:10.1016/j.biortech.2011.01.054.

Kundu MN (2019) The Speaking Tree: Socio-spiritual legacy of Swami Vivekananda. *Times of India*, July 4, 2019.

Kuswardhani N, Soni P, Shivakoti GP (2013) Comparative energy input-output and financial analyses of greenhouse and open field vegetables production in West Java, Indonesia. *Energy* 53: 83–92.

Labour Bureau (2017) Pocket Book of Labour Statistics, M/ Labour and Employment, Govt. of India, 283p, ISSN 0971 5398.

Labour Bureau, Ministry of Labour and Employment (2016) Report on Fifth Annual Employment-Unemployment Survey (2015–16). Chandigarh: Government of India.

Lacis AA, Hansen JE, Russel GL, Oinas V, Jonas J (2013) The role of long-lived greenhouse gases as principal LW control knob that governs the global surface temperature for past and future climate change. *Tellus B: Chemical and Physical Meteorology* 65(1): 19734. doi: 10.3402/tellusb.v65i0.19734.

Lal M (2003) Global climate change: India's monsoon and its variability. *J Environ Studies and Policy* 6: 1–34.

Laurance WF and Williamson GB (2001) Positive feedbacks among forest fragmentation, drought and climate change in the Amazon. *Conservation Biology* 15: 529–535.

Le Manach F, Abunge CA, McClanahan TR, Pauly D (2015) Tentative reconstruction of Kenya's marine fisheries catch, 1950–2010. pp. 37–51. In F Le Manach F and D Pauly (Eds.), *Fisheries Catch Reconstructions in the Western Indian Ocean, 1950–2010*. Fisheries Centre Research Reports 23(2). Fisheries Centre, University of British Columbia [ISSN 1198–6727].

Lemmen DS, Narhen FJ, James TS, Mercer C (2016) C.S.L. Editors (2016), *Canada's marine coast in a changing climate*. Government of Canada, Ottawa, ON.

Libes SM (2009) *Introduction to Marine Biogeochemistry*. 2nd edition. Amsterdam: Academic Press, Elsevier, 909pp.

Liu C (2010) How are China and the US building a clean-energy workforce? *ClimateWire*. www.scientificamerican.com/article/how-are-china-and-the-us/

LiVecchi AA, Copping D, Jenne A, Gorton R, Preus G, Gill R, Robichaud R, Green S, Geerlofs S, Gore D, Hume W, McShane C, Schmaus H, Spence H (2019). *Powering the Blue Economy; Exploring Opportunities for Marine Renewable Energy in Maritime Markets*. US Department of Energy, Office of Energy Efficiency and Renewable Energy. Washington, DC.

Lohani SP and Baral B (2012) Conceptual framework of low-carbon strategy for Nepal. *Low Carbon Economics* 2: 230–238.

MacDorman TL and Chircop A (2012) Canada's ocean policy framework: An overview. *Coastal Management* 40(2): 133–144.

Magesh R (2010) OTEC technology—A world of clean energy and water. Proceedings of the World Congress on Engineering 2010. World of Clean Energy and Water, London. www.iaeng.org/publication.

Maini H and Budhraja L (2018) *Ocean Based Blue Economy: An Insight into the SAGAR as the Last Growth Frontier*. NITI Aayog (National Institution for Transforming India) Internal Assessment, Government of India (www.niti.gov.in).

Mani S and Dhingra T (2013) Offshore wind energy policy for India—key factors to be considered. *Energy Policy* 56: 672–683.

Mann KH and Lazier RN (2006) *Dynamics of Marine Ecosystems: Biological-Physical Interactions in the Oceans*. Malden, MA: Wiley-Blackwell.

Marcu A, Stoefs W, Belis D, Katja T (2015) Country case study: Maldives for sustainable growth. Centre for European Policy Studies, Brussels, pp. 51.

Marete G (2018) Inside KPA plan for bigger vessels. *Business Daily* (Kenya), April 25. www.pressreader.com/ kenya/ business-daily-kenya /20180425/281947 428456553.

Maritime New Zealand (2017) The maritime sector of New Zealand's economy: Briefing to the incoming minister of transport. www.maritime.nz.govt.nz/about/documents/briefing-incoming -minister-2017.pdf.

Marketsandmarkets.com (2016) Omega-3 PUFA Market by Type (DHA, EPA, ALA), Application (Dietary Supplements, Functional Foods & Beverages, Pharmaceuticals, Infant Formula), Source (Marine, Plant), Sub-source), & Region – Global Forecasts to 2020. www.marketsandmarkets.com/Market-Reports/omega-3-omega-6–227.html.

Masson DG, Huvenne VAI, Stigter HC, de Wolff GA, Kiriakoulakis K, Arzola RG, Blackbird S (2010) Efficient burial of carbon in a submarine canyon. *Geology* 38: 831–834. doi:10.1130/g30895.1.

Masti CP (2018) Research for PECH Committee-Fisheries in Mauritania and European Union, European Parliament, European Union. http:// www.europarl.europa.u/ thinktank/ en/ document.html.

Matilal BK (1986) *Perception: An Essay on Classical Indian Theories of Knowledge*. Oxford: Oxford University Press, reprinted 2002.

Matilal BK (2017) *Epistemology, Logic, and Grammar in Indian Philosophical Analysis*. New Delhi: Oxford University Press.

Matsuda T (2020) Explaining Japan's post-Cold War security policy trajectory: Maritime realism. *Australian Journal of International Affairs*. doi.org/10.1080/10357718.2020.1782346.

McKenzie D and Sclater JG (1971) The evolution of the Indian Ocean since the Late Cretaceous. *Geophysical Journal International* 24(5): 437–528.

McPherson K (1984) Cultural exchange in the Indian Ocean Region. *Westerly* 29(4): 5–16.

MEC (2005) Ministry of Environment and Construction, Government of Maldives.

MENR (2000) Report of the Ministry of Environment and Natural Resources, Government of Sri Lanka, Colombo.

Messerli H, Weiss B, Kua J, Bakker M, Tomatis J, Rajeriarison P (2013a) *Indian Ocean Tourism: Regional Integration or Cooperation? (English)*. Washington, DC: World Bank. http:// documents.worldbank.org/ curated/ en/ 363671468001818766/ Indian- oceantourism- regional- integration- or- cooperation.

Messerli H, Weiss B, Kua J, Bakker M, Tomatis J, Rajeriarison P (2013b) *Indian Ocean Islands Tourism Sector Review: Madagascar (English)*. Washington, DC: World Bank. http://documents.worldbank.org/ curated/en/994701467992523618/ Indian-ocean-islands-tourism-sector-review-Madagascar.

Michael C and Moore S (2015) Developing Blue Economy in China and the United States, Centre for American Progress, www.americanprogress.org.

Ministry of Environment and Forest (2014) Gazette Notification, November 3. Bangladesh Govt. Press, Dhaka.

Ministry of Fisheries (2018) *Fisheries sector investment guide*, Ministry of Finance, Suva, Fiji.

Ministry of New and Renewable Energy (2011) Strategic plan for new and renewable energy sector for the period 2011–17, Feb. 2011, Govt of India, p. 85. https://mnre.gov.in/sites/default/files/uploads/strategic_plan_mnre_2011_17.pdf.

Ministry of Road Transport and Highways (2016) Indian Shipping Statistics 2015. New Delhi: Government of India. Accessed December 7. http://shipping.nic.in/showfile.php?lid=2280.

Ministry of Shipping (2011) Indian Maritime Agenda 2010–2020. New Delhi: Government of India.

Ministry of Statistics and Programme Implementation (2016) Indian States by GDP. Accessed December 5, 2016. http://statisticstimes.com/economy/gdp-of-indian-states.php.

Ministry of Tourism (2012) Statistiques Tourisme, Madagascar 1999 à 2012, Antananarivo.

Minshull TA, Lane CI, Collier JS, Whitmarsh RB (2008) The relationship between rifting and magmatism in the northeastern Arabian Sea. *National Geoscience* 1: 463–467.

Mirza MMQ (2002) Global warming and changes in the probability of the occurrence of floods in Bangladesh and implications. *Global Environmental Change* 12: 127–138.

Miththapala S (2008) *Mangroves*. Coastal Ecosystems Series Vol. 2, pp. 1–28. Ecosystems and Livelihoods Group Asia, IUCN, Colombo, Sri Lanka.

Mitroussi K (2013) Ship management: Contemporary developments and implications. *The Asian Journal of Shipping and Logistics* 29: 229–248.

Mittermeier RA, Turner WR, Larsen FW, Brooks TM, Gascon C (2011) Global biodiversity conservation: The critical role of hotspots. In FE Zachos and JC Habel (Eds.), *Biodiversity Hotspots*. Berlin, Heidelberg: Springer-Verlag.

Mittra S (2017) Blue economy: Beyond an economic proposition. *Observer Research Foundation* (173): 1–6.

MoFA (2014) Maritime Area of Bangladesh. Ministry of Foreign Affairs, Dhaka.

Mohan CR (2018) Raja Mandala: Securing the littoral. *The Indian Express*, June 5.

Mohan CR (2019) Why India should welcome Russia's presence in the Indian Ocean. *The Print*, November 26 (with Pia Krishnankutty and Kairvy Agarwal).

Mohanty SK, Dash P, Gupta A (2015) Unleashing the potential of blue economy. Policy Brief (Research and Information System for Developing Countries, India), 1: 1–8.

Mohanty SK, Dash P, Gupta A, Gaur P (2015) Prospects of Blue Economy in the Indian Ocean, Research and Information System for Developing Countries. New Delhi.

Mohanty SK et al. (2016) *Prospects of Blue Economy in the Indian Ocean*. New Delhi: Research and Information System for Developing Countries.

MOSPI (2019) Sustainable Development Goals, National Indicator Framework Baseline Report 2015–16, Ministry of Statistics and Programme Implementation, 492p.

Mukhopadhyay R (1998) Post-Cretaceous intraplate volcanism in the Central Indian Ocean Basin. *Marine Geology* 151: 135–142.

Mukhopadhyay R (2008) Interlinking of rivers: Are we on the right side? *Goa Today* XLIII(3): 36–38.

Mukhopadhyay R (2019) Global Warming: Formulating a Science Based Public Policy and Mitigation Strategy for South Asia. Ph.D. Thesis, Department of Public Administration, School of Social Sciences, Indira Gandhi National Open University, Delhi, 150pp.

Mukhopadhyay R Ghosh AK, Iyer SD (2008) *The Indian Ocean Nodule Field: Geology and Resource Potential*. 1st ed. Amsterdam: Elsevier.

Mukhopadhyay R, Ghosh AK, Iyer SD (2018a) *The Indian Ocean Nodule Field: Geology and Resource Potential*. Amsterdam: Elsevier.

Mukhopadhyay R, Karisiddaiah SM, Mukhopadhyay J (2018b) *Climate Change: Alternate Governance Policy for South Asia*. Amsterdam: Elsevier.

Mukhopadhyay R, Naik S, De Souza S, Dias O, Iyer SD, Ghosh AK (2019) The economics of mining seabed manganese nodules: A case study of the Indian Ocean nodule field. *Marine Georesources and Geotechnology* 37(7): 845–851.

Mukhopadhyay, R, George, P, Ranade, G (1997) Spreading rate dependent seafloor deformation in response to India-Eurasia collision: Results of a hydro-sweep survey in the Central Indian Ocean Basin. *Marine Geology* 140: 219–229.

Muller RD, Royer JY, Lawver LA (1993) Revised plate motions relative to the hotspots from combined Atlantic and Indian Ocean hotspot tracks. *Geology* 21: 275–278.

Murti Y and Agarwal T (2010) Marine derived pharmaceuticals: Development of natural health products from marine biodiversity. *International Journal of ChemTech Research* 2: 2198–2217.

Nabangchang O and Krairapanond N (2015) Ocean Economy and Ocean Health in Thailand. EAS Congress, Danang, Vietnam.

Nagao S (2018) The Growing Militarization of the Indian Ocean Power Game and Its Significance for Japan. The Sasakawa Peace Foundation (www.spf.org).

Naidu PD and Govil P (2010) New evidence on the sequence of deglacial warming in the tropical Indian Ocean. *J. Quaternary Sci.* 25: 1138–1143.

Nasiruddin (2015) Blue Economy for Bangladesh. https://mofl.portal.gov.bd/sites/default/files/files/mofl.portal.gov.bd/page/221b5a19_4052_4486_ae71_18f1ff6863c1/Blue%20economy%20for%20BD.pdf.

National Agency for Petroleum, Natural Gas and Biofuels (2018) Opportunities in the Brazilian Oil and Gas Industry, Rio de Janeiro, Brazil.

National Bureau of Statistics (2017) www.nbs.gov.sc/statistics/tourism.

National Fishery Authority, PNG (2013) PNG Marine Programme on coral reefs, Fisheries and Food Security, 2010–13, Department of Environment and Conservation.

National Fishery Authority, PNG (2019) www.fisheries.gov.pg/fisheriesindustry.

National Institute of Public Finance and Policy (2019) 43rd Annual Report, 2018–19. 104p.

National Marine Science Committee (2015) National Marine Science Plan 2015–2025: Driving the development of Australia's blue economy.

NCSI (2014) www.ncsi.gov.om/Pages/NCSI.aspx.

NDTV (2017) India, US, Japan Ready for Naval Drills in Indian Ocean, China Concerned. New Delhi Television News Desk, July 8.

Nellemann C, Corcoran E, Duarte CM, Valdrés L, De Young C, Fonseca L, Grimsditch G (2009) *Blue Carbon: The Role of Healthy Oceans in Binding Carbon. A Rapid Response Assessment*. Arendal, Norway: UNEP/GRID-Arendal. www.grida.no/publications/145.

New Zealand Govt. Portal, http://archive.stats.govt.nz/browse_for_stats/ environment/environmental -reporting-series/ environmental- indicators/Home/Marine/state-fish-stocks.aspx.

Nicolodi JL and Petermann RM (2010) Potential vulnerability of the Brazilian coastal line with environmental, social and technological aspects. *Pan American Journal of Aquatic Sciences* 5(2): 184–204.

Nihous G (2007) A preliminary assessment of ocean thermal energy conversion resources. *Journal of Energy Resources Technology* 129: 10–17.

NOAA (2014) Climate Report Jan 2014. National Center for Environmental Information, National Oceanic and Atmospheric Administration.

NOAA (2018) National Oceanic and Atmospheric Administration (NOAA) Report on the US Ocean and the Great Lakes Economy. NOAA office for coastal management (http://coast.noaa.gov/ digitalcoast/training/econreport.html).

NOEP (2016) State of US Ocean and Coastal Economies 2016 update, National Ocean Economic Programme, Center for Blue Economy.

Northcott M (2007) A Moral Climate: The Ethics of Global Warming (University of Edinburgh).

Obura D et al. (2017) *Reviving the Western Indian Ocean Economy: Actions for a Sustainable Future*. Gland, Switzerland: WWF International, 64pp.

Ocean Policy Statement (2002) The Indian Maritime Agenda 2010–2015, and the Latest Ensuring Secure Seas: Indian Maritime Security Strategy 2015.

Oceans Policy Science Advisory Group (2013) *Marine Nation 2025: Marine Science to Support Australia's Blue Economy*. Sydney: Government of Australia.

OECD (2016) *The Ocean Economy in 2030*. Paris: OECD Publishing. The Organisation for Economic Co-operation and Development, www.dx.doi.org/10.1787/9789264251724-en.

Ortuño Crespo G and Dunn DC (2016) Impacts of fisheries on open-ocean ecosystems. Nippon Foundation. www.nereusprogram.org/policy-brief-bbnj-impacts-of-fisheries/.

OXFAM International (2019) India: Extreme inequality in numbers. The Power of People against Poverty, www.oxfam.org/en/india.

Panda A and Parameswaran P (2018) Indian Ocean Update: The crisis in the Maldives and India's play at Duqm. *The Diplomat*, February 28. www.thediplomat.com.

Park DKS and Kildow DJT (2014) Rebuilding the classification system of the ocean economy. *Journal of Ocean and Coastal Economics* (1) https://doi.org/10.15351/2373–8456.1001.

Parthasarathi P and Reilio G (2014) The Indian Ocean in the long eighteenth century. *Eighteenth-Century Studies* 48(1): 1–19.

Patil PG, Virdin J, Colgan CS, Hussain MG, Failler P, Vegh T (2018) *Toward a Blue Economy: A Pathway for Sustainable Growth in Bangladesh*. Washington, DC: The World Bank Group, 93pp.

Patrik J and Ketker P (2017) Plastic Pollution and Our Oceans: What Everyone Should Know. https://camd.northeastern.edu /rugglesmedia/ 2017/02/08/ plastic- pollution- and-our-oceans-what-everyone-should-know/.

Patterson M, MacDonald GW, MacDonald N (2009) Ecosystem service appropriation of the Auckland region economy: An input–output analysis. *Regional Studies* 45(3): 333–350.

Paul B and Rashid H (1993) Flood damage to rice crop in Bangladesh. *Geogr Rev* 83(2): 151–159, pdf/Flood/statusreport.pdf.

Pauli G (2010) The Blue Economy: 10 Years, 100 Innovations, 100 Million Jobs. Report to the Club of Rome, Paradigm Publications, 309pp. ISBN: 9780912111902.

Peart R (2005) How New Zealand's coastal and marine management compares with international best practice, paper presented at Seachange05: Managing our coastal waters and oceans, Auckland (www.eds.org.nz/assets/Publications/Seachange).

Pelc R and Fujita RM (2002) Renewable energy from the ocean. *Marine Policy* 26: 471–479.

PEMSEA (2018) Proceeding of Blue Economy Forum, Bangkok, 2017, Partnerships in Environmental Management for the Sea of East Asia (PEMSEA), The Philippines.

Pendleton L, Donato DC, Murray BC, Crooks S, Jenkins WA, Sifleet S, Craft C, Fourqurean JW, Kauffman JB (2012) Estimating global "Blue Carbon" emissions from conversion and degradation of vegetated coastal ecosystems. *PLoS ONE*, 7:e43542. doi:10.1371/journal.pone.0043542. PMC 3433453. PMID 22962585.

Pinfold G (2009) Economic impact of marine related activities in Canada. *Statistical and Economic Analysis Series* 1–1(1): iii–125. http://waves-vagues.dfo-mpo.gc.ca/Library/338949.pdf.

Planning Commission (2016) State-wise YoY Growth Rate by Industry of Origin (2004–05 Prices). Accessed December 5, http://planningcommission.nic.in/data/datatable/data_2312/.

Planning Commission of India (2013) Press Note on Poverty Estimates, 2011–12. New Delhi: Government of India.

Poon, A (1989) Competitive strategies for new tourism. In C. Cooper (Ed.), *Progress in Tourism Recreation and Hospitality Management* (Vol. 1, pp. 91–102). London: Belhaven Press—in Lew (2008) Long tail tourism: New geographies for marketing niche tourism products. *Journal of Travel & Tourism Marketing* 25(3–4): 412.

Popescu I and Ogushi T (2013) Fisheries in Japan, Directorate General for Internal Policies, European Parliament.

Prange SR (2008) Scholars and the sea: A historiography of the Indian Ocean. *History Compass* 6(5): 1382–1393.

Press Information Bureau (2016) Economic Survey 2015–16: Services Sector remains the Key Driver of Economic Growth contributing almost 66.1% in 2015–16. Ministry of Finance. New Delhi: Government of India.

Prime Minister's Office (2013) *The Ocean Economy: A Roadmap for Mauritius.* Mauritius: Government of Mauritius.

Proceedings on Blue Economy Forum (2017) Indian Ocean Rim Association. Bangkok, Thailand.

Raewyn P (2015) How New Zealand's coastal and marine management compares with international best practice, paper presented at Seachange05: Managing our coastal waters and oceans, Auckland, 21–22.

Raghavan PS (2019) Keynote Address by Ambassador P.S. Raghavan, Chairman, National Security Advisory Board at Inaugural Session of the Indian Ocean Dialogue-6 (IOD VI), December 13, 2019, Indian Council of World Affairs, Sapru House, New Delhi.

Rajan R and Banerjee A (2019) Eight things India must do in 2019: The economic challenges we face and the reforms we need to carry out now. Edit-Page, *The Times of India*, January 1.

Rajan S (2018) The legal continental shelf: Geosciences at sea with UNCLOS. *J. Geol. Soc. India* 92: 131–133. doi:10.1007/s12594-018-0970-2.

Ramanathan V, Carmichael G (2008) Global and regional climate changes due to black carbon. *Nature Geoscience* 1, 221–227.

Ramanathan V, Crutzen PJ, Lelieveld J, Valero FPJ and 36 others (2001) Indian Ocean Experiment: An integrated analysis of the climate forcing and effects of the great Indo-Asian haze. *Journal of Geophysical Research* 106(D22): 28371–28398.

Ramanathan V, Ramana MV, Roberts G, Kim D, Corrigan C and 2 others (2007) Warming trend in Asia amplified by brown cloud solar absorption. *Nature* 448: 575–578.

Ranasinghe DMKH (2010) Climate Change mitigation—Sri Lanka's perspectives. In: Proceedings of the 15th International Forestry and Environment Symposium. November 26–27, University of Jayewardenepura, Sri Lanka, 290–296.

Raut A (2006) Climate impacts on Nepal. *Tiempo* 60: 3–5.

Regard V, Hatzfeld D, Molinaro M, Aubourg C, Bayer R, Bellier O, Yamini-Fard F, Peyret M, Abbassi M (2010) The transition between Makran subduction and the Zagros collision: Recent advances in its structure and active deformation. Geological Society, London, Special Publications 330(1): 43–64. doi:10.1144/SP330.4.

Ren W, Wang Q, Ji, J (2018) Research on China's marine economic growth pattern: An empirical analysis of China's eleven coastal regions. *Marine Policy* 87: 158–166. https://doi.org/10.1016/j.marpol.2017.10.021.

Report of EC (European Commission) (2012) Blue Growth—Opportunities for marine and maritime sustainable growth. Publications Office of the European Union. 13pp. doi:10.2771/43949

Report of Indian Ocean Tuna Commission (2018) Status Summary for Species of Tuna and Tuna-Like Species under the IOTC Mandate, as well as other Species Impacted by IOTC Fisheries. www.iotc.org/science/status-summary-species-tuna-and-tuna-species-under-iotc-mandate-well-other-species-impacted-iotc.

Report of Nordic Atlantic Cooperation (2018) Nordic Council of Ministers, FAO, The Commonwealth Secretariat, Faroe Islands, Final Report from the Large Ocean Nations Forum on Blue Growth, Malta, October 2017 (Torshavn: NORA, 2018).

Report on Energy Policy of the Republic of Seychelles 2010–2030 (2015) Document available at: www.iea.org/policiesandmeasures/pams/seychelles.

Reuters (2018) Indonesia, India plan to develop strategic Indian Ocean port. Report by Agustinus Beo Da Costa, World News, May 30, 2018.

Review of Maritime Transport (2018) UNCTAD/RMT/2017, 102pp.

Ripton JT (2015) The pollution crisis in sea of Japan. *Renewable Energy World* (www. reneableenergyworld.com/ugc/articles/2015/11/the-pollution-crisis-in-japan).

Rockström S and 26 others (2009) Report of the Stockholm Resilience Centre.

Roy A (2019) Blue economy in the Indian Ocean: Governance perspectives for sustainable development in the region. Observer Research Foundation Occasional Paper No. 181.

Royer JY and 12 others (1992) Indian Ocean plate reconstruction since the last Jurassic. In: Duncan RA, Rea DK, Kidd RB, vonRad U, Weissel JK (Eds.) *A Synthesis of Scientific Drilling Results from the Indian Ocean.* Am. Geophy. Union, Geophysical Monograph 70, 375–471.

RSA Government Portal (2018) Republic of South Africa. www.gov.za/blue economy.

Rubin CM, Horton BP, Sieh K, Pilarczyk JE, Daly P, Ismail N, Parnell AC (2017) Highly variable recurrence of tsunamis in the 7,400 years before the 2004 Indian Ocean tsunami. *Nature Communications* 8: 160–190.

Russell M (2014) Four ways climate change causes world hunger. Devex: International development news (www.devex.com).

SAARC (2016) South Asian Association for Regional Cooperation Web Portal 2016.

SAARC Technical Report (2010) Regional Study on Greenhouse effect and its impact on the region, 16 pdf Files, Environment & Climate Change. South Asian Association for Regional Cooperation.

Sabharwal M (2019) Why NYAY won't end poverty: Promise of income without work represents a panicked pessimism about India and her people. Edit-Page, *The Times of India,* April 23.

SAGAR as the Last Growth Frontier. https://niti.gov.in/writereaddata/Govt. of India Portal

Sakhuja V (2010) Security threats and challenges to maritime supply chains. Disarmament Forum, www.peacepalacelibrary.nl/ebooks/files/UNIDIR_pdf-art2959.pdf.

Sakhuja V (2011) Asian Maritime Power in the 21st Century: Strategic Transactions: China, India and Southeast Asia. ISEAS (January 15, 2011), ISBN-13: 978-9814311090, 380p.

Sannasiraj SA (2019) Ocean energy potential in India: Assessment and harvesting. In AS Raju (Ed.), *Blue Economy of India: Emerging Trends.* New Delhi: Studera Press, pp.79–82.

Santhana G (2003) Tourism development in coastal areas—Brazil: Economic demand and environmental issues. *Journal of Coastal Research,* Special issue No. 35: 85–93.

Saravanan A and Debnath D (2013) Patenting trends in marine biodiversity: Issues and challenges. *Pharma Utility* 7: 1–13.

Satpathy B, Muniapan B, Dass M (2013) UNESCAP's characteristics of good governance from the philosophy of Bhagavad-Gita and its contemporary relevance in the Indian context. *International Journal of Indian Culture and Business Management* 7(2): 192–212.

Schott FA, Xie SP, McCreary Jr. JP (2009) Indian Ocean Circulation and Climate Variability. *Reviews of Geophysics* 47: RG1002, 46p.

Science Daily (2018) New technologies in the ocean energy sector. European Commission Joint Research Centre, October 30, 2018. www.sciencedaily.com /releases/2018/ 10/ 181030174952.htm.

Scott D, Hall CM, Gossling S (2016) A review of the IPCC Fifth Assessment and implications for tourism sector climate resilience and decarbonisation. *Journal of Sustainable Tourism* 24(1): 8–30.

Senapati S and Gupta V (2014) Climate change and coastal ecosystem in India: Issues in perspectives. *International Journal of Environmental Sciences* 5(3).

SEPA (2019) www.worldwatch.org/sepa-chinas-marine-environment-faces-irreversible-damage).

Seraval TA and Alves FL (2011) International trends in ocean and coastal management in Brazil. *Journal of Coastal Research*, Special Issue 64, Proceedings of 11th International Coastal Symposium ICS2011, 1258–1262.

Sharma R (2015) Environmental issues of deep sea mining. *Proceedia Earth and Planetary Science* 11: 204–211.

Shindell D, and 23 others (2012) Simultaneously mitigating near-term climate change and improving human health and food security. *Science* 335: 183–189.

SIDS (2014) Third International Cooperation on Small Island Developing States, Samoa 2014, 21stAfrican Union Summit 2013.

Sikri V (2015) Gender, community and violence: Changing mindsets for empowering the women of South Asia. Conference co-organized by SWAN and Jamia Millia Islamia University in New Delhi on April 15–16, pp. 1–9.

Singh A (2016) India's Maritime Security Perspective (January 12, 2016). RUMLAE Research Paper No. 16-13. https://ssrn.com/abstract=2726411.

Singh N (2019) Purpose + profit: Capitalism is getting a makeover: New age companies are trying to do well by doing good. *Times of India*, September 29, 2019.

Sinha A, Stott I, Berkelhammer M (2011) A global context for mega-droughts in monsoon Asia during the past millennium. *Quaternary Science Review* 30: 47–62.

Sivakumar MVK, Stefanski R (2011) Climate change and food security in South Asia. In R Lal et al. (Eds.), *Climate Change and Food Security in South Asia*. Amsterdam: Springer Science + Business Media.

Smithers Group (2015) The Future of Marine Biotechnology for Industrial Applications to 2025. www.smithersrapra.com/products/market-reports/biomaterials/the-future-of-marine-biotechnology-for-industria.

Somaratne S, Dhanapala AH (1996) Potential impact of global climate change on forest in Sri Lanka. In L Erda, W Bolhofer, S Huq, S Lenhart, SK Mukherjee, JB Smith, J Wisniewski (Eds.), *Climate Change Variability and Adaptation in Asia and the Pacific*. Dordrecht: Kluwer, pp. 129–135.

Sousounis K (2016) How is climate change a feminist issue? www.girlup.org.

Spalding M, Kainuma M, Collins L (2010) *World Atlas of Mangroves*. London: Earthscan.

Spalding MD, Brumbaugh RD, Landis E (2016) *Atlas of Ocean Wealth*. Arlington, VA: The Nature Conservancy.

Sridhar KS (2010) Carbon emissions, climate change and impacts in Indian cities. In *India Infrastructure Development in a Sustainable Low Carbon Economy· Road Ahead for India*. New Delhi: Oxford University Press, pp. 345–354.

Statistics Japan (2019) *Statistical Handbook of Japan*, 2019, Statistics Bureau, Ministry of Internal Affairs and Communication, Japan.

Statistics New Zealand (2016) New Zealand's marine economy: 2007–13, www.stats.govt.nz.

Statistics New Zealand (2018) stats.govt.nz/indicators/marine economy, New Zealand Government.

STB (2011) *Seychelles 2012–2020 Tourism Master Plan*, Beau Vallon.

Stiffler C (2014) Colorado's Genuine Progress Indicator (GPI): A Comprehensive Metric of Economic Well-being in Colorado from 1960–2011, Colorado Fiscal Institute, 63p, www.coloradofiscal.org.

Strauss S (2012) Climate-Wire Climate Change. doi:10.1002/wcc.181.

Sukumaran PV, Unnikrishnan E, Gangadharan AV, Zaheer B, Abdulla NM, Kumaran K, Ramachandran KV, Hegde SV, Maran N, Bhat KK, Rao MK, Dinesh AC, Jayaprakash J, Praveen Kumar P, Shareef NM, Gopalan CC (2010) Marine sand resources in the south-west continental shelf of India. *Ind. J. Geo-Mar. Sci.* 39: 572–578.

Sultana H and Ali N (2006) Vulnerability of wheat production in different climatic zones of Pakistan under climate change scenarios using CSM-CERES-Wheat Model. Paper presented in the Second International Young Scientists' Global Change Conference, Beijing, November 7–9, 2006, organized by START (the global change System for Analysis, Research and Training) and the China Meteorological Administration.

Sun C, Li X, Zou W, Wang S, Wang Z (2018) Chinese marine economy development: Dynamic evolution and spatial difference. *Chinese Geographical Science* 28(1): 111–126. https://doi.org/10.1007/s11769-017-0912-8.

Sustainable Development Knowledge Platform (2016). Sustainable Development Goal 14. Accessed December 2, 2016, https://sustainabledevelopment.un.org/sdg14.

Tagore RN (1913) *Sadhana: The Realization of Life.* Calcutta: Viswa Bharati Prakashan.

Taneja A (2019) Integration of Karma, Bhakti and Jnana. Speaking Tree, *The Times of India*, February 5.

Tanner T, Mitchell T, Polack E, Guenther B (2009) Urban Governance for Adaptation: Assessing Climate Change Resilience in Ten Asian Cities. Institute of Development Studies, UK, Working paper 315, 47pp.

Tasir H (2018) Militarization of Indian Ocean: Implications for regional security. *The CSS Point*, April 10, 2018, www.globalvillagespace.com/.

Terashima H (2012) Japan's ocean policymaking. *Coastal Management* 40(2): 172–182. https://doi.org/10.1080/08920753.2012.652518.

The Telegraph (2019) The Telegraph, Calcutta Edition, August 22, 2019.

Third International Cooperation on Small Island Developing States, Samoa (2014) 21st African Union Summit 2013, Post-2015 Development Conference 2015, IOR's Blue Economy Conference 2015.

TNA (2016) Technology Needs Assessment Report: Climate Change mitigation, Government of Pakistan, M/Climate Change Islamabad. Joint report by Global Environment Facility and United Nation Environment Program. Asian Institute of Technology, Bangkok, pp 112.

TOI (2015) Livestock causes 15% of all emissions worldwide. *The Times of India*, December 16.

TOI (2019a) The Times of India Report of Wetlands of India, September 7.

TOI (2019b) Piracy and maritime safety. *The Times of India*, March 9.

TOI (2019c) Why a wallet full of cash is more likely to be returned than an empty one. *The Times of India*, July 5.

Torres T, Karger F, Keyes D, Thornton, H, Lather M, Alshariff K (2015) Whither the US National Ocean Policy Implementation Plan? *Marine Policy* 53: 198–222.

UK Government Office for Science (2018) *Foresight Future of the Sea.* 128. www.gov.uk/government/publications/future-of-the-sea--2.

UK Marine Industries Alliance (2011) A strategy for growth for UK Marine Industries, Global Excellence and Innovation, UK Marine Alliance.com

UNCSDG (2017) United Nations Conference to Support the Implementation of Sustainable Development Goal 14: Conserve and sustainably use the oceans, seas and marine resources for sustainable development. New York, June 5–9.

UNCTAD (2016) Review of Maritime Transport 2016, United Nations Conference of Trade and Development UNCTAD/RMT/2016.

UNCTAD (2019) Maritime Profile: Japan, United Nations Conference for Trade and Development, United Nations.

UNCTAD Maritime Profile (2019) https://unctadstat.unctad.org/CountryProfile / MaritimeProfile/en-GB/144/index.html.

UNDESA (2017) United Nations Department of Economic and Social Affairs, Population Division Report www.un.org/development/desa/en/.

UNEP (2016) United Nations Environment Program Assembly Convening in Nairobi: Governments Agree to 25 Landmark Resolutions to Drive Sustainability Agenda and Paris Climate Agreement, May 28. www.unep.org/newscentre/Default.asp x?DocumentID=27074&ArticleID=36197&l=e.

UNFCCC (2015) Paris Agreement. COP-21, http://unfccc.int/files/essential_ background/convention/application/pdf/english_paris_agreement.pdf.

UN-HDI (2019) Beyond income, beyond averages, beyond today—Inequalities in human development in the 21st Century. Human Development Index Ranking, United Nations Development Program, www.hdr.undp.org.

United Nations (2012) Blue Economy Concept Paper. Rio de Janeiro: United Nations.

US Census Bureau Shares (2019) www.census.gov/newsroom/press-releases/2019/paid-media-strategy.html.

USAID (2015) United States AID Agency Financial Report, 168pp.

U.S. Census Bureau Shares (2019) U.S. Census Bureau Shares 2020 Census Paid Media Strategy. https://www.census.gov/newsroom/press-releases/2019/paid-media-strategy.html.

USEPA (2016) United States Environmental Protection Agency. Climate change indicators in the United States Atmospheric concentration of greenhouse gases, pp. 1–13, www.epa.gov/climate -indicators.

USGCRP (2014) Our Changing Planet, Indicators of Change, Adapting to Change. Third National Climate Assessment, the Authoritative and Comprehensive Report on Climate Change and Its Impacts in the United States. US Global Change Research Program www.globalchange.gov/.

USGHGIR (2011) *United States Greenhouse Gas Inventory Report.* Washington, DC: US Environmental Protection Agency.

Vimal Kumar KG, Dinesh Kumar PK, Smitha BR, Habeeb Rahman H, Josia J, Muraleedharan KR, Sanjeevan VN, Achuthankutty CT (2008) Hydrographic characterization of southeast Arabian Sea during the wane of southwest monsoon and spring intermonsoon. *Environment Monitoring Assessment* 140: 231–247. https://doi.org/10.1007/s10661-007-9863-3

Visweswaran K (2004) Gendered states: Rethinking culture as a site of South Asian human rights work. *Human Rights Quarterly* 26: 483–511.

Vyas M (2019) Jobs are human capital: Quality of Indian jobs is falling, there are fewer rewards for getting an education today. Edit-Page, *The Times of India*, April 11.

Wagner C (2016) The role of India and China in South Asia. *Journal of Strategic Analysis* 40(4): 307–320.

Wai Ming To and Lee P (2018) China's maritime economic development: A review, the future trend, and sustainability implications. *Sustainability* 10: 4844.

Wairimu, E and Khainga D (2017) Kenya's Agenda in Developing the Blue Economy. http://kippra.or.ke/kenyas-agenda-in-developing-the-blue-economy.

Wang, Xiaohui (2016) The ocean economic statistical system of China and understanding of the Blue Economy. Journal of Ocean and Coastal Economics 2(2): 1–32. https://doi.org/10.15351/2373-8456.1055.

Ward J (2017) The Emerging Geopolitics of the Indian Ocean Region. Asia Pacific Bulletin, East-West Centre, Washington DC, 386, 2pp.

Wartsila (2018) Navigating through China's blue economy. https://www.wartsila.com/twentyfour7/innovation/navigating-through-China's-Blueeconomy.

Watanabe TN (2019) Japan's Rationale for the Free and Open Indo-Pacific Strategy [1]. International Information Network Analysis. The Sasakawa Peace Foundation. October 30, 2019. www.spf.org.

Wenhen R, Qi Wang, Jianyui Jo (2018) Research in China's maritime economic growth pattern: An empirical analysis of China's 11 coastal regions. *Marine Policy* 87:156–168.

Wertheim, S (2019) Can we stop a cold war with China? *The New York Times*, June 10, 2019.

White DG (2012) *Yoga in Practice*. Princeton, NJ: Princeton University Press, 416pp.

Wickremesinghe, R (2016) Inaugural Address delivered by Prime Minister of Sri Lanka at the Indian Ocean Conference on September 1, 2016, Shangri La Hotel, Singapore, Global Power Transition and the Indian Ocean. Colombo: Prime Minister's Office of the Democratic Socialistic Republic of Sri Lanka. www.pmoffice.gov.lk/ download / press /D00000000050_EN.pdf?p=7.

Wignaraja G (2019) Implications of a New Normal Indian Ocean Economy for Sri Lanka. ISAS Working paper. No. 317. pp. 1–18.

Wijeratne MA (1996) Tea: Plucking strategies. *Planters Chronicle* September: 443.

Wildlife Conservation Society (2015) Wildlife Conservation Society's Bangladesh Cetacean Project Proposal to establish a marine protected area in the Swatch-of-No-Ground submarine canyon and surrounding coastal waters in the Bay of Bengal. www.cbd.int/doc/meetings/mar/ebsaws-2015-01/other/ebsaws-2015-01-gobi-submission6-en.pdf.

Williams J, Fiona S, Hugh D, Konard H (2017) The economic contribution of commercial fishing to the New Zealand economy. BERL, www.berl.co.nz.

Wippel S (2013) Oman and the Indian Ocean Rim: Economic Integration Across Conventional Meta-Regions. In S Wippel (Ed.), *Regionalizing Oman: Political, Economic and Social Dynamics*. Dordrecht: Springer, pp. 159–183.

World Atlas of Coral Reefs, UNEP (WCMC). coral.unep.ch/atlaspr.htm.

World Bank (1972) *Bangladesh – Reconstructing the Economy (Vol. 2): Notes on the Revival of Economic Activity in the Major Sectors (English)*. South Asia series; no. SA 35. Washington, DC: World Bank. http://documents.worldbank.org/curated/en/540811468207864462/Notes-on-the-revival-of-economic-activity-in-the-major-sectors.

World Bank (2012) Connecting to Compete – Trade Logistics in the Global Economy. Washington, DC 20433.

World Bank (2013) Fish to 2030: Prospects for Fisheries and Aquaculture. Agriculture and environmental services discussion paper; no. 3. Washington, DC: World Bank. https://openknowledge.worldbank.org/handle/10986/17579.

World Bank (2014a) Climate-Smart Development: Adding Up the Benefits of Actions That Help Build Prosperity, End Poverty and Combat Climate Change.

World Bank (2014b) Does employment generation really matter for poverty reduction? Policy Research Working PAPER 4432 (authored by Gutierrez C, Orecchia C, Paci P, Serneels P), pp. 1–44.

World Bank (2017) World Bank indicators, with Millennium Project compilation and forecast.

World Bank Report (2016a) Oceans 2030: Blue Economy development Framework. www.worldbank.org/oceans.

World Bank Report (2016b) Blue Economy Development Framework. Growing the Blue Economy to Combat Poverty and Accelerate Prosperity. http://pubdocs.worldbank.org/en/446441473349079068/AMCOECC-Blue-Economy-Development-Framework.pdf.

World Bank Report (2017) Third South West Indian Ocean Fisheries Governance and Shared Growth Project (SWIOFish3) (P155642), p. 4.

World Economic Forum Report (2017) The Global Competitiveness Report. Accessed on July 2018, www.weforum.org/reports/the-global competitiveness-report-2017–2018.

World Ocean Council (2018) Ocean/maritime clusters: Leadership and collaboration for ocean sustainable development and implementing the sustainable development goals, World Ocean Council White Paper.

World Watch (2019) www.worldwatch.org/sepa-chinas-marine-environment-faces-irreversible-damage.

WRI (2015) World Resources Institute, Washington DC, 64pp.

WTTC (2012) *Travel & Tourism Economic Impact 2012: Seychelles*. London: WTTC.

WWF (2016) *Oceans Facts and Futures Report*.

WWI (2013) *State of the World 2013: Is Sustainability Still Possible*. Washington, DC: World Watch Institute.

Wyrtki K (1973) Physical oceanography of the Indian Ocean. In B Zeitschel and SA Gerlach (Eds.), *The Biology of the Indian Ocean*. Ecological Studies (Analysis and Synthesis), 3. Berlin; Heidelberg: Springer.

Xiao G, Liu W, Xu Q, Sun Z, Wang J (2005) Effects of temperature increase and elevated CO_2 concentration, with supplemental irrigation, on the yield of rain-fed spring wheat in a semiarid region of China. *Agriculture and Water Management* 74(3): 243–255.

XinhuaNet (2018) China Focus: China's maritime economy expands by 7.5 per cent in last five years (www.xinhuanet.com/english/2018-01/21/c_136913316.htm).

Yang B and Yue L (2018) Japan's marine economic development and competition with China. *CCAMLR Science* 25(2): 83–95.

Yeoman R and Akehurst G (2017) Cruise tourism's contribution to New Zealand's economy, me consulting, www.me.co.nz.

Zao K (2012) China's ocean policy making: Practice and lessons. *Coastal Management* 40(2): 145–160.

Zao R (2014) Defining and quantifying China's ocean economy. *Marine Policy* 43: 164–173.

Zhang W, Li Y-X, Tam C, Bougouffa S, Wang R, Pei B, Chiang H, Leung P, Lu Y, Sun J, Fu H, Bajic VB, Liu H, Webster NS, Qian P-Y (2019) Marine biofilms constitute a bank of hidden microbial diversity and functional potential. *Nature Communication*,10, Article No. 517.

Zhang YG, Dong LJ, Yang J, Wang SY, Song XR (2004) Sustainable development of marine economy in China. *Chinese Geographical Science* 14(1): 308–313.

Zou K (2012) China's ocean policymaking: Practice and lessons. *Coastal Management* 40(2): 145–160. https://doi.org/10.1080/08920753.2012.652514.

Index

aerosols 117, 118

Andaman & Nicobar Island 11, 14, 28, 40, 42, 49, 126–7, 138–9, 145

Andhra Pradesh 33, 47, 253–4

Antarctica 4, 13, 14

aplasia 33

aquaculture 2, 5, 6, 21, 25, 27–9, 31, 32, 42, 54, 58, 61, 64, 65, 67, 68, 70, 71, 73, 75–8, 80–6, 90, 91, 93, 98, 100, 103, 105, 108–10, 120, 121, 150, 153–5, 168–71, 177, 186, 190, 198, 200, 202, 239–40

aquaponics 201

Asian Development Bank 106, 190, 234–5, 239, 245

Asian Infrastructure Investment Bank (AIIB) 144

Association of South East Asian Nations (ASEAN) 141, 144, 149, 150–1, 163, 165, 215

Assumption Island 139–40

Australia: challenges, success and economic potential 87; resources, strategy and activities 86

bacteria 29, 33, 35, 276

Bangladesh: challenges, success and economic potential 76; resources, strategy and activities 75

Bay of Bengal Initiative for Multi-Sectoral Technical and Economic Cooperation (BIMSTEC) 84, 86, 104

Bay of Bengal Large Marine Ecosystem (BOBLME) 32, 237

beach nourishment 78, 94

biodiversity: conservation strategy 86, 104, 184; loss 108, 153–4, 25; potential 12

Blue: bond 67, 187, 247, 267; carbon xv, 7, 52–4, 71, 80–2, 150, 153, 184, 186, 190, 250, 266, 277; growth 166, 268, 278; planet 266

Blue Economy: activities 58, 62–3, 71, 76, 90, 93, 96, 98, 100, 103, 105–10, 151, 154, 186–7, 190, 253, 267; concept xv, 1, 9, 58, 83, 128, 208, 243, 246, 280; capacity building 185; potential 4, 250, 268; sectors 67, 70, 83–5, 107, 154–5, 177, 253, 266; vision 24 263–70

Blue audit 187; education 29, 186, 258, 269; skill development 4, 46, 138, 186, 222, 257; marine digital database 185; marine financing 187; marine spatial planning 68, 153, 166, 185, 239, 252

Brazil xiv, 103–4, 144, 158, 241, 250, 271, 277, 279–80; activities 103; environment conservation 104

Brazil-Russia-India-China-South Africa bloc (BRISC) 144, 145

BTVG Doctrine xv, 26, 211–3, 246, 267; agriculture 236; coastal zone 238; communication 242; culture 214; disaster 240; demography 218; economy 224; economic integration 234; education 222; ethical blue economy 211, 213, 243–7; ethics 215; gender 216; governance 228; health 222; hunger 241; judiciary 223; local administration 235; participative governance 234; population 218; regional cooperation 234; resource management 235; structural transformation 218; system engineering 214; technology 240; tourism 241; water 235; yoga 217

bunkering 48, 65, 68

Canada 3, 25, 43, 44, 48; activities 98; environment & conservation policy 100

capacity Building 140, 146, 184

carbon sequestration 6, 24, 54, 184, 211, 266

cargo handling 68, 71, 90, 177, 204

Caribbean 28, 49, 63, 131

catamarans 27, 31

Chabahar 139, 146

Chavakkad 38

China 20, 22, 28, 31, 33, 43, 89–93, 132–8; activities 89–90; environment & conservation policy 92; geopolitics & OBOR 130–7, 143–6, 160, 165, 267

China-Pakistan Economic Corridor (CPEC) 136, 145

choke points 11, 38, 114, 129–30, 134, 178–9

climate change 24, 54, 86, 114–28; catalysing 35; coastal ecosystem 24, 64, 86, 115, 239; collaboration 244; energy & commerce 41, 108; ethical blue economy 214, 216, 244, 267; geopolitics 128, 144; greenhouse gases 48, 54, 64; impact 114–5, 123–4, 128; monsoon 39; ocean 250; threats 266

cloud bursts 125

CMFRI 29, 30

Coastal Aquaculture Authority (CAA) 32

coast, coastal waters 35, 74, 124, 170

Cochin 38, 49

commercial fishing 8, 22, 29, 61, 93, 94, 103, 155, 169

Printed and bound by CPI Group (UK) Ltd, Croydon, CR0 4YY

24/10/2024

01778303-0003